Autonomous Search

Youssef Hamadi · Eric Monfroy · Frédéric Saubion

Editors

Autonomous Search

 Springer

Editors
Dr. Youssef Hamadi
Microsoft Research
Cambridge, UK

LIX
École Polytechnique
Palaiseau, France
youssefh@microsoft.com

Prof. Frédéric Saubion
Faculté des Sciences, LERIA
Université d'Angers
Angers, France
frederic.saubion@univ-angers.fr

Prof. Eric Monfroy
Departamento de Informática
Universidad Técnica Federico Santa María
Valparaíso, Chile
eric.monfroy@inf.utfsm.cl

LINA, UFR de Sciences et Techniques
Université de Nantes
Nantes, France
eric.monfroy@univ-nantes.fr

ISBN 978-3-642-21433-2   e-ISBN 978-3-642-21434-9
DOI 10.1007/978-3-642-21434-9
Springer Heidelberg Dordrecht London New York

Library of Congress Control Number: 2011941788

ACM Classification (1998): G.2, F.2, I.2

Printed on acid-free paper

Springer is part of Springer Science+Business Media (www.springer.com)

Acknowledgements

We would like to thank all the people who have contributed to this book. We are extremely grateful to the external reviewers for their detailed and constructive suggestions on the successive versions of the manuscript.

Youssef Hamadi, Eric Monfroy, and Frédéric Saubion

Contents

List of Contributors

Youssef Hamadi
Microsoft Research, Cambridge, CB3 0FB, UK,
LIX, École Polytechnique, F-91128 Palaiseau, France,
e-mail: youssefh@microsoft.com

Eric Monfroy
Universidad Santa María, Valparaíso, Chile,
e-mail: Eric.Monfroy@inf.utfsm.cl
and LINA, Université de Nantes, France,
e-mail: Eric.Monfroy@univ-nantes.fr

Frédéric Saubion
LERIA, Université d'Angers, France,
e-mail: Frederic.Saubion@univ-angers.fr

A. E. Eiben
Vrije Universiteit Amsterdam,
e-mail: gusz@cs.vu.nl

S. K. Smit
Vrije Universiteit Amsterdam,
e-mail: sksmit@cs.vu.nl

Holger H. Hoos
Department of Computer Science, University of British Columbia,
2366 Main Mall, Vancouver, BC, V6T 1Z4, Canada,
e-mail: hoos@cs.ubc.ca

Derek Bridge
Cork Constraint Computation Centre, University College Cork, Ireland,
e-mail: d.bridge@4c.ucc.ie

Eoin O'Mahony
Cork Constraint Computation Centre, University College Cork, Ireland,
e-mail: e.omahony@4c.ucc.ie

Barry O'Sullivan
Cork Constraint Computation Centre, University College Cork, Ireland,
e-mail: b.osullivan@4c.ucc.ie

Susan L. Epstein
Department of Computer Science, Hunter College and The Graduate Center
of The City University of New York, New York, New York,
e-mail: susan.epstein@hunter.cuny.edu

Smiljana Petrovic
Department of Computer Science, Iona College, New Rochelle, New York,
e-mail: spetrovic@iona.edu

Roberto Battiti
DISI – Dipartimento di Ingegneria e Scienza dell'Informazione,
Università di Trento, Italy,
e-mail: battiti@disi.unitn.it

Paolo Campigotto
DISI – Dipartimento di Ingegneria e Scienza dell'Informazione,
Università di Trento, Italy,
e-mail: campigotto@disi.unitn.it

Jorge Maturana
Instituto de Informática, Universidad Austral de Chile, Valdivia, Chile,
e-mail: jorge.maturana@inf.uach.cl

Álvaro Fialho
Microsoft Research – INRIA Joint Centre, Orsay, France,
e-mail: alvaro.fialho@inria.fr

Marc Schoenauer
Project-Team TAO, INRIA Saclay – Île-de-France and LRI (UMR CNRS 8623),
Orsay, France and Microsoft Research – INRIA Joint Centre, Orsay, France,
e-mail: marc.schoenauer@inria.fr

Frédéric Lardeux
LERIA, Université d'Angers, Angers, France,
e-mail: lardeux@info.univ-angers.fr

Michèle Sebag
Project-Team TAO, LRI (UMR CNRS 8623) and INRIA Saclay – Île-de-France,
Orsay, France and Microsoft Research – INRIA Joint Centre, Orsay, France,
e-mail: michele.sebag@inria.fr

Thomas Stützle,
IRIDIA, CoDE, Université Libre de Bruxelles, Brussels, Belgium,
e-mail: stuetzle@ulb.ac.be

Manuel López-Ibáñez
IRIDIA, CoDE, Université Libre de Bruxelles, Brussels, Belgium,
e-mail: manuel.lopez-ibanez@ulb.ac.be

Marco Montes de Oca
IRIDIA, CoDE, Université Libre de Bruxelles, Brussels, Belgium,
e-mail: mmontes@ulb.ac.be

Mauro Birattari
IRIDIA, CoDE, Université Libre de Bruxelles, Brussels, Belgium,
e-mail: mbiro@ulb.ac.be

Marco Dorigo
IRIDIA, CoDE, Université Libre de Bruxelles, Brussels, Belgium,
e-mail: mdorigo@ulb.ac.be

Michael Maur
Fachbereich Rechts- und Wirtschaftswissenschaften, TU Darmstadt,
Darmstadt, Germany,
e-mail: maur@stud.tu-darmstadt.de

Paola Pellegrini
Dipartimento di Matematica Applicata, Università Ca' Foscari Venezia,
Venezia, Italia,
e-mail: paolap@unive.it

Alejandro Arbelaez
Microsoft-INRIA joint lab, Orsay, France,
e-mail: alejandro.arbelaez@inria.fr

Said Jabbour
Université Lille-Nord de France, CRIL – CNRS UMR 8188,
Lens, France,
e-mail: jabbour@cril.fr

Lakhdar Sais
Université Lille-Nord de France, CRIL – CNRS UMR 8188,
Lens, France,
e-mail: sais@cril.fr

Marek Petrik
Department of Computer Science, University of Massachusetts Amherst,
Amherst, MA, USA,
e-mail: petrik@cs.umass.edu

Shlomo Zilberstein
Department of Computer Science, University of Massachusetts Amherst,
Amherst, MA, USA,
e-mail: shlomo@cs.umass.edu

Chapter 1
An Introduction to Autonomous Search

Youssef Hamadi, Eric Monfroy, and Frédéric Saubion

1.1 Introduction

In the foreword of his seminal book on artificial intelligence and on problem solving by man and machine [24], Jean-Louis Laurière pointed out that one of the main goals of research was to understand how a system (man or machine) may learn, analyze knowledge, transpose and generalize it in order to face real situations and finally solve problems. Therefore, the ultimate goal of this pioneer AI scientist was to build systems that were able to autonomously solve problems, and to acquire and even discover new knowledge, which could be reused later.

Finding the most suitable algorithm and its correct setting for solving a given problem is the holy grail of many computer science researchers and practitioners. This problem was investigated many years ago [31] as the *algorithm selection problem*. The proposed abstract model suggested extracting features in order to characterize the problem, searching for a suitable algorithm in the space of available algorithms and then evaluating its performance with respect to a set of measures. These considerations are still valid and can indeed be considered at least from two complementary points of view:

- Selecting solving techniques or algorithms from a set of possible available techniques;
- Tuning an algorithm with respect to a given instance of a problem.

Youssef Hamadi
Microsoft Research, Cambridge, CB3 0FB, UK
LIX, École Polytechnique, F-91128 Palaiseau, France
e-mail: youssefh@microsoft.com

Eric Monfroy
Universidad Santa María, Valparaíso, Chile, e-mail: Eric.Monfroy@inf.utfsm.cl and
LINA, Université de Nantes, France, e-mail: Eric.Monfroy@univ-nantes.fr

Frédéric Saubion
LERIA, Université d'Angers, France, e-mail: Frederic.Saubion@univ-angers.fr

Y. Hamadi et al. (eds.), *Autonomous Search*,
DOI 10.1007/978-3-642-21434-9_1,
© Springer-Verlag Berlin Heidelberg 2011

To address these issues, existing works often include tools from different computer science and mathematics areas, especially from machine learning and statistics. Moreover, many approaches have been developed to answer the algorithm selection problem in various fields, as described in the survey of K. Smith-Miles [33]. This present book focuses on the restriction of this general question to the fields of constraint satisfaction, constraint solving, and optimization problems.

Of course, this quest appears also related to the No Free Lunch theorems for optimization [34], which argue against the existence of a universal multipurpose solver. This is precisely the challenge to more autonomous and adaptive solving algorithms in opening up to Free Lunch theorems for large classes of problems, as recently suggested by R. Poli and M. Graff [30].

In the context of optimization problems, as we will see later, several terminologies are used to characterize those systems that include specific techniques in order to provide more intelligent solving facilities to the user: adaptive algorithms, Reactive Search, portfolio algorithms. We propose here unifying these approaches under the generic term Autonomous Search that reflects the ultimate goal of providing an easy-to-use interface for end users, who could provide a possibly informal description of their problem and obtain solutions with minimum interaction and technical knowledge. Such an ideal system is depicted in Figure 1.1.

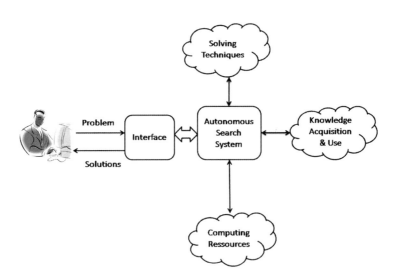

Fig. 1.1: Towards Autonomous Solvers

An Autonomous Search system should have the ability to advantageously modify its internal components when exposed to changing external forces and opportunities. It corresponds to a particular case of adaptive systems, with the objective of improving its problem-solving performance by adapting its search strategy to the problem at hand. Internal components correspond to the various algorithms involved in the search process: heuristics, inference mechanisms, etc. External forces correspond to the evolving information collected during this search process: search landscape analysis (quality, diversity, entropy, etc), external knowledge (prediction models, rules, etc) and so on. This information can be either directly extracted from the problem or indirectly computed through the perceived efficiency of the algorithm's components. Examples of collected information include the size of the search space (exact or estimated) and the number of subproblems. Computed information includes the discriminating degree of heuristics, the pruning capacity of inference techniques, and so on. Information can also refer to the computational environment, which can often vary, e.g., number of CPU cores.

Many works are scattered over various research areas and we see benefits in providing an overview of this new trend in problem solving. Nevertheless, Autonomous Search is particularly relevant to the constraint programming community, where many works have been conducted to improve the efficiency of constraint solvers and, more generally, of solving methods for handling combinatorial optimization problems. These improvements often rely on new heuristics, parameters or hybridizations of solving techniques, and, therefore, solvers are becoming more and more difficult to design and manage. They may involve *off-line* and *online* processes such as *tuning* (i.e., adjustment of parameters and heuristics before solving) and *control*. Note that previous taxonomies and studies have already been proposed for specific families of solving techniques and for specific solving purposes [13, 12, 14, 16, 8].

From a general point of view, directly related to the algorithm selection problem, a first idea is to manage a portfolio of algorithms and to use learning techniques in order to build prediction models for choosing the suitable techniques according to a set of features of the incoming instance to solve. This approach has been used successfully for solving various families of problems [21, 35, 16].

This general problem of finding the best configuration in a search space of heuristic algorithms is also related to the generic notion of Hyper-heuristics [7, 6, 10]. Hyper-heuristics are methods that aim at automating the process of selecting, combining, generating, or adapting several simpler heuristics (or components of such heuristics) to efficiently solve computational search problems. Hyper-heuristics are also defined as "heuristics to choose heuristics" [9] or "heuristics to generate heuristics" [1]. This idea was pioneered in the early 1960's with the combination of scheduling rules [15, 11]. Hyper-heuristics that manage a set of given basic search heuristics by means of search strategies or other parameters have been widely used for solving combinatorial problems (see Burke et al. [7] for a recent survey).

Nevertheless, the algorithm can be huge and one often has to fix the basic solving principles and act over a set of parameters. When considering parameter setting, the space of possible algorithms is the set of possible configurations of a given algorithmic scheme induced by the possible values of its parameters that control its computational behavior. Parameter tuning of evolutionary algorithms has been investigated for many years (we refer the reader to the book [25] for a recent survey). Adaptive control strategies were also proposed for other solving approaches such as local search [19, 29]. Powerful off-line techniques are now available for tuning parameters [22, 20, 27, 26].

Another important research community that focuses on very related problems has been established under the name *Reactive Search* by R. Battiti et al. [5, 3]. After focusing on local search with the seminal works on reactive tabu [4] and adaptive simulated annealing [23], this community is now growing through the dedicated conference LION [2].

It clearly appears that all these approaches share common principles and purposes and have been developed in different but connected communities. Their foundations rely on the fact that, since the solving techniques and search heuristics are more and more sophisticated and the problems structures more and more intricate, the choice and the correct setting of a solving algorithm is becoming an intractable task for most users. Therefore, there is a rising need for an alternative problem-solving framework.

In 2007, we organized the first workshop on Autonomous Search in Providence, RI, USA, in order to present relevant works aimed at building more intelligent solvers for the constraint programming community [17]. We tried to describe more conceptually the concept of Autonomous Search from the previously described related works [18]. The purpose of this book is to provide a clear overview of recent advances in autonomous tools for optimization and constraint satisfaction problems. In order to be as exhaustive as possible, keeping the focus on constrained problem solving, this book includes ten chapters that cover different solving techniques from metaheuristics to tree-based search and that illustrate how these solving techniques may benefit from intelligent tools by improving their efficiency and adaptability to solve the problems at hand.

1.2 What Is an Autonomous Solver?

Before presenting the chapters, we would like to quickly review some general principles and definitions that have guided us in the organization of this book. Therefore, we first describe the architecture of a solver, trying to be general enough to cover various solving techniques and paradigms. From this architecture, we will review different points of view concerning the different ways in which different features can be included and used to attain autonomy.

1.2.1 Architecture of the Solver

Optimization or constraint satisfaction problems may involve variables that take their values over various domains (integers, real numbers, Boolean, etc). In fact, solving such problems is the main interest of different but complementary communities in computer science: operations research, global optimization, mathematical programming, constraint programming, artificial intelligence, and so on. Among the different underlying paradigms that are associated with these research areas, we may try to identify common principles shared by the resulting solving algorithms and techniques that can be used for the ultimate solving purpose. Solvers could be viewed as a general skeleton whose components are selected according to the problem or class of problems to be solved. On the one hand, one has to choose the components of the solver and on the other hand one has to configure how these internal components are used during the solving process. Of course, a solving algorithm is designed according to the internal model, which defines the search space and uses a function to evaluate the elements of the search space. All these components can be subjected to various parameters that define their behavior. A given parametrization defines thus what we call a configuration of the solver. At this level, a control layer can be introduced, especially in an autonomous solver, to manage the previous components and modify the configuration of the solver during the solving process. The general solver architecture is illustrated by Figure 1.2.

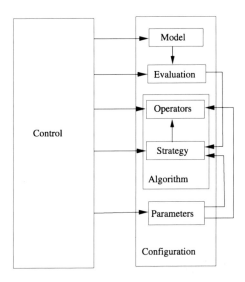

Fig. 1.2: The general architecture of a solver

1.2.2 Configuration of the Solver

In evolutionary computing, parameter setting [25] constitutes a major issue and, in the taxonomy proposed by Eiben et al. [13], methods are classified depending on whether they attempt to set parameters before the run (tuning) or during the run (control). The goal of parameter tuning is to obtain parameter values that could be useful over a wide range of instances. Such results require a large number of experimental evaluations and are generally based on empirical observations. Parameter control is divided into three branches according to the degree of autonomy of the strategies. Control is deterministic when parameters are changed according to a previously established schedule, adaptive when parameters are modified according to rules that take into account the state of the search, and self-adaptive when parameters are encoded into individuals in order to evolve conjointly with the other variables of the problem. In [32], Eiben and Smit discuss the differences between numeric and symbolic parameters. In [28], symbolic parameters are called components, and the operators of an evolutionary algorithm are configurable components of the solver that implement solving techniques.

In [3], Reactive Search is characterized by the integration of machine learning techniques into search heuristics. A classification of the source of information that is used by the algorithm is proposed for distinguishing between problem-dependent information, task-dependent information, and local properties.

In their survey [8], Burke et al. propose a classification of hyper-heuristics defined as "search methods or learning mechanisms for selecting or generating heuristics to solve computational search problems". As already mentioned, this classification also distinguishes between two dimensions: the different sources of feedback information and the nature of the heuristics' search space. The feedback, when used, corresponds here to the information that is learned while solving (on-line) or while using a set of training instances (off-line). The authors identify two families of low-level heuristics: construction heuristics (used to incrementally build a solution) and perturbation heuristics (used to iteratively improve a starting solution). The hyper-heuristics level can use heuristics selection methodologies that produce combinations of preexisting low-level heuristics, or heuristics generation methodologies that generate new heuristics from basic blocks of low level heuristics.

Another interesting classification is proposed in [16], in which Gagliolo et al. are interested in the algorithm selection problem and describe the different selection techniques according to the following points of views. Different dimensions are identified with respect to this algorithm selection problem (e.g., the generality of the selection process and its reactivity).

From these considerations, we may now provide a general overview of the chapters of this book with respect to the different aspects of Autonomous Search that they handle.

1.3 Outline and Overview of the Book

According to the existing taxonomies and different points of view that have been described above, it appears quite natural to classify the chapters of this book with respect to how they support autonomy from the user's point of view: off-line configuration of a solver or on-line control of a solver. The book also investigates future directions for Autonomous Search systems by considering new solving paradigms and application domains.

1.3.1 Off-line Configuration

Focusing first on the tuning of the parameters of a given solver, Chapter 2 addresses the problem of parameter tuning for evolutionary algorithms, which is clearly an important issue for the algorithms that most of the time involve many parameters whose impact is difficult to forecast. The chapter aims for a methodological approach to tuning and reviews automated tuning algorithms.

In a similar and complementary context, Chapter 3 provides a review of more automated configuration tools for solvers, focusing on local search-based tuning algorithms that explore the space of parameters. It also reviews other parameter tuning methods such as racing techniques.

As mentioned above, another way to address the algorithm selection problem is to consider a portfolio of solvers and to provide an intelligent oracle to select them with regard to the problems to solve at hand. In this context, Chapter 4 proposes an original use of learning processes in case-based reasoning applied to constraint programming and describes CPHYDRA, a case-based portfolio constraint solver. It illustrates how the user can take advantage of previously solved problems to solve similar cases. A general learning and reuse model is proposed to automate this whole process.

Turning to a more hyper-heuristics point of view, Chapter 5 describes, in the context of a complete tree based constraint solver, how combinations of low-level search heuristics can be learned from examples. The choice of the most suitable heuristics (i.e., variable/value choices) is a well-known difficult problem in constraint programming. In this chapter, a general scheme is proposed for designing an autonomous learner that is able to monitor its own performance and generate training examples to improve its quality. This chapter describes the Adaptive Constraint Engine, ACE, which constitutes a complete project that aims at allowing the user to benefit from an automated configuration of the basic branching heuristics of a complete constraint solver before running it on the instances she wants to solve efficiently.

1.3.2 On-line Control

Chapter 6 is a survey of Reactive Search, an important research trend mentioned above, and describes how reinforcement learning can be efficiently included in stochastic local search algorithms to improve their performance. This active community has developed reactive and adaptive search procedures in which, for instance, prohibition mechanisms or noise rates are dynamically adjusted along the search.

In a way complementary that of Chapter 6, Chapter 7 presents how learning mechanisms can be used in evolutionary algorithms to automatically select the suitable operators during the search process. This chapter also explores how operators can be generated on the fly according to the search needs. Therefore, the proposed techniques allow the user to handle two types of parameters: numerical parameters that adjust the behaviour of the solver and the solving components themselves, which can be modified along the search process.

In Chapter 8, the adaptive control of parameters in ant colony optimization, a broadly used solving paradigm for constraint optimization problems, is reviewed. This chapter presents a full empirical study of parameter setting for ACO and proposes an adaptation scheme for improving performance by controlling dynamic variations of the parameters.

1.3.3 New Directions and Applications

Following these reviews of autonomous online control of search processes in various solving paradigms, Chapter 9 considers new directions for building modern constraint solvers. Continuous search is thus a new concept that consists not only solving problems at hand using learned knowledge but also continuously improving the learning model.

Thanks to the generalization of multi-core processors, distributed solvers will certainly be an important issue in the next few years. Chapter 10 describes an adaptive solving policy for a distributed SAT solver, based on conflict-driven clause learning, the key technique of any modern SAT solver. This chapter also demonstrates that online control techniques can be very efficient and competitive for solving difficult combinatorial problems.

Last but not least, exploring another application domain, Chapter 11 illustrates how an autonomous solving process can be fully developed for a general class of problems by proposing learning techniques for planning, clearly an important research area in artificial intelligence. This chapter investigates how on-line and off-line techniques can be effectively used for learning heuristics and solving general-purpose planning problems.

1.4 Guideline for Readers

In this book, our purpose is to provide a wide panel of representative works in the scope of Autonomous Search. Therefore, different readers will be interested in focusing on diffrent areas, according to their scientific interests or main goals. To this aim, we propose hereafter some typical reading paths:

- Tuning parameters of an algorithm: Chapters 2 and 3.
- Autonomous complete (tree-based) constraint solvers: Chapter 4, 5, 9, and 10.
- Control in Metaheuristics: Chapters 6, 7, and 8.
- Managing solving heuristics: Chapters 4, 5, and 11.

Of course, the chapters are independent and can be read in any order, with many other possible combinations of interest.

References

[1] Bader-El-Den, M., Poli, R.: Generating SAT local-search heuristics using a GP hyper-heuristic Framework. In: 8th International Conference, Evolution Artificielle, EA 2007. Revised Selected Papers, Springer, no. 4926 in Lecture Notes in Computer Science, pp, 37–49 (2008)

[2] Battiti, R., Brunato, M. (eds), Learning and Intelligent Optimization second international conference, LION 2007 II. Selected Papers, Lecture Notes in Computer Science, vol., 5313. Springer (2008)

[3] Battiti, R., Brunato, M.: *Handbook of metaheuristics* (2nd edition), Gendreau, M., Potvin, J. Y. (eds), Springer, chap, Reactive search optimization: learning while optimizing (2009)

[4] Battiti, R., Tecchiolli, G.: The reactive tabu search. INFORMS Journal on Computing 6(2):126–140 (1994)

[5] Battiti, R., Brunato, M., Mascia, F.: Reactive search and intelligent optimization, Operations research/Computer Science Interfaces, vol., 45. Springer (2008)

[6] Burke, E., Kendall, G., Newall, J., Hart, E., Ross, P., Schulenburg, S.: *Handbook of Meta-heuristics*, Kluwer, chap., Hyper-heuristics: an emerging direction in modern search technology, pp, 457–474 (2003)

[7] Burke, E., Hyde, M., Kendall, G., Ochoa, G., Ozcan, E., Qu, R. A.: Survey of hyper-heuristics. Tech. Rep. Technical Report No. NOTTCS-TR-SUB-0906241418-2747, School of Computer Science and Information Technology, University of Nottingham, Computer Science (2009)

[8] Burke, E., Hyde, M., Kendall, G., Ochoa, G., Ozcan, E., Woodward, J.: *Handbook of metaheuristics* (2nd edition), Gendreau, M., Potvin, J. Y. (eds), Springer, chap., A classification of hyper-heuristics approaches (2009)

[9] Cowling, P., Soubeiga, E.: Neighborhood structures for personnel scheduling: a summit meeting scheduling problem (abstract). In: Burke, E., Erben, W. (eds), Proceedings of the 3rd International Conference on the Practice and Theory of Automated Timetabling (2000)

[10] Cowling, P., Kendall, G., Soubeiga, E.: Hyperheuristics: a tool for rapid prototyping in scheduling and optimisation. In: Applications of Evolutionary Computing, EvoWorkshops 2002: EvoCOP, EvoIASP, EvoSTIM/EvoPLAN, Springer, Lecture Notes in Computer Science, vol., 2279, pp, 1–10 (2002)

[11] Crowston, W., Glover, F., Thompson, G., Trawick, J.: Probabilistic and para-
     metric learning combinations of local job shop scheduling rules. Tech. rep.,
     ONR Research Memorandum No. 117, GSIA, Carnegie-Mellon University,
     Pittsburg, PA (1963)
[12] De Jong, K.: Parameter setting in EAs: a 30 year perspective. In: [25], pp, 1–18
     (2007)
[13] Eiben, A. E., Hinterding, R., Michalewicz, Z.: Parameter control in evolution-
     ary algorithms. IEEE Trans Evolutionary Computation 3(2):124–141 (1999)
[14] Eiben, A. E., Michalewicz, Z., Schoenauer, M., Smith, J. E.: Parameter control
     in evolutionary algorithms. In: [25], pp, 19–46 (2007)
[15] Fisher, H., Thompson, L.: *Industrial Scheduling*, Prentice Hall, chap., Proba-
     bilistic learning combinations of local job-shop scheduling rules (1963)
[16] Gagliolo, M., Schmidhuber, J.: Algorithm selection as a bandit problem with
     unbounded losses. Tech. rep., Tech. report IDSIA - 07 - 08 (2008)
[17] Hamadi, Y., Monfroy, E., Saubion, F.: Special issue on Autonomous
     Search. Constraint Programming Letters 4, URL http://www.
     constraint-programming-letters.org/ (2008)
[18] Hamadi, Y., Monfroy, E., Saubion, F.: *Hybrid Optimization: The Ten Years of
     CPAIOR*, Springer, chap., What is autonomous search? (2010)
[19] Hoos, H.: An adaptive noise mechanism for WalkSAT. In: AAAI/IAAI, pp,
     655–660 (2002)
[20] Hutter, F.: Automating the configuration of algorithms for solving hard compu-
     tational problems. Ph.D. thesis, Department of Computer Science, University
     of British Columbia (2009)
[21] Hutter, F., Hamadi, Y., Hoos, H., Brown, K. L.: Performance prediction and
     automated tuning of randomized and parametric algorithms. In: Twelfth In-
     ternational Conference on Principles and Practice of Constraint Programming
     CP'06 (2006)
[22] Hutter, F., Hoos, H., Stützle, T.: Automatic algorithm configuration based on
     local search. In: Proc. of the Twenty-Second Conference on Artifical Intelli-
     gence (AAAI '07), pp, 1152–1157 (2007)
[23] Ingber, L.: Very fast simulated re-annealing. Mathematical Computer Mod-
     elling 12(8):967–973 (1989)
[24] Laurière, J. L.: *Intelligence Artificielle, Résolution de problèmes par l'Homme
     et la machine*. Eyrolles (1986)
[25] Lobo, F., Lima, C., Michalewicz, Z. (eds), Parameter setting in evolutionary
     algorithms, Studies in Computational Intelligence, vol., 54. Springer (2007)
[26] Nannen, V., Eiben, A. E.: A method for parameter calibration and relevance
     estimation in evolutionary algorithms. In: Genetic and Evolutionary Compu-
     tation Conference, GECCO 2006, Proceedings, ACM, pp, 183–190 (2006)
[27] Nannen, V., Eiben, A. E.: Relevance estimation and value calibration of evo-
     lutionary algorithm parameters. In: IJCAI 2007, Proceedings of the 20th Inter-
     national Joint Conference on Artificial Intelligence, pp, 975–980 (2007)
[28] Nannen, V., Smit, S., Eiben, A. E.: Costs and benefits of tuning parameters of
     evolutionary algorithms. In: Parallel Problem Solving from Nature, PPSN X,

10th International Conference, Springer, Lecture Notes in Computer Science, vol., 5199, pp, 528–538 (2008)

[29] Patterson, D., Kautz, H.: Auto-Walksat: a self-tuning implementation of walksat. Electronic Notes in Discrete Mathematics 9:360–368 (2001)

[30] Poli, R., Graff, M.: There is a free lunch for hyper-heuristics, genetic programming and computer scientists. In: EuroGP, Springer, Lecture Notes in Computer Science, vol., 5481, pp, 195–207 (2009)

[31] Rice, J.: The algorithm selection problem. Tech. Rep. CSD-TR 152, Computer Science Department, Purdue University (1975)

[32] Smit, S., Eiben, A. E.: Comparing parameter tuning methods for evolutionary algorithms. In: Proceedings of the 2009 IEEE Congress on Evolutionary Computation (CEC 2009), IEEE, pp, 399–406 (2009)

[33] Smith-Miles, K.: Cross-disciplinary perspectives on meta-learning for algorithm selection. ACM Computing Surveys 41(1):1–25 (2008)

[34] Wolpert, D. H., Macready, W. G.: No free lunch theorems for optimization. IEEE Transactions on Evolutionary Computation 1(1):67–82 (1997)

[35] Xu, L., Hutter, F., Hoos, H., Leyton-Brown, K.: SATzilla: Portfolio-based algorithm selection for SAT. Journal of Artificial Intelligence Research 32:565–606 (2008)

Part I
Off-line Configuration

Chapter 2
Evolutionary Algorithm Parameters and Methods to Tune Them

A. E. Eiben and S. K. Smit

2.1 Background and Objectives

Finding appropriate parameter values for evolutionary algorithms (EA) is one of the persisting grand challenges of the evolutionary computing (EC) field. In general, EC researchers and practitioners all acknowledge that good parameter values are essential for good EA performance. However, very little effort is spent on studying the effect of EA parameters on EA performance and on tuning them. In practice, parameter values are mostly selected by conventions (mutation rate should be low), ad hoc choices (why not use uniform crossover?), and experimental comparisons on a limited scale (testing combinations of three different crossover rates and three different mutation rates). Hence, there is a striking gap between the widely acknowledged importance of good parameter values and the widely exhibited ignorance concerning principled approaches to tuning EA parameters.

To this end, it is important to recall that the problem of setting EA parameters is commonly divided into two cases, parameter tuning and parameter control [14]. In case of parameter control, the parameter values are changing during an EA run. In this case one needs initial parameter values and suitable control strategies, which run;in turn can be deterministic, adaptive, or self-adaptive. Parameter tuning is easier in the sense that the parameter values are not changing during a run, hence only a single value per parameter is required. Nevertheless, even the problem of tuning an EA for a given application is hard because of the large number of options and the limited knowledge about the effect of EA parameters on EA performance.

Given this background, we can regard the primary focus of this chapter as being parameter tuning. Our main message is that the technical conditions for choosing good parameter values are given and the technology is easily available. As it

A. E. Eiben
Vrije Universiteit Amsterdam, e-mail: gusz@cs.vu.nl

S. K. Smit
Vrije Universiteit Amsterdam, e-mail: sksmit@cs.vu.nl

Y. Hamadi et al. (eds.), *Autonomous Search*,
DOI 10.1007/978-3-642-21434-9_2,
© Springer-Verlag Berlin Heidelberg 2011

happens, there exist various algorithms that can be used for tuning EA parameters. Using such tuning algorithms (tuners, for short, in the sequel) implies significant performance improvements and the computational costs are moderate. Hence, changing the present practice and using tuning algorithms widely would lead to improvements on a massive scale: large performance gains for a large group of researchers and practitioners.

The overall aim of this chapter is to offer a thorough treatment of EA parameters and algorithms to tune them. This aim can be broken down into a number of technical objectives:

1. To discuss the notion of EA parameters and its relationship with the concepts of EAs and EA instances.
2. To consider the most important aspects of the parameter tuning problem.
3. To give an overview of existing parameter tuning methods.
4. To elaborate on the methodological issues involved here and provide recommendations for further research.

2.2 Evolutionary Algorithms, Parameters, Algorithm Instances

Evolutionary algorithms form a class of heuristic search methods based on a particular algorithmic framework whose main components are the variation operators (mutation and recombination – a.k.a. crossover) and the selection operators (parent selection and survivor selection); cf. [17]. The general evolutionary algorithm framework is shown in Figure 2.1.

A decision to use an evolutionary algorithm to solve some problem implies that the user or algorithm designer adopts the main design decisions that led to the general evolutionary algorithm framework and only needs to specify "a few" details. In the sequel we use the term *parameters* to denote these details.

Using this terminology, designing an EA for a given application amounts to selecting good values for the parameters. For instance, the definition of an EA might include setting the parameter crossoveroperator to onepoint, the parameter crossoverrate to 0.5, and the parameter populationsize to 100. In principle, this is a sound naming convention, but intuitively there is a difference between choosing a good crossover operator from a given list of three operators and choosing a good value for the related crossover rate $p_c \in [0, 1]$. This difference can be formalized if we distinguish between parameters by their domains. The parameter crossoveroperator has a finite domain with no sensible distance metric or ordering, e.g., {onepoint, uniform, averaging}, whereas the domain of the parameter p_c is a subset of $\mathbb{R}$ with the natural structure of real numbers. This difference is essential for searchability. For parameters with a domain that has a distance metric, or is at least partially ordered, one can use heuristic search and optimization methods to find optimal values. For the first type of parameter this is not possible because the domain has no exploitable structure. The only option in this case is sampling.

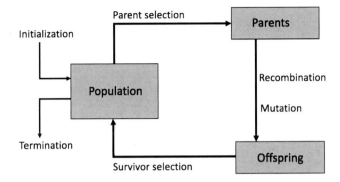

Fig. 2.1: General framework of an evolutionary algorithm.

Parameter with an unordered domain	Parameter with an ordered domain
qualitative	quantitative
symbolic	numeric
categorical	numerical
structural	behavioral
component	parameter
nominal	ordinal
categorical	ordered

Table 2.1: Pairs of terms used in the literature to distinguish between two types of parameters (variables)

The difference between the two types of parameters has already been noted in evolutionary computing, but different authors use different naming conventions. For instance, [5] uses the terms *qualitative* and *quantitative* parameters, [37] distinguishes between *symbolic* and *numeric* parameters, [10] calls them *categorical* and *numerical*, while [23] refers to *structural* and *behavioral* parameters. [30] calls unstructured parameters *components* and the elements of their domains, operators, and a parameter is instantiated by a value while a component is instantiated by allocating an operator to it. In the context of statistics and data mining one distinguishes between two types of variables (rather than parameters) depending on the presence of an ordered structure, but a universal terminology is lacking here too. Commonly used names are *nominal* vs. *ordinal* and *categorical* vs. *ordered* variables. Table 2.1 provides a quick overview of these options in general; Table 2.2 offers an EA-specific illustration with commonly used parameters in both categories.

From now on we will use the terms *qualitative parameter* and *quantitative parameter*. For both types of parameters the elements of the parameter's domain are called *parameter values* and we *instantiate* a parameter by allocating a value to it. In practice, quantitative parameters are mostly numerical values, e.g., the parame-

	EA_1	EA_2	EA_3
qualitative parameters			
Representation	bitstring	bitstring	real-valued
Recombination	1-point	1-point	averaging
Mutation	bit-flip	bit-flip	Gaussian $N(0,\sigma)$
Parent selection	tournament	tournament	uniform random
Survivor selection	generational	generational	(μ,λ)
quantitative parameters			
p_m	0.01	0.1	0.05
σ	n.a.	n.a	0.1
p_c	0.5	0.7	0.7
μ	100	100	10
λ	n.a.	n.a	70
tournament size	2	4	n.a.

Table 2.2: Three EA instances specified by the qualitative parameters representation, recombination, mutation, parent selection, survivor selection, and the quantitative parameters mutation rate (p_m), mutation step size (σ), crossover rate (p_c), population size (μ), offspring size (λ), and tournament size. In our terminology, the EA instances in columns EA_1 and EA_2 are just variants of the same EA. The EA instance in column EA_3 belongs to a different EA

ter crossover rate uses values from the interval $[0,1]$, and qualitative parameters are often symbolic, e.g., crossoveroperator. However, in general, quantitative parameters and numerical parameters are not the same, because it is possible to have an ordering on a set of symbolic values. For instance, one could impose an alphabetical ordering on the set of colors $\{blue, green, red, yellow\}$.

It is important to note that the number of parameters of EAs is not specified in general. Depending on design choices one might obtain different numbers of parameters. For instance, instantiating the qualitative parameter parentselection by tournament implies a new quantitative parameter tournamentsize. However, choosing roulettewheel does not add any parameters. This example also shows that there can be a hierarchy among parameters. Namely, qualitative parameters may have quantitative parameters "under them". For an unambiguous treatment we can call such parameters *sub-parameters*, always belonging to a qualitative parameter.

Distinguishing between qualitative and quantitative parameters naturally leads to distinguishing between two levels when designing a specific EA for a given problem. In the sequel, we perceive qualitative parameters as high-level ones that define an evolutionary algorithm, and look at quantitative parameters as low-level ones that define a variant of this EA. Table 2.2 illustrates this.

Following this naming convention an evolutionary algorithm is a partially specified algorithm where the values to instantiate qualitative parameters are defined, but the quantitative parameters are not. Hence, we consider two EAs to be different if they differ in one of their qualitative parameters, e.g., use different mutation operators. If the values for all parameters are specified, we obtain *an EA instance*. This terminology enables precise formulations while it requires care in phrasing.

Clearly, this distinction between EAs and EA instances is similar to distinguishing between problems and problem instances. For example, the abbreviation TSP represents the set of all possible problem configurations of the traveling salesman problem, whereas a TSP instance is one specific problem, e.g., ten specific cities with a given distance matrix D. If rigorous terminology is required then the right phrasing is "to apply an EA instance to a problem instance". However, such rigor is not always required, and formally inaccurate but understandable phrases like "to apply an EA to a problem" are fully acceptable in practice.

2.3 Algorithm Design and Parameter Tuning

In the broad sense, algorithm design includes all decisions needed to specify an algorithm (instance) for solving a given problem (instance). Throughout this chapter we treat parameter tuning as a special case of algorithm design. The principal challenge for evolutionary algorithm designers is that the design details, i.e., parameter values, largely influence the performance of the algorithm. An EA with good parameter values can be orders of magnitude better than one with poorly chosen parameter values. Hence, algorithm design in general, and EA design in particular, is an optimization problem.

To obtain a detailed view on this issue we distinguish between three layers: application layer, algorithm layer, and design layer; see Figure 2.2. As this figure indicates, the whole scheme can be divided into two optimization problems. The lower part of this three-tier hierarchy consists of a problem (e.g., the traveling salesman problem)on the application layer and an EA (e.g., a genetic algorithm) on the algorithm layer to find an optimal solution for the problem. Simply put, the EA is iteratively generating candidate solutions (e.g., permutations of city names), seeking one with maximal quality. The upper part of the hierarchy contains a tuning method to find optimal parameter values for the EA on the algorithm layer. Similarly to the lower part, the tuning method is iteratively generating parameter vectors seeking one with maximal quality, where the quality of a given parameter vector is based on the performance of the EA using its values. To avoid confusion we use distinct terms to designate the quality function of these optimization problems. In line with the usual EC terminology we use the term fitness for the quality of candidate solutions on the lower level, and the term *utility* to denote the quality of parameter vectors. Table 2.3 provides a quick overview of the related vocabulary.

With this nomenclature we can formalize the problem to be solved by the algorithm designer if we denote the qualitative parameters and their domains by $q_1, \ldots, q_m$ and $Q_1, \ldots, Q_m$, and likewise use the notations $r_1, \ldots, r_n$ and $R_1, \ldots, R_n$ for the quantitative parameters.[1] The problem of parameter tuning can then be seen as a search problem $\langle S, u \rangle$ in the parameter space

---

[1] Observe that due to the possible presence of sub-parameters the number of quantitative parameters n depends on the instantiations of $q_1, \ldots q_m$. This makes the notation somewhat inaccurate, but we use it for sake of simplicity.

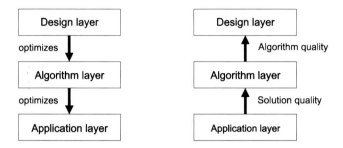

Fig. 2.2: The left diagram depicts the control flow through the three main layers in the hierarchy of parameter tuning, while the right diagram shows the information flow

	Problem solving	Parameter tuning
Method at work	evolutionary algorithm	tuning procedure
Search space	solution vectors	parameter vectors
Quality	fitness	utility
Assessment	evaluation	testing

Table 2.3: Vocabulary distinguishing the main entities in the context of problem solving (lower two blocks in Figure 2.2) and algorithm design, and parameter tuning (upper two blocks in Figure 2.2).

$$S = Q_1 \times \cdots \times Q_m \times R_1 \times \cdots \times R_n \qquad (2.1)$$

using a utility function u, where the utility $u(\bar{p})$ of a given parameter vector $\bar{p} \in S$ reflects the performance of $EA(\bar{p})$, i.e., the evolutionary algorithm instance using the values of $\bar{p}$, on the given problem instance(s). Solutions of the parameter tuning problem can then be defined as parameter vectors with maximum utility. Given this conceptual framework it is easy to distinguish between the so-called "structural" and "parametric" tunings [9] in a formal way: structural tuning takes place in the space $S = Q_1 \times \cdots \times Q_m$, while parametric tuning refers to searching through $S = R_1 \times \cdots \times R_n$.

Now we can define the *utility landscape* as an abstract landscape where the locations are the parameter vectors of an EA and the height reflects utility. It is obvious that fitness landscapes - commonly used in EC - have a lot in common with utility landscapes as introduced here. To be specific, in both cases we have a search space (candidate solutions vs. parameter vectors), a quality measure (fitness vs. utility) that is conceptualized as 'height', and a method to assess the quality of a point in the search space (fitness evaluation vs. utility testing). Finally, we have a search method (an evolutionary algorithm vs. a tuning procedure) that is seeking a point with maximum height.

Despite the obvious analogies between the upper and the lower halves of Figure 2.2, there are some differences we want to note here. First of all, fitness values are most often deterministic – depending, of course, on the problem instance to be solved. However, the utility values are always stochastic, because they reflect the performance of an EA, which is a stochastic search method. The inherently stochastic nature of utility values implies particular algorithmic and methodological challenges that will be discussed later. Second, the notion of fitness is usually strongly related to the objective function of the problem on the application layer and differences between suitable fitness functions mostly concern arithmetic details. The notion of utility, however, depends on the problem instance(s) to be solved and the performance metrics used to define how good an EA is. In the next secion we will have a closer look at these aspects.

2.4 Utility, Algorithm Performance, Test Functions

As stated above, solutions of the parameter tuning problem are parameter vectors with maximum utility, where utility is based on some definition of EA performance and some objective functions or problem instances.

In general, there are two atomic performance measures for EAs: one regarding solution quality and one regarding algorithm speed or search effort. The latter is interpreted here in the broad sense, as referring to any sensible measure of computational effort for evolutionary algorithms, e.g., the number of fitness evaluations, CPU time, and wall clock time.[2] There are different combinations of fitness and time that can be used to define algorithm performance in one single run. For instance:

- Given a maximum running time (computational effort), algorithm performance is defined as the best fitness at termination.
- Given a minimum fitness level, algorithm performance is defined as the running time (computational effort) needed to reach it.
- Given a maximum running time (computational effort) and a minimum fitness level, algorithm performance is defined through the Boolean notion of success: a run succeeds if the given fitness is reached within the given time; otherwise it fails.

Obviously, by the stochastic nature of EAs multiple runs on the same problem are necessary to get a good estimate of performance. Aggregating the measures mentioned above over a number of runs we obtain the performance metrics commonly used in evolutionary computing, (cf. [17, Chapter 14]):

- MBF (mean best fitness),
- AES (average number of evaluations to solution),
- SR (success rate),

respectively. When designing a good EA, one may tune it on either of these performance measures, or a combination of them.

Further to the given performance metrics, utility is also determined by the problem instance(s) or objective functions for which the EA is being tuned. In the sim-

---

[2] Please refer to [15] for more details on measuring EA search effort.

plest case, we are tuning our EA on one function f. Then the utility of a parameter vector $\bar{p}$ is measured by the EA performance on f. In this case, tuning delivers a *specialist*, that is, an EA that is very good at solving one problem instance, the function f, with no claims or indications regarding its performance on other problem instances. This can be a satisfactory result if one is only interested in solving f. However, algorithm designers in general, and evolutionary computing experts in particular, are often interested in so-called robust EA instances, that is, in EA instances that work well on many objective functions. In terms of parameters, this requires "robust parameter settings". To this end, test suites consisting of many test functions are used to test and evaluate algorithms. For instance, a specific set $\{f_1, \ldots, f_n\}$ is used to support claims that the given algorithm is good on a "wide range of problems". Tuning an EA on a set of functions delivers a *generalist*, that is, an EA that is good at solving various problem instances. Obviously, a true generalist would perform well on all possible functions. However, this is impossible by the no-free-lunch theorem [36]. Therefore, the quest for generalist EAs is practically limited to less general claims that still raise serious methodology issues, as discussed in [15].

Technically, tuning an EA on a collection of functions $\{f_1, \ldots, f_n\}$ means that the utility is not a single number, but a vector of utilities corresponding to each of the test functions. Hence, finding a good generalist is a multi-objective problem for which each test function is one objective. The current parameter tuning algorithms can only deal with tuning problems for which the utilities can be compared directly; therefore a method is needed to aggregate the utility vectors into one scalar number. A straightforward solution to this can be obtained by averaging over the given test suite. For a precise definition we need to extend the notation of the utility function such that it shows the given objective function f. Then $u_f(\bar{p})$ reflects the performance of the evolutionary algorithm instance $EA(\bar{p})$ on f, and the utility of $\bar{p}$ on $F = \{f_1, \ldots, f_z\}$ can be defined as the average of the utilities $u_{f_1}(\bar{p}), \ldots, u_{f_z}(\bar{p})$:

$$u_F(\bar{p}) = \frac{1}{z} \cdot \sum_{i=1}^{z} u_{f_i}(\bar{p}).$$

Obviously, instead of a simple arithmetic average, one could use weighted averages or more advanced multi-objective aggregation mechanisms, for instance, based on lexicographic orderings.

Summarizing, the core of our terminology is as follows:
1. Solution vectors have fitness values, based on the objective function related to the given problem instance to be solved.
2. EA instances have performance values, based on information regarding fitness and running time on one or more problem instances (objective functions).
3. Parameter vectors have utility values, defined by the performance of the corresponding EA instance and the problem instances (objective functions) used for tuning.

Figure 2.3 illustrates the matter in a graphical form. It shows that the solutions of a tuning problem depend on the problem(s) to be solved, the EA used, and the utility

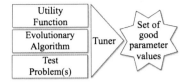

Fig. 2.3: Generic scheme of parameter tuning showing how good parameter values depend on four factors: the problem(s) to be solved, the EA used, the utility function, and the tuner itself

function. Adding the tuner to the equation we obtain this picture showing how a set of good parameter values obtained through tuning depends on four factors.

2.5 Algorithmic Approaches to Parameter Tuning

In this section we review algorithmic approaches to solving the parameter tuning problem. In essence, all of these methods work by the generate and test principle, i.e., by generating parameter vectors and testing them to establish their utility.

2.5.1 Iterative and Non-Iterative Tuners

The tuners we are to review in the following are variants of this generic scheme, which can be divided into two main categories:
1. *iterative* and
2. *non-iterative*.

All non-iterative tuners execute the GENERATE step only once, during initialization, thus creating a fixed set of vectors. Each of those vectors is then tested during the TEST phase to find the best vector in the given set. Hence, one could say that non-iterative tuners follow the INITIALIZE -and- TEST template. Initialization can be done by random sampling, generating a systematic grid, or some space-filling set of vectors.

The second category of tuners is formed by iterative methods that do not fix the set of vectors during initialization, but start with a small initial set and create new vectors iteratively during execution.

2.5.2 Single and Multistage Procedures

Given the stochastic nature of EAs, a number of tests is necessary for a reliable estimate of utility. Following [8] and [5], we distinguish between

1. *single-stage* and
2. *multistage procedures.*

Single-stage procedures perform the same number of tests for each given vector, while multistage procedures use a more sophisticated strategy. In general, they augment the TEST step with a SELECT step, where only promising vectors are selected for further testing and those with low performance are deliberately ignored.

As mentioned in Section 2.2, some methods are only applicable to quantitative parameters. Sophisticated tuners, such as SPOT [3], however, can be used for quantitative, qualitative or even mixed parameter spaces.

2.5.3 Measuring Search Effort

In the sequel, we present an overview of parameter tuning algorithms. The arranging principle we use is based on the search effort perspective. Obviously, good tuning algorithms try to find a good parameter vector with the least possible effort. In general, the total search effort can be expressed as $A \times B \times C$, where

1. A is the number of parameter vectors to be tested by the tuner.
2. B is the number of tests, e.g., EA runs, per parameter vector needed to establish its utility. The product $A \times B$ represents the total number of algorithm runs used by the tuners. Note, $B = 1$ for deterministic algorithms which optimize deterministic problems.
3. C is the number of function evaluations performed in one run of the EA. This is related to the estimate of the performance, because we are trying to estimate the performance of an algorithm based on a small number of function evaluations only, and not after the complete run is finished.

Based on this perspective, we summarize existing methods in four categories: tuners that try to allocate search efforts optimally by saving on A, B, and C, respectively, and those that try to allocate search efforts optimally by saving on A and B.

2.5.4 Classification of Tuning Methods

2.5.4.1 Using a Small Number of Parameter Vectors (A)

In this group we find tuners that are trying to allocate search efforts efficiently by cleverly generating parameter vectors. Strictly speaking, they might not always minimize A, but try to "optimize the spending", that is, get the most out of testing A parameter vectors. Such tuners are usually iterative methods, although there are exceptions, e.g., Latin square [25] and Taguchi orthogonal arrays [35]. The idea behind most tuners in this category is to start with a relatively small set of vectors and iteratively generate new sets in a clever way, i.e., such that new vectors are likely to be good. These tuners are only appropriate for quantitative parameters.

Finding a good set of parameter values for EAs is the kind of search problem that EAs are good at. It is therefore possible to use an EA (on the design layer) for optimizing the parameters of an EA (on the algorithm layer). The EA on the design layer is called a *meta-EA* and in general any kind of evolutionary algorithm, e.g., *genetic algorithms* (GAs), *evolution strategies* (ESs), *evolutionary programming* (EP), *differential evolution* (DE), *particle swarm optimization* (PSO), and *estimation of distribution algorithms* (EDAs), could be used for this purpose. However, existing literature only reports on using GAs, ESs, and EDAs as meta-EAs. We assume that readers of this chapter are familiar with the basics of evolutionary algorithms and need no introduction to GAs or ESs. Therefore, here we only discuss the generic template of meta-EA applications, which is the following. The individuals used in the meta-EA encode a parameter vector $\bar{p}$ of (quantitative) parameter values of the baseline EA. This encoding depends on the type of EA on the design layer, and may differ for a meta-GA, and a meta-ES, or meta-EDA. To evaluate the utility of such a vector $\bar{p}$, the baseline EA is run B times using the given parameter values in $\bar{p}$, thereby performing B tests. The utility of $\bar{p}$ is determined based on these runs/tests, and the actual definition of EA performance. Using this utility information, the meta-EA can perform selection, recombination, and mutation of parameter vectors in the usual way. As early as 1978, Mercer and Sampson [24] elaborated on this idea by their meta-GA (called meta-plan), but due to the large computational costs, their research was very limited. Greffenstette [18] was the first to conduct more extensive experiments with meta-EAs and showed the effectiveness of the approach. The use of meta-ES has been reported by Greffenstette [18] and Herdy [20] with mixed results.

The meta-GA and meta-ES approaches do not differ significantly; the GA or ES is used as an optimizer and any preference between them is solely based on their optimization power and/or speed. The approach based on ideas from EDAs is different in this respect. While EAs only try to find good points in a search space belonging to a certain problem, EDAs try to estimate the distribution of promising values over the search space. Consequently, they provide more information than GAs or an ESs, because the evolved distributions disclose a lot about the parameters, in particular about the utility landscape, and can be used to get insight into the sensitivity and relevance of the different parameters. Hence, the choice for using a meta-EDA can be motivated by the additional information it provides w.r.t. a meta-GA or a meta-ES. The *Relevance Estimation and VAlue Calibration* method (REVAC) offered by Nannen and Eiben is based on this idea. REVAC has been introduced in [29] and further polished in [27, 28] and [26]. In [30] and [34] it is demonstrated how REVAC can be used to indicate the costs and benefits of tuning each parameter –something a meta-GA or a meta-ES cannot do. In the process of searching for good parameter values REVAC implicitly creates probability distributions (one probability distribution per parameter) in such a way that parameter values that proved to be good in former trials have a higher probability than poor ones. Initially, all distributions represent a uniform random variable and after each new test they are updated based on the new information. After the termination of the tuning process, i.e., the stopping of REVAC, these distributions can be retrieved and analyzed, showing not only the range of promising parameter values, but also disclosing information about the relevance of each parameter. For a discussion of REVAC's probability distributions, entropy, and parameter relevance we refer you to [34].

2.5.4.2 Using a Small Number of Tests (B)

The methods in this group try to reduce B, i.e., the number of tests per parameter vector. The essential dilemma here is that fewer tests yield fewer reliable estimates of utility and if the number of tests is too low, then the utilities of two parameter vectors might not be statistically distinguishable. More tests can improve (sharpen) the stability of the expected utility to a level such that the superiority of one vector over another can be safely concluded, but more tests come with a price of longer runtime. The methods in this group use the same trick to deal with this problem: performing only a few tests per parameter vector initially and increasing this number to the minimum level that is enough to obtain statistically sound comparisons between the given parameter vectors. Such methods are known as *statistical screening, ranking, and selection*. Some of these methods are designed to select a single best solution, others are designed to screen a set of solutions by choosing a (random size) subset of the solutions containing the best one. Several of these approaches are based on multistage procedures such as sequential sampling [8, 11].

Maron et al. [22] introduced the term *racing* for a basic ranking and selection scheme and applied several methods, based on this scheme, to EAs. These non-iterative methods assume that a set $P = \{\bar{p}_1, \ldots, \bar{p}_A\}$ of parameter vectors is given and a maximum number of tests, B, per parameter vector is specified. Then they work according to the following procedure.

1. Test all $\bar{p} \in P$ once.
2. Determine $\bar{p}_{best}$ with highest (average) utility.
3. Determine P' as the set of parameters vectors whose utility is not significantly worse than that of $\bar{p}_{best}$.
4. $P = P'$.
5. If $size(P) > 1$ and the number of tests for each $\bar{p} \in P$ is smaller than B, goto 1.

The main difference between the existing racing methods can be found in step 3 as there are several statistical tests[3] that can indicate significant differences in utility:

- Analysis of Variance (ANOVA), cf. [32].
- Kruskal-Wallis (Rank Based Analysis of Variance).
- Hoeffding's bound, cf. [22].
- Unpaired Student T-Test, cf. [22].

All these methods are multistage procedures and they can be used for both quantitative and qualitative parameters; see also the discussion in [9]. Recently, Balaprakash et al. [1] introduced iterative racing procedures. Birattari et al. [10] present an overview of state-of-the-art racing procedures.

---

[3] Not to be confused with tests of parameter vectors; cf. Table 2.3.

2.5.4.3 Using a Small Number of Parameter Vectors and Tests (*A* and *B*)

There are currently four approaches that try to combine the main ideas from both categories. Yuan et al. [37] showed how *racing* [22] can be combined with meta-EAs in order to reduce both *A* and *B*. They propose two different approaches. In both methods, they use the standard meta-EA approach, but introduce a racing technique into the evaluation of parameter vectors.

In the first method, the quantitative (numerical) parameters are represented by a vector and are evolved using a meta-$(1+\lambda)$ ES. At each generation, a set of λ new vectors is created using a Gaussian distribution centered at the current best vector. Racing is then used to determine which vector has the highest utility. By using racing, not all of the λ individuals have to be evaluated *B* times to determine the current best vector. It is expected that many of the vectors are eliminated after just a few tests saving a large portion of computational resources.

Their second method uses a population of parameter vectors containing the quantitative (numerical) parameters, which is evolved using selection, recombination, and mutation. However, the utility of a vector of numerical parameters is evaluated not only within one single EA instance, but on all possible combinations of qualitative parameters. The utility of the vector is equal to the performance of the best combination. Racing is then used to determine which combination is most successfully using the vector of numerical parameters. By employing racing, not every combination has to be evaluated *B* times. This saves computational resources, while the search space of both quantitative and qualitative parameters can still be explored.

Introducing the *sequential parameter optimization toolbox* (SPOT), Bartz-Beielstein et al. [3] outlined another approach which uses a small number of parameter vectors *A* and tests *B* by means of an iterative procedure. SPOT starts with an initial set of parameter vectors usually laid out as a space filling design, e.g., *Latin hybercube design* (LHD). These are tested B_0 times to estimate their expected utility. Based on the results, a prediction model is fitted to represent an approximation of the utility landscape (additionally, the often used kriging component enables estimating the error at any point in the model). Then *l* new parameter vectors are generated and their expected utility is estimated using the model, i.e., without additional runs of the EA. This step employs predictions and error estimations within the expected improvement heuristic to find points with good potential to improve over the currently existing ones. The most promising points are then tested *B* times. If none of those points results in a better expected utility, *B* is increased. To establish fairness, the number of reevaluations of the current best and new design points is the same. As the estimates of the current vectors are sharpened, utility prediction gets more accurate over time and "lucky samples" from the start are dropped. Note that the set of parameter vectors employed after the initial step is usually very small, often in the range of one to four vectors, and is held constant in size.

In order to explore the search space relatively quickly, *B* is initialized to a small value. If a promising area is found, the method focuses on improving the model and the best vector using more accurate utility estimates by increasing *B*. Although in

[3] a kriging enhanced regression model is used to predict utilities, it is in principle a general framework suited for a large range of modeling techniques: see, e.g. [4].

Note that SPOT may also be combined with racing-like statistical techniques for keeping B low where possible. In [7], random permutation tests have been employed to decide if a newly tested configuration shall be repeated as often as the current best one or if it can safely be regarded as inferior. Lasarczyk [21] implemented Chen's *optimal computing budget allocation* (OCBA) [12] into SPOT.

The fourth approach that explicitly aims at using a small number of parameter vectors and tests is REVAC++, the most recent variant of REVAC. The name RE-VAC++ is motivated by the fact that the method can be simply described as REVAC + racing + sharpening, where racing and sharpening are separate methods that serve as possible add-ons to REVAC, or in fact any search-based tuner. This version of REVAC has been introduced in [33], where the advantageous effects of the racing and sharpening extensions have been also shown.

2.5.4.4 Using a Small Number of Function Evaluations (C)

Optimizing A and/or B are the most commonly used methods to optimally allocate search efforts, but in principle the number of fitness evaluations per EA run (C) could also be optimized. Each decrease in EA run-time is multiplied by both A and B; hence such runtime reduction methods can have a high impact on the total search effort. However, at the time of writing this chapter, we are not aware of any methods proposed for this purpose.

In theory, methods to reduce the number of evaluations per EA run could be based on the idea of terminating an EA run if there is an indication that the parameter vector used by the given EA will not reach some utility threshold, e.g., the average utility, or the utility of another, known parameter-vector such as the current best or current worst one. Such an indication can be statistically or theoretically grounded, but also based on heuristics. This concept is very similar to racing, but instead of decreasing the number of tests (B), it decreases the number of evaluations (C). In general, such a method could deal with both qualitative and quantitative parameters.

As mentioned above, we could not find any general parameter tuning methods of this type in the literature. However, we found one paper whose main concept would fit under the umbrella of saving on C whenever possible. This paper of Harik et al. is concerned with population sizes only, as all other EA parameters are fixed and only population sizes are varied and compared [19]. The essence of the method is to run multiple EAs in parallel, each with a different population size, compare the running EAs after a fixed number of fitness evaluations, and terminate those that use a population size with an expectedly poor performance (i.e., utility). The expectations regarding the utilities of different population sizes are based on a simple heuristic observing that small populations always have a head-start, because of the fast convergence to a (local) optimum. This fact motivates the assumption that once a large population EA overtakes a small population EA (using the same number of fitness evaluations), the small population EA will always remain behind. Hence, when an EA with a small population size starts performing worse than EAs with a large population size it can be terminated in an early stage, thus reducing search effort. The method in [19] can only use a fitness-based notion of utility, i.e., not speed of success rate information, and the heuristic is only applicable to population sizes. Further research is needed to develop heuristics for parameters regarding variation and selection operators.

2.6 Successful Case Studies on Tuning Evolutionary Algorithms

In recent years, the number of well-documented case studies on parameter tuning have been increasing. Depending on the type of tuning method, the focus varies from increasing algorithm performance to decreasing effort. Examples of the latter are the experiments of Clune et al., who tune a traditional GA on the counting ones and a four-bit deceptive trap problem [13]. Using the DAGA2 algorithm, they found parameter values that performed only slightly worse than the parameters found using extensive experimentation by experts. They concluded that 'In situations in which the human investment required to set up runs is more precious than performance, the meta-GA could be preferred'.

The experiments of Yuan and Gallagher in [37] are similar, but with a slightly different focus. They compared the tuning effort of their Racing-GA with the effort needed to test all possible combinations with parameter values, rather than manual labor. They tuned a range of algorithms to the 100-bit One-Max problem. They reported a 90% decrease in effort related to the brute force method with a similar precision.

Nannen et al. illustrate the relation between tuning costs and performance in [30]. In contrast to [13], tuning effort is not compared with human effort, but measured as the computational effort to reach a certain performance. The study reports on tuning a large collection of EAs on the Rastrigin problem. The results show that the amount of tuning needed to reach a given performance differs greatly per EA. Additionally, the best EA depends on the amount of tuning effort that is allowed. Another interesting outcome indicates that the tuning costs mainly depend on the overall setup of the Evolutionary Algorithm, rather than the number of free parameters.

There are two studies we know of reporting on a significant performance improvement over that of the 'common wisdom' parameter values found using an automated tuning algorithm. Bartz-Beielstein and Markon [6] showed how an Evolution Strategy and Simulated Annealing algorithm can be tuned using DACE and regression analysis on a real-world problem concerning elevator group control. They reported that standard parameterizations of search algorithms might be improved significantly. Smit and Eiben [34] compare three different tuning algorithms with and without extra add-ons to enhance performance. The problem at the application layer was a the Rastrigin function, which was solved using a simple GA. They concluded that tuning parameters pays off in terms of EA performance. All nine tuner instances managed to find a parameter vector that greatly outperformed the best guess of a human user. As the authors summarize: "no matter what tuner algorithm you use, you will likely get a much better EA than relying on your intuition and the usual parameter setting conventions".

Tuning evolutionary algorithms can be considered as a special case of tuning metaheuristics. Birattari's book, cf. [9], gives a good treatment from a machine learning perspective, including several case studies, for instance, on tuning iterated local search and ant colony optimization by the F-Race tuner.

2.7 Considerations for Tuning EAs

In this section we discuss some general issues of EA tuning as we advocate it here and provide some guidelines for tuning EAs.

To begin, let us address concerns about the benefits of the whole tuning approach for real-world applications due to the computational overhead costs. For a detailed answer it is helpful to distinguish between two types of applications, one-off problems and repetitive problems; cf. [17, Chapter 14]. In the case of one-off problems, one has to solve a given problem instance only once and typically the solution quality is much more important than the total computational effort. Optimizing the road network around a new airport is an example of such problems. In such cases it is a sound strategy to spend the available time on running the EA several times – with possibly long run times and different settings – and keep the best solution found without inferring anything about good parameters. After all, this problem instance will not be faced again, so information about the best EA parameters for solving it is not really relevant. In these cases, i.e., one-off problems, the added value of tuning is questionable. Applying some parameter control technique to adjust the parameter values on the fly, while solving the problem, is a better approach here. In the case of repetitive problems, one has to solve many different instances of a given problem, where it is assumed that the different instances are not too different. Think, for example, of routing vehicles of a parcel delivery service in a city. In this case it is beneficial to tune the EA to the given problem, because the computational overhead costs of tuning only occur once, while the advantages of using the tuned EA are enjoyed repeatedly. Given these considerations, it it interesting to note that academic research into heuristic algorithms is more akin to solving repetitive problems than to one-off applications. This is especially true for research communities that use benchmark test suites and/or problem instance generators.

Given the above considerations, we can argue that algorithmic tuning of EA parameters is beneficial for at least two types of users, practitioners facing real-world applications with a repetitive character and academics willing to advance the development of evolutionary algorithms in general. To these groups our first and foremost recommendation is

> Do tune your EA with a tuner algorithm (and report the tuning efforts alongside the performance results).

This suggestion may sound trivial, but considering the current practice in evolutionary computing it is far from being superfluous. As we argue in this chapter, there are enough good tuners available, the efforts of tuning are limited, and the gains can be substantial. A typical use case is the publication of a research paper showing that a newly invented EA (NI-EA) is better than some carefully chosen benchmark EA (BM-EA). In this case, tuning NI-EA can make the difference between comparable-and-sometimes-better and convincingly-better in favor of NI-EA.

Let us note that using parameter tuners in performance comparisons leads to the methodological question about whether the benchmark algorithm should be tuned too. One argument for not tuning is that a benchmark may be interpreted as a fixed and invariable algorithm that provides the same challenge to every comparison using it. This may sound unreasonable, giving an unfair advantage to the new algorithm in the comparison. However, the current EC research and publication practice ignores the tuning issue completely, hiding the tuning efforts, and thus hiding the possible unfairness of a comparison between a tuned NI-EA and an untuned BM-EA. In this respect, using a tuner for NI-EA and reporting the tuning efforts alongside the performance results is more illustrative than the contents of the majority of publications at present. A typical claim in a paper following the usual practice sounds like

"NI-EA is better than BM-EA,"

with possible refinements concerning the number or type of test functions where this holds. Using tuners and reporting the tuning effort would improve the present practice by making things public and transparent. A typical claim of the new style would sound like

"Spending effort X on tuning NI-EA, we obtain an instance that is better than (the untuned) BM-EA."

To this end, the tuning effort could be measured by the number of parameter vectors tested (A), the number of utility tests executed $(A \times B)$[4], the computing time spent on running the tuner, and so on.

Alternatively, one could reinterpret the concept of a benchmark EA from as is (including the values of its parameters) to as it could be (if its parameters were tuned). This stance would imply that both NI-EA and BM-EA need to be tuned for a fair comparison. A typical claim of this style would sound like

"The best instance of NI-EA is better than the best instance of BM-EA, where for both EAs the best instance is the one obtained through spending effort X on tuning."

Note that this new philosophy eliminates the rather accidental parameter values as a factor when comparing EAs and focuses on the type of EA operators (recombination, mutation, etc.) instead. In other words, we would obtain statements about EAs rather than EA instances.

An important caveat with respect to tuning EAs is the fact that the set of important parameters is often less clear than it may seem. At first sight, it is easy to identify the

---

[4] If the number of tests per parameter vector is not constant, then $A \times B$ is not the total number of utility tests executed; but for the present argument this is not relevant.

parameters of a given EA, but a closer look *at its code* may disclose that the working of the algorithm depends on several constants hidden in the code. These "magic constants" are determined by the algorithm designer/programmer, but are not made visible as independent parameters. For instance, the conceptual description of the well-known G-CMA-ES will indicate the population size μ and the offspring size λ as parameters with corresponding default values, such as $\lambda = 4 + 3 \cdot log_2(n)$ and $\mu = \frac{\lambda}{2}$, including the magic constants 4, 3, and 2. These constants can be designated as independent parameters of the given EA, thereby making them subject to tuning. The advantage of this practice is clear; it provides additional knobs to tune, potentially increasing the maximum achievable performance. If your game is to obtain a very well performing EA for some application, then you should definitely consider this option. For academic comparisons, it may raise questions regarding the fairness of comparison similar to those discussed above regarding benchmarking. The main question here is whether it is correct to compare the given EA using some good values[5] for its original parameters with a new variant of it using tuned values for those parameters and tuned values for some of its magic constants.

Finally, note that the adaptability of EAs or related algorithms to a specific problem is different. We have seen this in the documented results from [30], as explained in Section 2.6: The default parameters of algorithm A lead to very good performance, whether or not the ones of algorithm B do. Tuning is thus much more important for B. An attempt to measure the adaptablity by means of the *empirical tuning potential* (ETP) is given in [31]. However, a lower adaptability may stem from two sources. Either the algorithm is harder to tune and requires more effort while tuning, or the obtainable best parameter vectors are not much better than the default values. Running tuners with increased number of tests T may pay off by giving insight here.

2.8 Conclusions and Outlook

In this chapter we discussed the notions of EA parameters, elaborated on the issue of tuning EA parameters and reviewed several algorithmic approaches to address it. Our main message can be summarized as follows: Contrary to what contemporary practice would suggest, *there are* good tuning methods that enable EA users to boost EA performance at moderate cost through finding good parameter values tailored to the given problem(s).

Regarding the future, we are moderately optimistic. We foresee a change in attitude on a large scale, fueled by the easily accessible and easy-to-use tuners. The most prominent change we expect is the wide spread use of tuners in scientific publications as well as in applications. To be specific, we expect that evolutionary computing researchers and practitioners will spend an extra day on tuning their EA before creating the figures for an experimental comparison or de-

---

[5] The definition of "good" is irrelevant here. The algorithm could be hand-tuned, algorithmically tuned, commonly used, whatever.

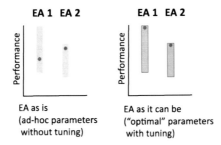

Fig. 2.4: The effect of parameter tuning on comparing EAs. Left: the traditional situation, where the reported EA performance is an "accidental" point on the scale ranging from the worst to the best performance (as determined by the parameter values used). Right: the improved situation, where the reported EA performance is a near-optimal point on this scale, belonging to the tuned instance. This indicates the full potential of the given EA, i.e., how good it can be when using the right parameter values

ploying their EA-based optimizer module within a decision support system. Given that the costs of this extra effort are limited, but the gains can be substantial, it is hard to imagine a reason for not undertaking it. To support this development, let us point to freely available software from `http://sourceforge.net/projects/tuning/`. This Web site contains the Matlab codes of the tuning algorithms we have been using over the last couple of years. Additionally, `http://sourceforge.net/projects/mobat/` provides the Meta-Heuristic Optimizer Benchmark and Analysis Toolbox. It can be used to configure, benchmark and analyze Evolutionary Algorithms and other Meta-Heuristic Optimizers. MOBAT is written in Java and optimized for parallel processing. Alternatively, SPOT can be obtained from `www.gm.fh-koeln.de/campus/personen/lehrende/thomas.bartz-beielstein/00489/`.

A change of practice as foreseen here could also imply a change in methodology in the long term, illustrated in Figure 2.4. In today's research practice, comparisons of EAs are typically based on ad hoc parameter values. Hence, the experiments show a comparison between two rather arbitrary *EA instances* and tell nothing about the full potential of the given EAs. Using tuner algorithms could fix this deficiency and eliminate much of the noise introduced by ad hoc parameter values, thus supporting comparisons of evolutionary algorithms (i.e., configurations of the qualitative parameters), rather than evolutionary algorithm instances.

References

[1] Balaprakash, P., Birattari, M., and Stützle, T.: Improvement strategies for the F-race algorithm: Sampling design and iterative refinement. In: *Hybrid Meta-heuristics*, pp., 108–122 (2007)

[2] Bartz-Beielstein, T., Chiarandini, M., Paquete, L., and Preuss, M., editors, *Empirical methods for the analysis of optimization algorithms*. Springer (2010)

[3] Bartz-Beielstein, T., Parsopoulos, K.E., and Vrahatis, M. N.: Analysis of particle swarm optimization using computational statistics. In: Chalkis, editor, *Proceedings of the International Conference of Numerical Analysis and Applied Mathematics (ICNAAM 2004)*, pp., 34–37 (2004)

[4] Bartz-Beielstein, T.: Experimental analysis of evolution strategies: overview and comprehensive introduction. Technical Report Reihe CI 157/03, SFB 531, Universität Dortmund, Dortmund, Germany (2003)

[5] Bartz-Beielstein, T.: *Experimental research in evolutionary computation—the new experimentalism*. Natural Computing Series. Springer, Berlin, Heidelberg, New York (2006)

[6] Bartz-Beielstein, T., and Markon, S.: Tuning search algorithms for real-world applications: A regression tree based approach. Technical Report of the Collaborative Research Centre 531 Computational Intelligence CI-172/04, University of Dortmund, March (2004)

[7] Bartz-Beielstein, T., and Preuss, M.: Considerations of budget allocation for sequential parameter optimization (SPO). In: Paquete, L., et al., editors, *Workshop on Empirical Methods for the Analysis of Algorithms, Proceedings*, pp., 35–40, Reykjavik, Iceland (2006)

[8] Bechhofer, R. E., Dunnett, W. C., Goldsman, D. M., and Hartmann, M.: A comparison of the performances of procedures for selecting the normal population having the largest mean when populations have a common unknown variance. *Communications in Statistics*, B19:971–1006 (1990)

[9] Birattari, M.: *Tuning metaheuristics*. Springer, Berlin, Heidelberg, New York (2005)

[10] Birattari, M., Yuan, Z., Balaprakash, T., and Stützle, T.: F-frace and iterated F-frace: An overview. In: Bartz-Beielstein, T., et al. [2]

[11] Branke, J., Chick, S. E., and Schmidt, C.: New developments in ranking and selection: An empirical comparison of the three main approaches. In: *WSC '05: Proceedings of the 37th Conference on Winter simulation*, pp., 708–717. Winter Simulation Conference (2005)

[12] Chen, J. E., Chen, C. H., and Kelton, D. W.: Optimal computing budget allocation of indifference-zone-selection procedures. Working paper, taken from http://www.cba.uc.edu/faculty/keltonwd (2003)

[13] Clune, J., Goings, S., Punch, B., and Goodman, E.: Investigations in meta-GAs: Panaceas or pipe dreams? In: *GECCO '05: Proceedings of the 2005 Workshops on Genetic and Evolutionary Computation*, pp., 235–241, ACM (2005)

[14] Eiben, A. E., Hinterding, R., and Michalewicz, Z.: Parameter control in evolutionary algorithms. *IEEE Transactions on Evolutionary Computation*, 3(2):124–141 (1999)

[15] Eiben, A. E., and Jelasity, M.: A critical note on experimental research methodology in EC. In: *Proceedings of the 2002 Congress on Evolutionary Computation (CEC'2002)*, pp., 582–587. IEEE Press, Piscataway, NJ (2002)

[16] Eiben, A. E., and Smith, J. E.: *Introduction to Evolutionary Computation*. Natural Computing Series. Springer (2003)

[17] Eiben, A. E., and Smith, J. E.: *Introduction to evolutionary computing*. Springer, Berlin, Heidelberg, New York (2003)

[18] Greffenstette, J. J.: Optimisation of control parameters for genetic algorithms. In: *IEEE Transactions on Systems, Man and Cybernetics*, volume 16, pp., 122–128 (1986)

[19] Harik, G. R., and Lobo, F. G.: A parameter-less genetic algorithm. In: Banzhaf, W., et al., editors, Proceedings of the Genetic and Evolutionary Computation Conference GECCO-99, San Francisco, CA. Morgan Kaufmann, pp., 258–265 (1999)

[20] Herdy, M.: Reproductive isolation as strategy parameter in hierarichally organized evolution strategies. In: Männer, R., and Manderick, B., editors, *Proceedings of the 2nd Conference on Parallel Problem Solving from Nature*, pp., 209–209. North-Holland, Amsterdam (1992)

[21] Lasarczyk, C. W. G.: *Genetische programmierung einer algorithmischen chemie*. Ph.D. thesis, Technische Universität Dortmund (2007)

[22] Maron, O., and Moore, A.: The racing algorithm: model selection for lazy learners. In: *Artificial Intelligence Review*, volume 11, pp., 193–225 (1997)

[23] Maturana, J., Lardeux, F., and Saubion, F.: Autonomous operator management for evolutionary algorithms. Journal of Heuristics **16**, 881–909 (2010)

[24] Mercer, R. E., and Sampson, J. R.: Adaptive search using a reproductive metaplan. *Kybernetes*, 7:215–228 (1978)

[25] Myers, R., and Hancock, E. R.: Empirical modelling of genetic algorithms. *Evolutionary Computation*, 9(4):461–493 (2001)

[26] Nannen, V.: Evolutionary agent-based policy analysis in Ddnamic Environments. Ph.D. thesis, Free University Amsterdam (2009)

[27] Nannen, V., and Eiben, A. E.: Efficient relevance estimation and value calibration of evolutionary algorithm parameters. In: *IEEE Congress on Evolutionary Computation*, pp., 103–110. IEEE (2007)

[28] Nannen, V. and Eiben, A. E.: Relevance estimation and value calibration of evolutionary algorithm parameters. In: Veloso, M., editor, *IJCAI 2007, Proceedings of the 20th International Joint Conference on Artificial Intelligence*, pp., 1034–1039 (2007)

[29] Nannen, V. and Eiben, A. E.: A method for parameter calibration and relevance estimation in evolutionary algorithms. In: Keijzer, M. ,editor, *Proceedings of the Genetic and Evolutionary Computation Conference (GECCO-2006)*, pp., 183–190. Morgan Kaufmann, San Francisco (2006)

[30] Nannen, V., Smit, S. K., and Eiben, A. E.: Costs and benefits of tuning parameters of evolutionary algorithms. In: Rudolph, G., Jansen, T., Lucas, S., Poloni, C., and Beume,N. , editors, *PPSN*, volume 5199 of *Lecture Notes in Computer Science*, pp., 528–538. Springer (2008)

[31] Preuss, M.: Adaptability of algorithms for real-valued optimization. In: Giacobini, M., et al., editors, *Applications of Evolutionary Computing, EvoWork-*

shops 2009. Proceedings, volume 5484 of Lecture Notes in Computer Science, pp., 665–674, Berlin, Springer (2009)

[32] Schaffer, J. D., Caruana, R. A., Eshelman, L. J., and Das, R.: A study of control parameters affecting online performance of genetic algorithms for function optimization. In: *Proceedings of the Third International conference on Genetic algorithms*, pp., 51–60,San Francisco, CA, USA, 1989. Morgan Kaufmann Publishers Inc. (1989)

[33] Smit, S. K., and Eiben, A. E.: Comparing parameter tuning methods for evolutionary algorithms. In: *Proceedings of the 2009 IEEE Congress on Evolutionary Computation*, pp., 399–406. IEEE Computational Intelligence Society, IEEE Press (2009)

[34] Smit, S. K., and Eiben, A. E.: Using entropy for parameter analysis of evolutionary algorithms. In: Bartz-Beielstein, T., et al. [2]

[35] Taguchi, G., and Yokoyama, T.: *Taguchi methods: design of experiments*. ASI Press (1993)

[36] Wolpert, D. H., and Macready, W. G.: No free lunch theorems for optimization. *IEEE Transaction on Evolutionary Computation*, 1(1):67–82 (1997)

[37] Yuan, B. and Gallagher, M.: Combining meta-EAs and racing for difficult EA parameter tuning tasks. In: Lobo, F. G., Lima, C. F., and Michalewicz, Z., editors, *Parameter Setting in Evolutionary Algorithms*, pp., 121–142. Springer (2007)

Chapter 3
Automated Algorithm Configuration and Parameter Tuning

Holger H. Hoos

3.1 Introduction

Computationally challenging problems arise in the context of many applications, and the ability to solve these as efficiently as possible is of great practical, and often also economic, importance. Examples of such problems include scheduling, time-tabling, resource allocation, production planning and optimisation, computer-aided design and software verification. Many of these problems are $\mathcal{NP}$-hard and considered computationally intractable, because there is no polynomial-time algorithm that can find solutions in the worst case (unless $\mathcal{NP} = \mathcal{P}$). However, by using carefully crafted heuristic techniques, it is often possible to solve practically relevant instances of these 'intractable' problems surprisingly effectively (see, e.g., 55, 3, 54) [1].

The practically observed efficacy of these heuristic mechanisms remains typically inaccessible to the analytical techniques used for proving theoretical complexity results, and therefore needs to be established empirically, on the basis of carefully designed computational experiments. In many cases, state-of-the-art performance is achieved using several heuristic mechanisms that interact in complex, non-intuitive ways. For example, a DPLL-style complete solver for SAT (a prototypical $\mathcal{NP}$-complete problem with important applications in the design of reliable soft- and hardware) may use different heuristics for selecting variables to be instantiated and the values first explored for these variables, as well as heuristic mechanisms for managing and using logical constraints derived from failed solution attempts. The activation, interaction and precise behaviour of those mechanisms is often controlled by parameters, and the settings of such parameters have a substantial impact on the

Holger H. Hoos
Department of Computer Science, University of British Columbia, 2366 Main Mall, Vancouver, BC, V6T 1Z4, Canada, e-mail: hoos@cs.ubc.ca

[1] We note that the use of heuristic techniques does not imply that the resulting algorithms are necessarily incomplete or do not have provable performance guarantees, but often results in empirical performance far better than the bounds guaranteed by rigorous theoretical analysis.

Y. Hamadi et al. (eds.), *Autonomous Search*,
DOI 10.1007/978-3-642-21434-9_3,
© Springer-Verlag Berlin Heidelberg 2011

efficacy with which a heuristic algorithm solves a given problem instance or class of problem instances. For example, the run-time of CPLEX 12.1 – a widely used, commercial solver for mixed integer programming problems – has recently been demonstrated to vary by up to a factor of over 50 with the settings of 76 user-accessible parameters [42].

A problem routinely encountered by designers as well as end users of parameterised algorithms is that of finding parameter settings (or *configurations*) for which the empirical performance on a given set of problem instances is optimised. Formally, this *algorithm configuration* or *parameter tuning* problem can be stated as follows:

Given
  - an algorithm A with parameters $p_1, \ldots, p_k$ that affect its behaviour,
  - a space C of configurations (i.e., parameter settings), where each configuration $c \in C$ specifies values for A's parameters such that A's behaviour on a given problem instance is completely specified (up to possible randomisation of A),
  - a set of problem instances I,
  - a performance metric m that measures the performance of A on instance set I for a given configuration c,

find a configuration $c^* \in C$ that results in optimal performance of A on I according to metric m.

In the context of this problem, the algorithm whose performance is to be optimised is often called the *target algorithm*, and we use $A(c)$ to denote target algorithm A under a specific configuration c. The set of values any given parameter p can take is called the *domain* of p. Depending on the given target algorithm, various types of parameters may occur. *Categorical parameters* have a finite, unordered set of discrete values; they are often used to select from a number of alternative mechanisms or components. Using *Boolean parameters*, heuristic mechanisms can be activated or deactived, while the behaviour and interaction of these mechanisms is often controlled by *integer-* and *real-valued parameters* (the former of which are a special cases of *ordinal parameters*, whose domains are discrete and ordered). *Conditional parameters* are only active when other parameters are set to particular values; they routinely arise in the context of mechanisms that are activated or selected using some parameter, and whose behaviour is then controlled by other parameters (where the latter parameters conditionally depend on the former). Sometimes, it is useful to place additional constraints on configurations, e.g., to exclude certain combinations of parameter values that would lead to ill-defined, incorrect or otherwise undesirable behaviour of a given target algorithm.

Clearly, the number and types of parameters, along with the occurrence of conditional parameters and constraints on configurations, determine the nature of the configuration space C and have profound implications on the methods to be used for searching performance-optimising configurations within that space. These methods range from well-known numerical optimisation procedures, such as the Nelder-Mead Simplex algorithm [49, 13] or the more recent gradient-free CMA-ES algorithm [25, 27, 26], to approaches based on experimental design methods (see, e.g., 11, 5, 1), response-surface models (see, e.g., 44, 7) and stochastic local search procedures (see, e.g., 36, 37).

In general, when configuring a specific target algorithm, it is desirable to find parameter configurations that work well on problem instances other than those in the given instance set I. To this end, care needs to be taken in selecting the instances in I to be representative of the kinds of instances to which the optimised target algorithm configuration is expected to be applied. Difficulties can arise when I is small, yet contains very different types of instances. To recognise situations in which a configured target algorithm, $A(c^*)$, fails to perform well when applied to instances other than those used in the configuration process, it is advisable to test it on a set of instances not contained in I; this can be done by including in I only part of the overall set of instances available, or by means of cross-validation.

It is also advisable to investigate performance variation of $A(c^*)$ over instance set I, since, depending on the performance metric m used for the configuration of A and differences between instances in I, the optimised configuration c^* might represent a trade-off between strong performance on some instances and weaker performance on others. In particular, when using a robust statistic, such as median run-time, as a performance metric, poor performance on large parts of a given instance set can result. To deal effectively with target algorithm runs in which no solution was produced (in particular, time-outs encountered when optimising run-time), it is often useful to use a performance metric based on penalised averaging, in which a fixed penalty is assigned to any unsuccessful run of A (see also 37).

In the existing literature, the terms *algorithm configuration* and *parameter tuning* are often used interchangeably. We prefer to use *parameter tuning* in the context of target algorithms with relatively few parameters with mostly real-valued domains, and *algorithm configuration* in the context of target algorithms with many categorical parameters. Following Hoos [28], we note that algorithm configuration problems arise when dealing with an algorithm schema that contains a number of instantiable components (typically, subprocedures or functions), along with a discrete set of concrete choices for each of these. While most standard numerical optimisation methods are not applicable to these types of algorithm configuration problems, F-Race [11, 5], Calibra [1] and ParamILS [36, 37] have been used successfully in this context. However, so far only ParamILS has been demonstrated to be able to deal with the vast design spaces resulting from schemata with many independently instantiable components (see, e.g., 45, 66), and promising results have been achieved by a genetic programming procedure applied to the configuration of local search algorithms for SAT [18, 19], as well as by a recent gender-based genetic algorithm [2].

In the remainder of this chapter, we discuss three classes of methods for solving algorithm configuration and parameter tuning problems. *Racing procedures* iteratively evaluate target algorithm configurations on problem instances from a given set and use statistical hypothesis tests to eliminate candidate configurations that are significantly outperformed by other configurations; *ParamILS* uses a powerful stochastic local search (SLS) method to search within potentially vast spaces of candidate configurations of a given algorithm; and sequential model-based optimisation (SMBO) methods build a response surface model that relates parameter settings to performance, and use this model to iteratively identify promising settings. We also give a brief overview of other algorithm configuration and parameter tuning procedures and comment on the applications in which various methods have proven to be effective.

While we deliberately limit the scope of our discussion to the algorithm configuration problem defined earlier in this section, we note that there are several conceptually closely related problems that arise in the computer-aided design of algorithms (see also 28): *per-instance algorithm selection methods* choose one of several target algorithms to be applied to a given problem instance based on properties of that instance determined just before attempting to solve it (see, e.g., 57, 46, 24, 67, 68); similarly, *per-instance algorithm configuration methods* use instance properties to determine the specific configuration of a parameterised target algorithm to be used for solving a given instance (see, e.g., 34). *Reactive search procedures, on-line algorithm control methods* and *adaptive operator selection techniques* switch between different algorithms, heuristic mechanisms and parameter configurations while running on a given problem instance (see, e.g., 14, 10, 16); and *dynamic algorithm portfolio approaches* repeatedly adjust the allocation of CPU shares between algorithms that are running concurrently on a given problem instance (see, e.g., 20).

Furthermore, we attempt neither to cover all algorithm configuration methods that can be found in the literature, nor to present all details of the procedures we describe; instead, we focus on three fundamental approaches of algorithm configuration methods and survey a number of prominent methods based on these, including the state-of-the-art procedures at the time of this writing. We briefly discuss selected applications of these procedures to illustrate their scope and performance, but we do not attempt to give a complete or detailed account of the empirical results found in the literature.

3.2 Racing Procedures

Given a number of candidate solvers for a given problem, the concept of racing is based on a simple yet compelling idea: sequentially evaluate the candidates on a series of benchmark instances and eliminate solvers as soon as they have fallen too far behind the current leader, i.e., the candidate with the overall best performance at a given stage of the race.

Racing procedures were originally introduced for solving model selection problems in machine learning. The first such technique, dubbed *Hoeffding Races* [48], was introduced in a supervised learning scenario, where a black-box learner is evaluated by measuring its error on a set of test instances. The key idea is to test a given set of models, one test instance at a time, and to discard models as soon as they are shown to perform significantly worse than the best ones. Performance is measured as error over all test instances evaluated so far, and models are eliminated from the race using non parametric bounds on the true error, determined based on Hoeffding's inequality (which gives an upper bound on the probability of the sum of random variables deviating from its expected value). More precisely, a model is discarded from the race if the lower bound on its true error (for a given confidence level $1 - \delta$) is worse than the upper bound on the error of the currently best model. As a result, the computational effort expended in evaluating models becomes increasingly focussed on promising candidates, and the best candidate models end up getting evaluated most thoroughly.

This idea can be easily transferred to the problem of selecting an algorithm from a set of candidates, where each candidate may correspond to a configuration of a parameterised algorithm [11]. In this context, candidate algorithms (or configurations) are evaluated on a given set of problem instances. As in the case of model selection, the race proceeds in steps, where in each step every candidate is evaluated on the same instance, taken from the given instance set, and candidates that performed significantly worse on the instances considered so far are eliminated from the race. (We note that the evaluation of candidates in each step can, in principle, be performed independently in parallel.)

This procedure requires that the set of candidate algorithms be finite and, since in the initial steps of a race all candidates will need to be evaluated, of somewhat reasonable size. Therefore, when applied to algorithm configuration or parameter tuning scenarios with continuous parameters, racing approaches need to make use of discretisation or sampling techniques. In the simplest case, all continuous parameters are discretised prior to starting the race. Alternatively, stages of sampling and racing can be interleaved, such that the candidate configurations being considered become increasingly concentrated around the best performing configurations.

In the following, we will first present the F-Race procedure of Birattari et al. [11] in more detail and outline its limitations. We will then discuss variations of F-Race that overcome those weaknesses [5, 12], and finally summarise some results achieved by these racing procedures in various algorithm configuration scenarios.

3.2.1 F-Race

The F-Race algorithm by Birattari et al. [11] closely follows the previously discussed racing procedure. Similarly to Hoeffding races, it uses a non parametric test as the basis for deciding which configurations to eliminate in any given step. However, rather than just performing pairwise comparisons with the currently best configuration (the so-called *incumbent*), F-Race first uses the rank-based Friedman test (also known as *Friedman two-way analysis of variance by ranks*) for ni independent s-variate random variables, where s is the number of configurations still in the race, and ni is the number of problem instances evaluated so far. The Friedman test assesses whether the s configurations show no significant performance differences on the ni given instances; if this null hypothesis is rejected, i.e., if there is evidence that some configurations perform better than others, a series of pairwise

procedure *F-Race*
    input *target algorithm A, set of configurations C, set of problem instances I,*
           performance metric m;
    parameters *integer ni_{min};*
    output *set of configurations C^*;*

    $C^* := C$; $ni := 0$;
    repeat
        randomly choose instance i from set I;
        run all configurations of A in C^* on i;
        $ni := ni + 1$;
        if $ni \geq ni_{min}$ **then**
            perform rank-based Friedman test on results for configurations in C^* on all instances
                in I evaluated so far;
            if test indicates significant performance differences **then**
                $c^* :=$ best configuration in C^* (according to m over instances evaluated so far);
                for all $c \in C^* \setminus \{c^*\}$ **do**
                    perform pairwise Friedman post hoc test on c and c^*;
                    if test indicates significant performance differences **then**
                        eliminate c from C^*;
                  end if;
                end for;
            end if;
        end if;
    until termination condition met;
    return C^*;
end *F-Race*

Fig. 3.1: Outline of F-Race for algorithm configuration (original version, according to 11). In typical applications, ni_{min} is set to values between 2 and 5; further details are explained in the text. When used on its own, the procedure would typically be modified to return $c^* \in C^*$ with the best performance (according to m) over all instances evaluated within the race

post hoc tests between the incumbent and all other configurations is performed. All configurations found to have performed significantly worse than the incumbent are eliminated from the race. An outline of the F-Race procedure for algorithm configuration, as introduced by [11], is shown in Figure 3.1; as mentioned by [5], runs on a fixed number of instances are performed before the Friedman test is first applied. The procedure is typically terminated either when only one configuration remains, or when a user-defined time budget has been exhausted.

The Friedman test involves ranking the performance results of each configuration on a given problem instance; in the case of ties, the average of the ranks that would have been assigned without ties is assigned to each tied value. The test then determines whether some configurations tend to be ranked better than others when considering the rankings for all instances considered in the race up to the given iteration. Following Birattari et al. [11], we note that performing the ranking separately for each problem instance amounts to a blocking strategy on instances. The use of this strategy effectively reduces the impact of noise effects that may arise from the performance variation observed over the given instances set for any configuration of the target algorithm under consideration; this can become critical when those performance variations are large, as has been observed for many algorithms for various hard combinatorial problems (see, e.g., 21, 29).

3.2.2 Sampling F-Race and Iterative F-Race

A major limitation of this basic version of F-Race stems from the fact that in the initial steps all given configurations have to be evaluated. This property of basic F-Race severely limits the size of the configuration spaces to which the procedure can be applied effectively – particularly when dealing with configuration spaces corresponding to so-called *full factorial designs*, which contain all combinations of values for a set of discrete (or discretised) parameters. Two more recent variants of F-Race, Sampling F-Race and Iterative F-Race, have been introduced to address this limitation [5]; both use the previously described F-Race procedure as a subroutine.

Sampling F-Race (short: *RSD/F-Race*) is based on the idea of using a sampling process to determine the initial set of configurations subsequently used in a standard F-Race. In RSD/F-Race, a fixed number r of samples is determined using a so-called *Random Sampling Design*, in which each configuration is drawn uniformly at random from the given configuration space C. (In the simplest case, where no conditional parameters or forbidden configurations exist, this can be done by sampling values for each parameter independently and uniformly at random from the respective domain.) As noted by Balaprakash et al. [5], the performance of this procedure depends substantially on r, the number of configurations sampled in relation to the size of the given configuration space.

A somewhat more effective approach for focussing a procedure based on F-Race on promising configurations is *Iterative F-Race* (short: *I/F-Race*). The key idea behind I/F-Race is the use of an iterative process, where in the first stage of each iteration configurations are sampled from a probabilistic model M, while in the second stage a standard F-Race is performed on the resulting sample, and the configurations surviving this race are used to define or update the model M used in the following iteration. (See Figure 3.2.)

The probabilistic model used in each iteration of I/F-Race consists of a series of probability distributions, $\mathcal{D}_1, \ldots, \mathcal{D}_s$, each of which is associated with one of s 'promising' parameter configurations, $c_1, \ldots, c_s$. Balaprakash et al. [5] consider only numerical parameters and define each distribution $\mathcal{D}_i$ to be a k-variate normal distribution $\mathcal{N}_i := \mathcal{N}(\mu_i, \Sigma_i)$ that is centred on configuration c_i, i.e., $\mu_i = c_i$. They further define the covariance between any two different parameters in a given $\mathcal{N}_i$ to be zero, such that $\mathcal{N}_i$ can be factored into k independent, univariate normal distributions. To start the process with an unbiased probabilistic model, in the first iteration of I/F-Race a single k-variate uniform distribution is used, which is defined as the product of the k independent uniform distributions over the ranges of each given parameter

procedure *I/F-Race*
    input *target algorithm A, set of configurations C, set of problem instances I,*
        performance metric m;
    output *set of configurations* C^*;

    initialise probabilistic model M;
    $C' := \emptyset$;   // *later*, C' *is the set of survivors from the previous F-Race*
    repeat
        based on model M, sample set of configurations $\widehat{C} \subseteq C$;
        perform F-Race on configurations in $\widehat{C} \cup C'$ to obtain set of configurations C^*;
        update probabilistic model M based on configurations in C^*;
        $C' := C^*$;
    until termination condition met;
    return $c^* \in C^*$ with best performance (according to m) over all instances evaluated;
end *I/F-Race*

Fig. 3.2: High-level outline of Iterated F-Race, as introduced by [5]; details are explained in the text. The most recent version of I/F-Race slightly deviates from this outline (see 12)

(we note that this can be seen as a degenerate case of the normal distributions used subsequently, in which the variance is infinite and truncation is applied).

In each iteration of I/F-Race, a certain number of configurations are sampled from the distributions $\mathcal{N}_1, \ldots, \mathcal{N}_s$. In the first iteration, this corresponds to sampling configurations uniformly at random from the given configuration space. In subsequent iterations, for each configuration to be sampled, first, one of the $\mathcal{N}_i$ is chosen using a rank-based probabilistic selection scheme based on the performance of the configuration c_i associated with $\mathcal{N}_i$ (for details, see 5), and then a configuration is sampled from this distribution. Values that are outside the range allowable for a given parameter are set to the closer of the two boundaries, and settings for parameters with integer domains are rounded to the nearest valid value. The number a of configurations sampled in each iteration depends on the number s of configurations that survived the F-Race in the previous iteration; Balaprakash et al. [5] keep the overall number of configurations considered in each iteration of I/F-Race constant at some value r, and therefore simply replace those configurations eliminated by F-Race with newly sampled ones (i.e., $a := r - s$, where in the first iteration, $s = 0$).

The resulting population of $a + s$ configurations is subjected to a standard F-Race; this race is terminated using a complex, disjunctive termination condition that involves a (lower) threshold on the number of surviving configurations as well as upper bounds on the computational budget (measured in target algorithm runs) and the number of problem instances considered [2]. Each of the F-Races conducted within I/F-Race uses a random permutation of the given instance set in order to

---

[2] The threshold mechanism ends the race as soon as the number of survivors has fallen below k, the number of target algorithm parameters.

avoid bias due to a particular instance ordering. The s configurations that survived the race (where the value of s depends on the part of the termination condition that determined the end of that race) induce the probabilistic model used in the following iteration of I/F-Race.

To increasingly focus the sampling process towards the most promising configurations, the standard deviations of the component distributions of the probabilistic models $\mathcal{N}_i$ are gradually decreased using a volume reduction technique. More precisely, after each iteration, the standard deviation vector σ_i of each distribution $\mathcal{N}_i$ is scaled by a factor $(1/r)^k$, where r is the total number of configurations entered into the F-Race, and k is the number of given parameters; this corresponds to a reduction of the total volume of the region bounded by $\mu_i \pm \sigma_i$ (over all k parameters) by a factor of r. At the beginning of I/F-Race, when configurations are sampled uniformly, the standard deviation values are (somewhat arbitrarily) set to half of the range of the respective parameter values.

I/F-Race, as specified by Balaprakash et al. [5], assumes that all parameters are numerical. This limitation is overcome in a later variant [12], which supports categorical parameters by sampling their values from discrete probability distributions that are updated by redistributing probability mass to values seen in good configurations, as determined by F-Race. This version of I/F-Race, which we call *I/F-Race-10* for clarity, also differs from the one described previously in several other aspects. Notably, the number of iterations in I/F-Race-10 is determined as $2 + \lfloor \log_2(k) + 0.5 \rfloor$, and the overall computational budget (i.e., number of target algorithm runs) is distributed equally over these iterations. Furthermore, the number r of configurations considered at iteration number t is set to $\lfloor b/(5+t) \rfloor$, where b is the computational budget available for that iteration; this leads to fewer configurations being considered in later iterations. The threshold on the number of survivors below which any given F-Race is terminated is also determined as $2 + \lfloor \log_2(k) + 0.5 \rfloor$. Finally, I/F-Race-10 handles conditional parameters by only sampling values for them when they are active, and by only updating the respective component of the model in situations where such parameters are active in a configuration surviving one of the subsidiary F-Races. (For further details, see 12.)

3.2.3 Applications

Balaprakash et al. [5] describe applications of F-Race, Sampling F-Race and Iterative F-Race to three high-performance stochastic local search algorithms: MAX-MIN Ant System for the TSP with six parameters [64], an estimation-based local search algorithm for the probabilistic TSP (PTSP) with three parameters [6], and a simulated annealing algorithm for vehicle routing with stochastic demands (VRP-SD) with four parameters [53]. The empirical results from these case studies indicate that both, Sampling F-Race and Iterative F-Race can find good configurations in spaces that are too big be handled effectively by F-Race, and that Iterative F-Race tends to give better results than Sampling F-Race, especially when applied to more difficult configuration problems. Both, the PTSP and the VRP-SD algorithms as configured by Iterative F-Race represented the state of the art in solving these problems at the time of this study.

More applications of F-Race have recently been summarised by Birattari et al. [12]. These include tuning the parameters of various meta-heuristic algorithms for university timetabling problems [58], of a control system for simple robots [52], and of a new state-of-the-art memetic algorithm for the linear ordering problem [61]. In all of these cases, the basic F-Race algorithm was applied to target algorithms with few parameters and rather small configuration spaces (48–144 configurations).

Yuan et al. [71] report an application of I/F-Race for tuning various heuristic algorithms for solving a locomotive scheduling problem provided by the German railway company, Deutsche Bahn. The target algorithms considered in this work had up to five parameters, mostly with continuous domains. The most complex application of I/F-Race reported by Birattari et al. [12] involves 12 parameters of the ACOTSP software, some of which conditionally depend on the values of others.

While these (and other) racing procedures have been demonstrated to be useful for accomplishing a broad range of parameter tuning tasks, it is somewhat unclear how well they perform when applied to target algorithms with many more parameters, and how effectively they can deal with the many categorical and conditional parameters arising in the context of more complex computer-aided algorithm design tasks, such as the ones considered by Hutter et al. [35], KhudaBukhsh et al. [45], Hutter et al. [42], and Tompkins and Hoos [66].

3.3 ParamILS

When manually solving algorithm configuration problems, practitioners typically start from some configuration (often default or arbitrarily chosen settings) and then attempt to achieve improvements by modifying one parameter value at a time. If such an attempt does not result in improved performance, the modification is rejected and the process continues from the previous configuration. This corresponds to an iterative first-improvement search in the space of configurations.

While the idea of performing local search in configuration space is appealing considering the success achieved by similar methods on other hard combinatorial problems, iterative improvement is a very simplistic method that is limited to finding local optima. The key idea behind ParamILS is to combine more powerful stochastic local search (SLS) methods with mechanisms aimed at exploiting specific properties of algorithm configuration problems. The way in which this is done does not rely on an attempt to construct or utilise a model of good parameter configurations or of the impact of configuations on target algorithm perfomance; therefore, ParamILS is a *model-free search procedure*.

3.3.1 The ParamILS Framework

At the core of the ParamILS framework for automated algorithm configuration [36, 37] lies *Iterated Local Search (ILS)*, a well-known and versatile stochastic local search method that has been applied with great success to a wide range of difficult combinatorial problems (see, e.g., 47, 30). ILS iteratively performs phases of simple local search designed to rapidly reach or approach a locally optimal solution to the given problem instance, interspersed with so-called perturbation phases, whose purpose is to effectively escape from local optima. Starting from a local optimum x, in each iteration one perturbation phase is performed, followed by a local search phase, with the aim of reaching (or approaching) a new local optimum x'. Then, a so-called acceptance criterion is used to decide whether to continue the search process from x' or whether to revert to the previous local optimum, x. Using this mechanism, ILS aims to solve a given problem instance by effectively exploring the space of its locally optimal solutions. At a lower level, ILS – like most SLS methods – visits (i.e., moves through) a series of candidate solutions such that at any given time there is a current candidate solution, while keeping track of the incumbent (i.e., the best solution encountered so far).

ParamILS uses this generic SLS method to search for high-performance configurations of a given algorithm as follows (see also Figure 3.3). The search process is initialised by considering a given configuration (which would typically be the given target algorithm's default configuration) as well as r further configurations that are chosen uniformly at random from the given configuration space. These $r + 1$ configurations are evaluated in a way that is specific to the given ParamILS variant, and the best-performing configuration is selected as the starting point for the iterated local search process. This initialisation mechanism can be seen as a combination of the intuitive choice of starting from a user-defined configuration (such as the target algorithm's default settings) and a simple experimental design technique, where the latter makes it possible to exploit situations where the former represents a poor choice for the given set of benchmark instances. Clearly, there is a trade-off between the effort spent on evaluating randomly sampled configurations at this point and the effort used in the subsequent iterated local search process. Hutter et al. [39] reported empirical results suggesting that $r = 10$ results in better performance than $r = 0$ and $r = 100$ across a number of configuration scenarios. However, we suspect that more sophisticated initialisation procedures, in particular ones based on racing or sequential model-based optimisation techniques, might result in even better performance.

The subsidiary local search procedure used in ParamILS is based on the one-exchange neighbourhood induced by arbitrary changes in the values of a single target algorithm parameter. ParamILS supports conditional parameters by pruning neighbourhoods such that changes in inactive parameters are excluded from consideration; it also supports exclusion of (complete or partial) configurations explicitly declared 'forbidden' by the user. Using the one-exchange neighbourhood, ParamILS performs iterative first-improvement search – an obvious choice, considering the computational cost of evaluating candidate configurations. We believe that larger neighbourhoods might prove effective in situations in which parame-

```
procedure ParamILS
    input target algorithm A, set of configurations C, set of problem instances I,
        performance metric m;
    parameters configuration c₀ ∈ C, integer r, integer s, probability pr;
    output configuration c*;

    c* := c₀;
    for i := 1 to r do
        draw c from C uniformly at random;
        assess c against c* based on performance of A on instances from I according to metric m;
        if c found to perform better than c* then
            c* := c;
        end if;
    end for;

    c := c* ;
    perform subsidiary local search on c;
    while termination condition not met do
        c' := c;
        perform s random perturbation steps on c'
        perform subsidiary local search on c';
        assess c' against c based on performance of A on instances from I according to metric m;
        if c' found to perform better than c then        // acceptance criterion
            update overall incumbent c*;
            c := c';
        end if;
        with probability pr do
            draw c from C uniformly at random;
        end with probability;
    end while;
    return c*;
end ParamILS
```

Fig. 3.3: High-level outline of ParamILS, as introduced by [36]; details are explained in the text

ter effects are correlated, as well as in conjunction with mechanisms that recognise and exploit such dependencies in parameter response. Furthermore, search strategies other than iterative first-improvement could be considered in variants of ParamILS that build and maintain reasonably accurate models of local parameter responses.

The perturbation procedure used in the ParamILS framework performs a fixed number, s, of steps chosen uniformly at random in the same one-exchange neighbourhood used during the local search phases. Computational experiments in which various fixed values of s as well as several multiples of the number of target algorithm parameters were considered suggest that relatively small perturbations (i.e., $s = 2$) are sufficient for obtaining good performance of the overall configuration procedure [39]. Considering the use of iterative first-improvement during the local search phases, this is not overly surprising; still, larger perturbations might be effective in combination with model-based techniques within the ParamILS framework in the future.

While various acceptance criteria have been used in the literature on ILS, ParamILS uses one of the simplest mechanisms: Between two given candidate configurations, it always chooses the one with better observed performance; ties are broken in favour of the configuration reached in the most recent local search phase. This results in an overall behaviour of the iterated local search process equivalent to that of an iterative first-improvement procedure searching the space of configurations reached by the subsidiary local search process. Considering once again the computational effort involved in each iteration of ParamILS, this is a natural choice; however, in cases where many iterations of ParamILS can be performed, and where the given configuration space contains attractive regions with many local minima, more complex acceptance criteria that provide additional search diversification (e.g., based on the Metropolis criterion) might prove useful.

In addition to the previously described perturbation procedure, ParamILS also uses another diversification mechanism: At the end of each iteration, with a fixed probability pr (by default set to 0.01), the current configuration is abandoned in favour of a new one that is chosen uniformly at random and serves as the starting point for the next iteration of the overall search process. This restart mechanism provides the basis for the probabilistic approximate completeness of FocusedILS, the more widely used of the two ParamILS variants discussed in the following. We believe that it also plays an important role towards achieving good performance in practice, although anecdotal empirical evidence suggests that additional diversification of the search process is required in order to eliminate occasionally occurring stagnation behaviour.

Finally, like the racing procedures discussed in the previous section, ParamILS performs blocking on problem instances, i.e., it ensures that comparisons between different configurations are always based on the same set of instances. This is important, since the intrinsic hardness of problem instances for any configuration of the given target algorithm may differ substantially. Furthermore, when used for optimising the performance of a randomised target algorithm A, ParamILS also blocks on the pseudo random number seeds used in each run of A; the main reason for this lies in our desire to avoid spurious performance differences in cases where the differences between two configurations have no impact on the behaviour of A.

3.3.2 BasicILS

The conceptually simplest way of assessing the performance of a configuration of a given target algorithm A is to perform a fixed number of runs of A. This is precisely what happens in BasicILS(N), where the user-defined parameter N specifies the number of target algorithm runs performed for each configuration to be assessed, using the same instances and pseudo random number seeds. Applied to a randomised target algorithm A, BasicILS(N) will only perform multiple runs per instance if N

exceeds the number of given problem instances; in this case, the list of runs performed is determined by a sequence of random permutations of the given set of instances, and the random number seed used in each run is determined uniformly at random.

This approach works well for configuration scenarios where a relatively small set of benchmark instances is representative of all instances of interest. Furthermore, the N target algorithm runs per configuration can be performed independently in parallel. As for all ParamILS variants – and, indeed, for any SLS algorithm – further parallelisation can be achieved by performing multiple runs of BasicILS(N) in parallel. Finally, in principle, it is possible to perform multiple parallel runs of the subsidiary local search in each iteration or to evaluate multiple neighbours of a configuration in each search step independently in parallel.

3.3.3 FocusedILS

One drawback of BasicILS is that it tends to make substantial effort evaluating poor configurations, especially when used to configure a given target algorithm for minimised run-time. The only way to reduce that effort is to choose a small number of runs, N; however, this can (and often does) result in poor generalisation of performance to problem instances other than those used during the configuration process. FocusedILS addresses this problem by initially evaluating configurations using few target algorithm runs and subsequently performing additional runs to obtain increasingly precise performance estimates for promising configurations. We note that the idea of focussing the computational effort in evaluating configurations on candidates that have already shown promising performance is exactly the same as that underlying the concept of racing. However, unlike the previously discussed racing procedures, FocusedILS determines promising configurations heuristically rather than using statistical tests.

The mechanism used by FocusedILS to assess configurations is based on the following concept of *domination*: Let c_1 and c_2 be configurations for which $N(c_1)$ and $N(c_2)$ target algorithm runs have been performed, respectively. As in the case of BasicILS, the runs performed for each configuration follow the same sequence of instances (and pseudo random number seeds). Then c_1 dominates c_2 if, and only if, $N(c_1) \geq N(c_2)$ and the performance estimate for c_1 based on its first $N(c_2)$ runs is at least as good as that for c_2 based on all of its $N(c_2)$ runs. This definition incorporates the previously discussed idea of blocking, as configurations are compared based on their performance on a common set of instances (and pseudo random number seeds).

Whenever FocusedILS decides that one configuration, c_1, performs better than another, c_2, it ensures that c_1 dominates c_2 by performing additional runs on either or both configurations. More precisely, when comparing two configurations, an additional run is first performed for the configuration whose performance estimate is based on fewer runs or, in the case of a tie, on both configurations. Then, as long as neither configuration dominates the other, further runs are performed based on the same criterion. Furthermore, when domination has been determined, FocusedILS performs additional runs for the winner of the comparison (ties are always broken in favour of more recently visited configurations). The number of these *bonus runs* is determined as the number of configurations visited since the last improving search step, i.e., since the last time a comparison between two configurations was decided in favour of the one that had been visited more recently. This mechanism ensures that the better a configuration appears to perform, the more thoroughly it is evaluated, especially in cases where a performance improvement is observed after a number of unsuccessful attempts.

As first established by Hutter et al. [36], FocusedILS has an appealing theoretical property: With increasing run-time, the probability of finding a configuration with globally optimal performance on the given set of benchmark instances approaches 1. This probabilistic approximate completeness (PAC) property follows from two key observations: Firstly, thanks to the previously mentioned probabilistic restart mechanism used in the ParamILS framework, over time any configuration from a finite configuration space is visited arbitrarily often. Secondly, as the number of visits to a given configuration increases, so does the number of target algorithm runs FocusedILS performs on it, which causes the probability of mistakenly failing to recognise that its true performance on the given instance set is better than that of any other configuration it is compared with approach 0. While this theoretical guarantee only concerns the behaviour of FocusedILS at the limit, as run-time approaches infinity, the mechanisms giving rise to it appear to be very effective in practice when dealing with surprisingly large configuration spaces (see, e.g., 35, 37, 42). Nevertheless, stagnation of the search process has been observed in several cases and is typically ameliorated by performing multiple runs of FocusedILS independently in parallel, from which a single winning configuration is determined based on the performance observed on the set of benchmark instances used in those runs. We expect that by replacing the simplistic probabilistic restart mechanism, and possibly modifying the mechanism used for allocating the additional target algorithm runs to be performed when assessing configurations, stagnation can be avoided or overcome more effectively.

3.3.4 Adaptive Capping

Both BasicILS and FocusedILS can be improved by limiting under certain conditions the time that is spent evaluating poorly performing configurations. The key idea is that when comparing two configurations c_1 and c_2, a situation may arise where, regardless of the results of any further runs, c_2 cannot match or exceed the performance of c_1 [37]. This is illustrated by the following example, taken from Hutter et al. [37]: Consider the use of BasicILS(100) for minimising the expected run-time of a given target algorithm on a set of 100 benchmark instances, where configuration c_1 has solved all 100 instances in a total of ten CPU seconds, and c_2 has run for the same ten CPU seconds on the first instances without solving them.

Clearly, we can safely terminate that latter run after $10 + \varepsilon$ CPU seconds (for some small time ε), since the average run-time of c_2 must exceed 0.1 CPU seconds, regardless of its performance in the remaining $N - 1$ runs, and therefore be worse than that of c_1.

Based on this insight, the *trajectory-preserving adaptive capping mechanism* of Hutter et al. [37] limits the effort spent on evaluating configurations based on comparing lower bounds on the performance of one configuration c_2 to upper bounds (or exact values) on that of another configuration c_1, based on the results of given sets of runs for c_1 and c_2. We note that this corresponds to the notion of racing, where each of the two configurations works independently through a given number of runs, but the race is terminated as soon as the winner can be determined with certainty. Apart from the potential for savings in running time, the use of trajectory-preserving capping does not change the behaviour of ParamILS.

A heuristic generalisation of this capping mechanism makes it possible to achieve even greater speedups, albeit at the price of possibly substantial changes to the search trajectory followed by the configuration procedure. The key idea behind this generalisation (dubbed *aggressive capping*) is to additionally bound the time allowed for evaluating configurations based on the performance observed for the current incumbent, i.e., the best-performing configuration encountered since the beginning of the ParamILS run. The additional bound is obtained by multiplying the performance estimate of the incumbent by a constant, *bm*, called the *bound multiplier*. Formally, for $bm = \infty$, the additional bound becomes inactive (assuming the performance measure is to be minimised), and the behaviour of trajectory-preserving capping is obtained. For $bm = 1$, on the other hand, a very aggressive heuristic is obtained, which limits the evaluation of any configuration to the time spent on evaluating the current incumbent. In practice, $bm = 2$ appears to result in good performance and is used as a default setting in ParamILS. Despite its heuristic nature, this modified capping mechanism preserves the PAC property of FocusedILS.

Although Hutter et al. [37] spelled out their adaptive capping mechanisms for the performance objective of minimising a target algorithm's mean run-time only, these mechanisms generalise to other objectives in a rather straightforward way (a discussion of capping in the context of minimising quantiles of run-time is found in Ch. 7 of [32]). We note, however, that – especially when several target algorithm runs are conducted in parallel – adaptive capping would be most effective in the case of run-time minimisation. Particularly substantial savings can be achieved during the assessment of the $r + 1$ configurations considered during initialisation of the search process, as well as towards the end of each local search phase. Finally, it should be noted that adaptive capping mechanisms can be used in the context of configuration procedures other than ParamILS; for example, Hutter et al. [37] mention substantial speedups achieved by using adaptive capping in combination with simple random sampling (the same procedure as that used during the initialisation of ParamILS).

3.3.5 Applications

ParamILS variants, and in particular FocusedILS, have been very successfully applied to a broad range of high-performance algorithms for several hard combinatorial problems. An early version of FocusedILS was used by Thachuk et al. [65] to configure a replica-exchange Monte Carlo (REMC) search procedure for the 2D and 3D HP protein structure prediction problems; the performance objective was to minimise the mean run-time for finding ground states for a given set of sequences in these abstract but prominent models of protein structure, and the resulting configurations of the REMC procedure represented a considerable improvement over the state of the art techbniques for solving these challenging problems.

FocusedILS has also been used in a series of studies leading to considerable advances in the state-of-the-art in solving the satisfiability problem in propositional logic, one of the most widely studied $\mathcal{NP}$-hard problems in computer science. Hutter et al. [35] applied this procedure to SPEAR, a complete, DPLL-type SAT solver with 26 parameters (ten of which are categorical), which jointly give rise to a total of $8.34 \cdot 10^{17}$ possible configurations. The design of SPEAR was influenced considerably by the availability of a powerful configuration tool such as FocusedILS, whose application ultimately produced configurations that solved a given set of SAT-encoded software verification problems about 100 times faster than previous state-of-the-art solvers for these types of SAT instances and won the first prize in the QF_BV category of the 2007 Satisfiability Modulo Theories (SMT) Competition.

KhudaBukhsh et al. [45] used FocusedILS to find performance-optimised instantiations of SATenstein-LS, a highly parametric framework for stochastic local search (SLS) algorithms for SAT. This framework was derived from components found in a broad range of high-performance SLS-based SAT solvers; its 41 parameters induce a configuration space of size $4.82 \cdot 10^{12}$. Using FocusedILS, performance improvements of up to three orders of magnitudes were achieved over the previous best-performing SLS algorithms for various types of SAT instances, for several of which SLS-based solvers are the most effective SAT algorithms overall. Several automatically determined configurations of SATenstein-LS were used in the most recent SATzilla solvers, which led the field in the 2009 SAT Competition, winning prizes in five of the nine main categories [69].

Very recently, Xu et al. [70] used FocusedILS in an iterative fashion to obtain sets of configurations of SATenstein-LS that were then used in combination with state-of-the-art per-instance algorithm selection techniques (here: SATzilla). In each iteration of the overall procedure, dubbed *Hydra*, FocusedILS was used to find configurations that would best complement a given portfolio-based per-instance algorithm selector. This approach resulted in a portfolio-based SAT solver that, while derived in a fully automated fashion from a single, highly parameterised algorithm, achieved state-of-the-art performance across a wide range of benchmark instances.

Tompkins and Hoos [66] applied FocusedILS to a new, flexible framework for SLS-based SAT solvers called VE-Sampler (which is conceptually orthogonal to the previously mentioned SATenstein-LS framework). VE-Sampler has a large number of categorical and conditional parameters, which jointly give rise to more than 10^{50}

distinct configurations, and using FocusedILS, configurations could be found that were shown to solve two well-known sets of SAT-encoded software verification problems between 3.6 and nine times faster than previous state-of-the-art SLS-based SAT solvers for these types of SAT instances.

Chiarandini et al. [15] used FocusedILS to configure a hybrid, modular SLS algorithm for a challenging university timetabling problem that subsequently placed third in Track 2 of the Second International Timetabling Competition (ITC 2007). The configuration space considered in this context was relatively small (seven parameters, 50,400 configurations), but the use of automated algorithm configuration made it possible to achieve close to state-of-the-art performance in very limited human development time and without relying on deep and extensive domain expertise. In subsequent work, Fawcett et al. [17] expanded the design space by parameterising additional parts of the solver, and – using multiple rounds of FocusedILS with different performance objectives – obtained a configuration that was demonstrated to substantially improve upon the state of the art for solving the post-enrolment course timetabling problem considered in Track 2 of ITC 2007. The overall performance metric used in these studies (and in the competition) was solution quality achieved by the target solver after a fixed amount of time.

Recently, Hutter et al. [42] reported substantial improvements in the performance of several prominent solvers for mixed integer programming (MIP) problems, including the widely used industrial CPLEX solver. In the case of CPLEX, FocusedILS was used to configure 76 parameters, most of which are categorical (and some conditional), giving rise to a configuration space of size $1.9 \cdot 10^{47}$. Despite the fact that the default parameter settings for CPLEX are known to have been chosen carefully, based on a considerable amount of thorough experimentation, substantial performance improvements were obtained for many prominent types of MIP instances, in terms of both time required for finding optimal solutions (and proving-optimality) and the minimising of the solution quality (optimality gap) achieved within a fixed amount of time (speedup factors beween 1.98 and 52.3, and gap reduction factors between 1.26 and 8.65). Similarly impressive results were achieved for another commercial MIP solver, Gurobi, and a prominent open-source MIP solver, lpsolve.

Finally, Hutter et al. [39] reported a successful meta-configuration experiment, in which BasicILS was used to optimise the four parameters that control the behaviour of FocusedILS. BasicILS was chosen as the meta-configurator, since its runs can be parallelised in a straightforward manner. At least partly because of the enormous computational cost associated with this experiment (where each target algorithm run corresponds to solving an algorithm configuration task, and hence to executing many costly runs of the algorithm configured in that task, which itself solved instances of an $\mathcal{NP}$-hard problem such as SAT), only marginal improvements in the performance of the configurator, FocusedILS, on a number of previously studied configuration benchmarks could be achieved.

The target algorithms considered in most of these applications have continuous parameters, and up to this point ParamILS requires these parameters to be discretised. While in principle finding reasonable discretisations (i.e., ones whose use does not cause major losses in the performance of the configurations found by ParamILS) could be difficult, in most cases generic approaches such as even or geometric subdivisions of a given interval seem to give good results. Where this is not the case, multiple runs of the configuration procedure can be used to iteratively refine the domains of continuous parameters. The same approach can be used to extend domains in cases where parameter values in an optimised configuration lie at the boundary of their respective domains. Nevertheless, the development of ParamILS variants that natively deal with continuous parameters and support dynamic extensions of parameter domains remains an interesting direction for future work.

3.4 Sequential Model-Based Optimisation

A potential disadvantage of the model-free search approach underlying ParamILS is that it makes no explicit attempt to recognise and benefit from regularities in a given configuration space. The *model-based search paradigm* underlying the approach discussed in this section, on the other hand, uses the information gained from the configurations evaluated so far to build (and maintain) a model of the configuration space, based on which configurations are chosen to be evaluated in the future. We note that, as also pointed out by its authors, the Iterative F-Race procedure (I/F-Race) discussed in Section 3.2.2 of this chapter is a model-based configuration procedure in this sense. But unlike in I/F-Race, the models used by the methods discussed in the following capture directly the dependency of target algorithm performance on parameter settings. These *response surface models* can be used as surrogates for the actual parameter response of a given target algorithm and provide the basis for determining promising configurations at any stage of an iterative model-based search procedure; this generic approach is known as *sequential model-based optimisation (SMBO)* and can be seen as a special case of sequential analysis – a broader area within statistics that also comprises sequential hypothesis testing and so-called multi-armed bandits. An outline of SMBO for algorithm configuration is shown in Figure 3.4; in principle, performance measurements for multiple configurations can be performed independently in parallel.

The setting considered in almost all work on sequential model-based optimisation procedures up to this day is known as the *black-box optimisation problem*: Given an unknown function f and a space of possible inputs X, find an input $x \in X$ that optimises f based on measurements of f on a series of inputs. The function to be optimised, f, may be deterministic or stochastic; in the latter case, measurements are subject to random noise, and formally the values of f are random variables. Algorithm configuration can be seen as a special case of black-box optimisation, where the function to be optimised is the performance m of an algorithm A on a set of problem instances I. However, in contrast to algorithm configuration procedures such as FocusedILS or F-Race, black-box optimisation procedures do not take into account the fact that approximate measurements can be obtained at lower computational cost by running A on subsets of I, and that, by blocking on instances (and pseudo-random

procedure *SMBO*
    input *target algorithm A, set of configurations C, set of problem instances I,*
        performance metric m;
    output *configuration c^**;

    determine initial set of configurations $C_0 \subset C$;
    for all $c \in C_0$, measure performance of A on I according to metric m;
    build initial model M based on performance measurements for C_0;
    determine incumbent $c^* \in C_0$ for which best performance was observed or predicted;
    repeat
        based on model M, determine set of configurations $C' \subseteq C$;
        for all $c \in C'$, measure performance of A on I according to metric m;
        update model M based on performance measurements for C';
        update incumbent c^*;
    until termination condition met;
    return c^*;
end *SMBO*

Fig. 3.4: High-level outline of the general sequential model-based optimisation approach to automated algorithm configuration; model M is used to predict the performance of configurations that have not (yet) been evaluated, and set C' is typically chosen to contain configurations expected to perform well based on those predictions. Details of various algorithms following this approach are explained in the text

number seeds), performance measurements for different configurations can be compared more meaningfully; furthermore, they have no means of exploiting knowledge about the instances in I acquired from earlier target algorithm runs.

Because black-box function optimisation is somewhat more general than algorithm configuration, and methods for solving black-box functions are easily applicable to modelling and optimising the response of a wide range of systems, in the following we use standard terminology from the statistics literature on experimental design, in particular, *design point* for elements of the given input space X and *response* for values of the unknown function f. In the context of algorithm configuration, design points correspond to configurations of a given target algorithm A, and response values represent A's performance m on instance set I. A unified, more technical presentation of the methods covered in this section can be found in the dissertation of Hutter [32], and further details are provided in the original articles referenced throughout.

3.4.1 The EGO Algorithm

The efficient global optimisation (EGO) algorithm for black-box function optimisation by Jones et al. [44] uses a response surface model obtained via noise-free Gaussian process regression in combination with an expected improvement criterion for selecting the next configuration to be evaluated. The noise-free Gaussian process (GP) model utilised by EGO is also known as the *DACE model*, after its prominent use in earlier work by Sacks et al. [59]. It defines for every input x a random variable $\hat{F}(x)$ that characterises the uncertainty over the true response value $f(x)$ at point x.

The model-based optimisation process carried out by EGO starts with about $10 \cdot k$ design points determined using a k-dimensional space-filling Latin hypercube design (LHD). After measuring the response values for these values, the $2 \cdot k + 2$ parameters of a DACE model are fit to the pairs of design points and response values, using maximum likelihood estimates (as described by 44, this can be partially done in closed form). The resulting model is assessed by means of so-called *standard-ized cross-validated residuals*, which reflect the degree to which predictions made by the model agree with the observed response values on the design points used for constructing the model. If the model is deemed unsatisfactory, the response values may be transformed using a log- or inverse-transform (i.e., modified by applying the function $\ln y$ or $1/y$) and the model fitted again.

After a satisfactory initial model has been obtained, it is used in conjunction with an *expected improvement criterion* to determine a new design point to be evaluated. The expected improvement measure used in this context uses the current DACE model M to estimate the expected improvement over the best response value measured so far, f_{min}, at any given design point x, and is formally defined as $EI(x) := E[\max\{f_{min} - \hat{F}(x), 0\}]$, where $\hat{F}(x)$ is the random variable describing the response for a design point x according to model M. Using a closed-form expression for this measure given by Jones et al. [44] and a branch & bound search method (which can be enhanced heuristically), the EGO algorithm then determines a design point x' with maximal expected improvement $EI(x')$. If $EI(x')$ is less than 1% of the current incumbent, the procedure terminates. Otherwise, the response value $f(x')$ is measured, and the DACE model is refitted on the previous set of data extended by the pair $(x', f(x'))$, and a new iteration begins, in which the updated model is used to determine the next design point using the same process that yielded x'.

Note that in every iteration of this process, the DACE model has to be fitted, which involves a matrix inversion of cost $O(n^3)$, where n is the number of design points used. Depending on the cost of measuring the response value for a given design point, this may represent a substantial computational overhead. Furthermore, the noise-free Gaussian process model used in EGO cannot directly characterise the stochastic responses obtained when solving algorithm configuration problems involving randomised target algorithms.

3.4.2 Sequential Kriging Optimisation and Sequential Parameter Optimisation

We now discuss two black-box optimisation procedures that deal with stochastic responses, as encountered when modelling phenomena subject to observation noise or configuring randomised algorithms.

The first of these, known as *Sequential Kriging Optimisation* [31] estimates a Gaussian process (GP) model (also known as a *kriging model*) directly from samples (i.e., noisy measurements) of response values. Similarly to EGO, SKO starts with an LHD of $10 \cdot k$ design points, for which response values are sampled. To facilitate initial estimates of the observation noise, one additional sample is then drawn for each of the k best of these design points, after which a Gaussian process model M is fitted directly to the resulting set of $11 \cdot k$ pairs of design points and corresponding response values. The resulting model M is then assessed and possibly modified based on a transformation of the response, as in the EGO algorithm.

Next, an incumbent design point is determined based on model M by minimising the expression $\mu(x) + \sigma(x)$ using the Nelder-Mead Simplex algorithm [49], where $\mu(x)$ and $\sigma(x)$ are the mean and standard deviation predicted by M for input x, and the minimisation is over the design points used in constructing the model. This risk adverse strategy is less easily misled by inaccurate estimates of the mean response value than a minimisation of the predicted mean only. The next design point to be evaluated is determined based on model M using an augmented expected improvement measure, designed to steer the process away from design points with low predictive variance. This augmented expected improvement measure is formally defined as

$$EI'(x) := E[\max\{\hat{f}_{min} - \hat{F}(x), 0\}] \cdot \left(1 - \sigma_{\varepsilon}/\sqrt{s^2(x) - \sigma_{\varepsilon}^2}\right),$$

where $\hat{f}_{min}$ is the model's prediction for the current best input (as in EGO, obtained by considering all design points used for building the model), $\hat{F}(x)$ is the random variable describing the response for a design point x according to model M, σ_{ε} is the standard deviation of the measurement noise (assumed to be identical for all inputs), and $s^2(x)$ is the variance of the response $\hat{F}(x)$ given by the model at point x, where the second term in the product decreases as the predictions of M become more accurate. Based on the given model M, the next design point to be evaluated, x', is determined by maximising $EI'(x)$ using the Nelder-Mead Simplex algorithm [49]. Next, the model is refitted, taking into account x' and a response value sampled at x', and a new iteration begins, in which the updated model is used to determine the next design point using the same process that yielded x'. If the maximum $EI'(x)$ values from $d + 1$ successive iterations all fall below a user-defined threshold, the iterative sampling process is terminated. (This treshold can be specified as an absolute value or as a fraction of the difference between the largest and smallest observed response values.)

Unfortunately, SKO assumes that the variability of the response values at each design point is characterised by a Gaussian distribution, and that the standard deviations of those distributions are the same across the entire design space. Both of these assumptions are problematic in the context of configuring randomised algorithms, particularly when minimising run-time (see, e.g., 30). Furthermore, the time required for fitting the model in each iteration of SKO is cubic in the number of response values sampled, which can represent a substantial computational burden.

The *Sequential Parameter Optimisation (SPO)* procedure by Bartz-Beielstein et al. [8] follows a fundamentally different strategy to deal with noisy response measurements:[3] Rather than fitting a Gaussian process model directly to a set of sampled responses, for each design point x the measure to be optimised is estimated empirically based on all samples taken at x, and a noise-free Gaussian process model (like the one used in the EGO algorithm) is fitted to the resulting data. In contrast to the approach taken by SKO, this makes it possible to optimise arbitrary statistics of the noisy function values, i.e., in the case of algorithm configuration, of the given target algorithm's run-time distributions; examples of such statistics are the mean, median, and arbitrary quantiles, and also measures of variability, as well as combinations of measures of location and variability. Another advantage of this approach is its substantially lower computational complexity: While SKO requires time cubic in the number of function values sampled, SPO's run-time is only cubic in the number of distinct design points – typically a much lower number.

Like SKO and EGO, SPO uses a Latin hypercube design as a basis for constructing the initial model; however, SPO chooses d design points and samples r response values for each of these, where d and r are specified by the user (the default value of r is 2). Based on these samples, empirical estimates of the measure to be optimised are calculated for each design point, and the point with the best resulting value is chosen as the initial incumbent. Next, a noise-free Gaussian process model is fitted to the resulting set of d pairs of design points and empirical response statistics. This model, M, is sampled for $10,000$ design points chosen uniformly at random,[4] and the best j of these according to an expected improvement (EI) measure are selected for further evaluation, where j is a user-defined number with a default setting of 1. The EI measure used in this context is formally defined as

$$EI^2(x) := E[(f_{min} - \hat{F}(x))^2] = E^2[f_{min} - \hat{F}(x)] + \mathrm{Var}[f_{min} - \hat{F}(x)],$$

where f_{min} is the best value of the measure to be optimised observed so far, and $\hat{F}(x)$ is the distribution over the predictions obtained from model M at design point x. This EI measure has been introduced by Schonlau et al. [62] with the aim of encouraging the exploration of design points for which the current model produces highly uncertain predictions.

At each of the design points determined in this way, r new response values are measured. Furthermore, additional response values are measured for the current incumbent to ensure that it is evaluated based on as many samples as available for any of the new design points. Then, the best of all the design points considered so far, according to the given measure to be optimised, is selected as the new incumbent

---

[3] In the literature, the term *sequential parameter optimisation* is also used to refer to a broad methodological framework encompassing fully automated as well as interactive approaches for understanding and optimising an algorithm's performance in response to its parameter settings. Here, as in the work of Hutter et al. [38], we use the term more narrowly to refer to the fully automated SMBO procedures implemented in various versions of the Sequential Parameter Optimization Toolbox (SPOT) by [9].

[4] We note that the use of a space-filling design, such as an LHD, should in principle yield better results if implemented sufficiently efficiently.

(with ties broken uniformly at random). If the design point thus selected has been an incumbent at any point earlier in the search process, r is increased; in SPO version 0.3 [8], r is doubled, while in the newer version 0.4 [7], it is merely incremented by 1, and in both cases values of r are limited to a user-specified maximum value r_{max}. At this point, a new iteration of SPO begins, in which a noise-free GP is fitted on the augmented set of data.

3.4.3 Recent Variants of Sequential Parameter Optimisation: SPO$^+$ and TB-SPO

Based on a detailed investigation of the core components of the SPO algorithm, Hutter et al. [38] introduced a variant called SPO$^+$ that shows considerably more robust performance on standard benchmarks than the SPO 0.3 and SPO 0.4 algorithms described previously.

The main difference between SPO$^+$ and the previous SPO procedures lies in the way in which new design points are accepted as incumbents. Inspired by FocusedILS, SPO$^+$ uses a mechanism that never chooses a new incumbent $\hat{x}'$ without ensuring that at least as many responses have been sampled at $\hat{x}'$ as at any other design point $x \neq \hat{x}'$. To achieve this, for any challenger to the current incumbent $\hat{x}$, i.e., for any design point x' that appears to represent an improvement over $\hat{x}$ based on the samples taken so far, additional response values are sampled until either x' ceases to represent an improvement, or the number of response values sampled at x' reaches that taken at $\hat{x}$, with x' still winning the comparison with $\hat{x}$ based on the respective samples; only in the latter case does x' become the new incumbent, while in the former case it is dismissed, and as many additional response values are sampled for $\hat{x}$ as newly measured for x'.

The new response values determined for a challenger x' are sampled in batches, with the number of new samples taken doubling in each successive batch. As noted by Hutter et al. [38], using this mechanism, rejection of challengers is done in a rather aggressive, heuristic manner, and frequently occurs after only a single response value has been sampled at x' – long before a statistical test could conclude that the x' is worse than the current incumbent.

The *Time-Bounded Sequential Parameter Optimisation (TB-SPO)* algorithm by Hutter et al. [41] introduces a number of further modifications to the SMBO framework underlying the previously described SPO variants. In particular, in contrast to all SMBO procedures discussed so far, TB-SPO does not construct its initial model based on a large set of samples determined using a Latin hypercube design, but rather interleaves response measurements at randomly chosen points with ones taken at points that appear to be promising based on the current model. The initial model is based on a single sample only; when used for algorithm configuration, where the black-box function to be optimised represents the output of a parameterised algorithm, the default configuration for the algorithm is used as the design point at which this initial sample is taken. At any stage of the iterative model-based search process that follows, response values are sampled at a series of design points in which odd-numbered points are determined by optimising an expected improvement measure (as is done in SPO$^+$), while even-numbered points are sampled uniformly at random from the given design space. (Mechanisms that achieve a different balance between promising and randomly chosen design points could lead to better performance but have not been explored so far.)

The number of design points at which response values are sampled between any two updates to the model is determined based on the time t required for constructing a new model and the search for promising parameter settings; to be precise, after at least two design points have been evaluated, further points are considered until the time used for evaluating design points since the last model update exceeds a user-defined multiple (or fraction) of the overhead t.

Finally, in order to reduce the computational overhead incurred by the model construction process, TB-SPO uses an approximate version of the standard Gaussian process models found in the other SPO variants. This so-called *projected process (PP) approximation* is based on the idea of representing explicitly only a randomly sampled subset of the given data points (here: pairs of input and response values) when building the Gaussian process model; if this subset comprises s data points, while the complete set has n data points, the time complexity of fitting a GP model decreases from $O(n^3)$ to $O((s+n) \cdot s^2)$, while the time required for predicting a response value (mean and variance of the predictive distribution at a given design point) decreases from $O(n^2)$ to $O(s^2)$ [56]. In the context of an SMBO procedure, this will typically lead to substantial savings, since the number of data points available increases over time, and n can easily reach values of several thousand, while effective PP approximations can be based on constant-size subsets with s no larger than 300 [41]. (Details on other, minor differences between TB-SPO and SPO$^+$ can be found in [41].)

3.4.4 Applications

As mentioned earlier, sequential model-based optimisation methods have primarily been developed for the optimisation of black-box functions, but can obviously be applied to algorithm configuration problems by defining the function to be optimised to be the performance of a target observed algorithm applied to one or more benchmark instances. The design points are thus algorithm configurations, and the response values capture the performance of A on the given benchmark instance(s) according to some performance measure m. In principle, SMBO procedures like those described earlier in this section can be applied to optimise a target algorithm on a set I of benchmark instances by using a measure m that captures the performance on the entire set I; however, as discussed earlier for the case of BasicILS, this tends to quickly become impractical as the size of I grows. Therefore, the empirical evaluation of SMBO procedures tends to be focussed on performance optimisation on single benchmark instances. Furthermore, because of the nature of the response surface models used, SMBO methods are usually restricted to dealing with real- and integer-valued target algorithm parameters (although, very recently, [43] has introduced techniques that can handle categorical parameters).

Following an example from Bartz-Beielstein et al. [9], the SPO variants discussed in this section have been empirically evaluated using CMA-ES [25, 26, 27] – one of the best-performing gradient-free numerical optimisation procedures currently known – on several standard benchmark functions from the literature on gradient-free numerical optimisation (see, e.g., 26). The configuration space considered in these examples, which involve the convex Sphere function as well as the non-convex Ackley, Griewank and Rastrigin functions, is spanned by three real- and one integer-valued parameters of CMA-ES, and the performance measure was solution quality, achieved after a fixed number of evaluations of the respective benchmark function. The empirical results reported by Hutter et al. [38] for CMA-ES applied to the ten-dimensional instances of these functions indicate that SPO$^+$ tends to perform significantly better than SPO 0.3 and 0.4, which in turn appear to perform substantially better than SKO. In addition, Hutter et al. [38] considered the minimisation of the median number of search steps required by SAPS [33], a well-known stochastic local search algorithm for SAT, to solve a single benchmark instance obtained from encoding a widely studied quasi-group completion problem into SAT; in this case, four continuous parameters were optimised. The results from that experiment confirmed that SPO$^+$ tends to perform better than previous SPO variants and suggest that, at least on some configuration problems with a relatively modest number of predominently real-valued parameters, it can also yield slightly better results than FocusedILS when allowed the same number of target algorithm runs.

TB-SPO has been empirically compared to SPO$^+$ and FocusedILS on relatively simple algorithm configuration tasks involving the well-known SAT solver SAPS [33], with four continuous parameters, running on single SAT instances. In these experiments, TB-SPO was shown to perform significantly better than SPO$^+$ (sometimes achieving over 250-fold speedups), and moderately better than FocusedILS [41]. However, it is important to keep in mind that, unlike TB-SPO (and all other SMBO procedures covered in this section), FocusedILS explicitly deals with multiple problem instances, and can therefore be expected to perform substantially better on realistic algorithm configuration tasks. Furthermore, while SMBO procedures like TB-SPO do not require continuous algorithm parameters to be discretised, they presently cannot deal with conditional parameters, which are routinely encountered in the more challenging algorithm configuration tasks on which FocusedILS has been shown to be quite effective.

3.5 Other Approaches

In addition to the methods covered in the previous sections, there are many other procedures described in the literature that can, at least in principle, be applied to the algorithm configuration problem considered here.

Experimental design methods, such as full or fractional factorial designs, stratified random sampling, Latin hypercube designs and various other types of space-filling and uniform designs (see, e.g., 60), are applicable to algorithm configuration, but per se do not take into account one fundamental aspect of the problem: Namely, that we are interested in performance on multiple instances and have control over the number of instances used for evaluating any given configuration. Furthermore, when minimizing the run-time of the given target algorithm (a very common performance objective in algorithm configuration and parameter tuning), it is typically necessary to work with censored data from incomplete runs, and it can be beneficial to cut off runs early (as done by the adaptive capping strategy explained in Section 3.3.4). Perhaps more importantly, experimental design methods lack the heuristic guidance that is often crucial for searching large configuration spaces effectively.

Nevertheless, these simple design methods are sometimes used for the initialisation of more complex procedures (see Section 3.4 and [1], which will be discussed in slightly more detail later). There is also some evidence that in certain cases a method as simple as uniform random sampling, when augmented with adaptive capping or with the mechanism used by TB-SPO for evaluating configurations, can be quite effective (see the recent work [41]).

In principle, gradient-free numerical optimisation methods are directly applicable to parameter tuning problems, provided that all parameters are real-valued (and that there are no parameter dependencies, such as conditional parameters). Prominent and relatively recent methods that appear to be particularly suitable in this context are the covariance matrix adaptation evolution strategy (CMA-ES) by Hansen and Ostermeier [27] and the mesh adaptive direct search (MADS) algorithms by Audet and Orban [4]. Similarly, it is possible to use gradient-based numerical optimisation procedures – in particular, quasi-Newton methods such as the Broyden-Fletcher-Goldfarb-Shanno (BFGS) algorithm (see, e.g., 51) – in conjunction with suitable methods for estimating or approximating gradient information. However, in order to be applied in the context of configuring target algorithms with categorical and conditional parameters, these methods would require non-obvious modifications; we also expect that in order to be reasonably effective, they would need to be augmented with mechanisms for dealing with multiple problem instances and capped (or censored) runs. The same holds for standard methods for solving stochastic optimisation problems (see, e.g., 63).

The CALIBRA algorithm by Adenso-Diaz and Laguna [1], on the other hand, has been specifically designed for parameter tuning tasks. It uses a specific type of fractional factorial design from the well-known Taguchi methodology in combination with multiple runs of a local search procedure that gradually refines the region of interest. Unfortunately, CALIBRA can handle no more than five parameters, all of which need to be ordinal (the limitation to five parameters stems from the specific fractional designs used at the core of the procedure).

The CLASS system by Fukunaga [18, 19] is based on a genetic programming approach; it has been specifically built for searching a potentially unbounded space of the heuristic variable selection method used in an SLS-based SAT solver. Like most genetic programming approaches, CLASS closely links the specification of the configuration space and the evolutionary algorithm used for exploring this space.

The Composer system developed by Gratch and Dejong [23] is based on an iterative improvement procedure not unlike that used in ParamILS; this procedure is conceptually related to racing techniques in that it moves to a new configuration only after gathering sufficient statistical evidence to conclude that this new configuration performs significantly better than the current one. In a prominent application, Gratch and Chien [22] used the Composer system to optimise five parameters of an algorithm for scheduling communication between a spacecraft and a set of ground-based antennas.

Ansótegui et al. [2] recently developed a gender-based genetic algorithm for solving algorithm configuration problems. Their GGA procedure supports categorical, ordinal and real-valued parameters; it also allows its user to express independencies between parameter effects by means of so-called variable trees – a concept that could be of particular interest in the context of algorithm configuration problems where such independencies are known by construction, or heuristic methods are are available for detecting (approximate) independencies. Although there is some evidence that GGA can solve some moderately difficult configuration problems more effectively than FocusedILS without capping [2], it appears to be unable to reach the performance of FocusedILS version 2.3 with aggressive capping on the most challenging configurations problems [40]. Unfortunately, GGA also offers less flexibility than FocusedILS in terms of the performance metric to be optimised. More algorithm configuration procedures based on evolutionary algorithms are covered in Chapter 2 of this book.

Finally, work originating from the Ph.D. project of Hutter [32] has recently overcome two major limitations of the sequential model-based optimisation methods discussed in Section 3.4 of this chapter by introducing a procedure that can handle categorical parameters while explicitly exploiting the fact that performance is evaluated on a set of problem instances. There is some evidence that this procedure, dubbed *Sequential Model-based Algorithm Configuration (SMAC)*, can, at least on some challenging configuration benchmarks, reach and sometimes exceed the performance of FocusedILS [43], and we are convinced that, at least in cases where the parameter response of a given target algorithm is reasonably regular and performance evaluations are very costly, such advanced SMBO methods hold great promise.

3.6 Conclusions and Future Work

Automated algorithm configuration and parameter tuning methods have been developed and used for more than a decade, and many of the fundamental techniques date back even further. However, it has only recently become possible to effectively solve complex configuration problems involving target algorithms with dozens of parameters, which are often categorical and conditional. This success is based in part on the increased availability of computational resources, but has mostly been enabled by methodological advances underlying recent configuration procedures.

Still, we see much room (and, indeed, need) for future work on automated algorithm configuration and parameter tuning methods. We believe that in developing such methods, the fundamental features underlying all three types of methods discussed in this chapter can play an important role, and that the best methods will employ combinations of these. We further believe that different configuration procedures will likely be most effective for solving different types of configuration problems (depending, in particular, on the number and type of target algorithm parameters, but also on regularities in the parameter response). Therefore, we see a need for research aiming to determine which configurator is most effective under which circumstances. In fact, we expect to find situations in which the sequential or iterative application of more than one configuration procedure turns out to be effective – for example, one could imagine applying FocusedILS to find promising configurations in vast, discrete configuration spaces, followed by a gradient-free numerical optimisation method, such as CMA-ES, for fine-tuning a small number of real-valued parameters.

Overall, we believe that algorithm configuration techniques, such as the ones discussed in this chapter, will play an increasingly crucial role in the development, evaluation and use of state-of-the-art algorithms for challenging computational problems, where the challenge could arise from high computational complexity (in particular, $\mathcal{NP}$-hardness) or from tight resource constraints (e.g., in real-time applications). Therefore, we see great value in the design and development of software frameworks that support the real-world application of various algorithm configuration and parameter tuning procedures. The High-Performance Algorithm Lab (HAL), recently introduced by Nell et al. [50], is a software environment designed to support a wide range of empirical analysis and design tasks encountered during the development, evaluation and application of high-performance algorithms for challenging computational problems, including algorithm configuration and parameter tuning. Environments such as HAL not only facilitate the application of automated algorithm configuration and parameter tuning procedures, but also their development, efficient implementation and empirical evaluation.

In closing, we note that the availability of powerful and effective algorithm configuration and parameter tuning procedures has a number of interesting consequences for the way in which high-performance algorithms are designed and used in practice. Firstly, for developers and end users, it is now possible to automatically optimise the performance of (highly) parameterised solvers specifically for certain classes of problem instances, leading to potentially much improved performance in real-world applications. Secondly, while on-line algorithm control mechanisms that adjust parameter settings during the run of a solver (as covered, for example, in Chapters 6, 7 and 8 of this book) can in principle lead to better performance than the (static) algorithm configuration procedures considered in this chapter, we expect these latter procedures to be very useful in the context of (statically) configuring the parameters and heuristic components that determine the behaviour of the on-line control mechanisms. Finally, during algorithm development, it is no longer necessary (or even desirable) to eliminate parameters and similar degrees of freedom, but instead, developers can focus more on developing ideas for realising certain heuristic mechanisms or components, while the precise instantiation can be left to automated configuration procedures [28]. We strongly believe that this last effect will lead to a fundamentally different and substantially more effective way of designing and implementing high-performance solvers for challenging computational problems.

Acknowledgements This chapter surveys and discusses to a large extent work carried out by my research group at UBC, primarily involving Frank Hutter, Kevin Leyton-Brown and Kevin Murphy, as well as Thomas Stützle at Université Libre de Bruxelles, to all of whom I am deeply grateful for their fruitful and ongoing collaboration. I gratefully acknowledge valuable comments by Frank Hutter, Thomas Stützle and Maverick Chan on earlier drafts of this chapter, and I thank the members of my research group for the many intellectually stimulating discussions that provide a fertile ground for much of our work on automated algorithm configuration and other topics in empirical algorithmics.

References

[1] Adenso-Diaz, B., Laguna, M.: Fine-tuning of algorithms using fractional experimental design and local search. Operations Research 54(1):99–114 (2006)

[2] Ansótegui, C., Sellmann, M., Tierney, K.: A gender-based genetic algorithm for the automatic configuration of algorithms. In: Proceedings of the 15th International Conference on Principles and Practice of Constraint Programming (CP 2009), pp. 142–157 (2009)

[3] Applegate, D. L., Bixby, R. E., Chvátal, V., Cook, W. J.: The Traveling Salesman Problem: A Computational Study. Princeton University Press (2006)

[4] Audet, C., Orban, D.: Finding optimal algorithmic parameters using the mesh adaptive direct search algorithm. SIAM Journal on Optimization 17(3):642–664 (2006)

[5] Balaprakash, P., Birattari, M., Stützle, T.: Improvement strategies for the F-Race algorithm: Sampling design and iterative refinement. In: Bartz-Beielstein, T., Blesa, M., Blum, C., Naujoks, B., Roli, A., Rudolph, G., Sampels, M. (eds) 4th International Workshop on Hybrid Metaheuristics, Proceedings, HM 2007, Springer Verlag, Berlin, Germany, Lecture Notes in Computer Science, vol. 4771, pp. 108–122 (2007)

[6] Balaprakash, P., Birattari, M., Stützle, T., Dorigo, M.: Estimation-based metaheuristics for the probabilistic traveling salesman problem. Computers & OR 37(11):1939–1951 (2010)

[7] Bartz-Beielstein, T.: Experimental Research in Evolutionary Computation: The New Experimentalism. Natural Computing Series, Springer Verlag, Berlin, Germany (2006)

[8] Bartz-Beielstein, T., Lasarczyk, C., Preuß, M.: Sequential parameter optimization. In: McKay, B., et al. (eds) Proceedings 2005 Congress on Evolutionary

Computation (CEC'05), Edinburgh, Scotland, IEEE Press, vol. 1, pp. 773–780 (2005)

[9] Bartz-Beielstein, T., Lasarczyk, C., Preuss, M.: Sequential parameter optimization toolbox, manual version 0.5, September 2008, available at http://www.gm.fh-koeln.de/imperia/md/content/personen/lehrende/bartz_beielstein_thomas/spotdoc.pdf (2008)

[10] Battiti, R., Brunato, M., Mascia, F.: Reactive Search and Intelligent Optimization. Operations Research/Computer Science Interfaces, Springer Verlag (2008)

[11] Birattari, M., Stützle, T., Paquete, L., Varrentrapp,K.: A racing algorithm for configuring metaheuristics. In: GECCO '02: Proceedings of the Genetic and Evolutionary Computation Conference, pp. 11–18 (2002)

[12] Birattari, M., Yuan, Z., Balaprakash, P., Stützle, T.: F-race and Iterated F-Race: An overview. In: Bartz-Beielstein, T., Chiarandini, M., Paquete, L., Preuss, M. (eds) Experimental Methods for the Analysis of Optimization Algorithms, Springer, Berlin, Germany, pp. 311–336 (2010)

[13] Bűrmen, Á., Puhan, J., Tuma, T.: Grid restrained Nelder-Mead algorithm. Computational Optimization and Applications 34(3):359–375 (2006)

[14] Carchrae, T., Beck, J.: Applying machine learning to low knowledge control of optimization algorithms. Computational Intelligence 21(4):373–387 (2005)

[15] Chiarandini, M., Fawcett, C., Hoos, H.: A modular multiphase heuristic solver for post enrollment course timetabling (extended abstract). In: Proceedings of the 7th International Conference on the Practice and Theory of Automated Timetabling PATAT (2008)

[16] Da Costa, L., Fialho, Á., Schoenauer, M., Sebag, M.: Adaptive Operator Selection with Dynamic Multi-Armed Bandits. In: Proceedings of the 10th Annual Conference on Genetic and Evolutionary Computation (GECCO'08), pp. 913–920 (2008)

[17] Fawcett, C., Hoos, H., Chiarandini, M.: An automatically configured modular algorithm for post enrollment course timetabling. Tech. Rep. TR-2009-15, University of British Columbia, Department of Computer Science (2009)

[18] Fukunaga, A. S.: Automated discovery of composite SAT variable-selection heuristics. In: Proceedings of the 18th National Conference on Artificial Intelligence (AAAI-02), pp. 641–648 (2002)

[19] Fukunaga, A. S.: Evolving local search heuristics for SAT using genetic programming. In: Genetic and Evolutionary Computation – GECCO-2004, Part II, Springer-Verlag, Seattle, WA, USA, Lecture Notes in Computer Science, vol. 3103, pp. 483–494 (2004)

[20] Gagliolo, M., Schmidhuber, J.: Dynamic algorithm portfolios. In: Amato, C., Bernstein, D., Zilberstein, S. (eds) Proceedings of the 9th International Symposium on Artificial Intelligence and Mathematics (AI-MATH-06) (2006)

[21] Gomes, C. P., Selman, B., Crato, N., Kautz, H.: Heavy-tailed phenomena in satisfiability and constraint satisfaction problems. Journal of Automated Reasoning 24(1-2):67–100 (2000)

[22] Gratch, J., Chien, S. A.: Adaptive problem-solving for large-scale scheduling problems: A case study. Journal of Artificial Intelligence Research 4:365–396 (1996)

[23] Gratch, J., Dejong, G.: Composer: A probabilistic solution to the utility problem in speed-up learning. In: Rosenbloom, P., Szolovits, P. (eds) Proceedings of the 10th National Conference on Artificial Intelligence (AAAI-92), AAAI Press / The MIT Press, Menlo Park, CA, USA, pp. 235–240 (1992)

[24] Guerri, A., Milano, M.: Learning techniques for automatic algorithm portfolio selection. In: Proceedings of the 16th European Conference on Artificial Intelligence (ECAI 2004), pp. 475–479 (2004)

[25] Hansen, N.: The CMA evolution strategy: A comparing review. In: Lozano, J., Larranaga, P., Inza, I., Bengoetxea, E. (eds) Towards a new evolutionary computation. Advances on estimation of distribution algorithms, Springer, pp. 75–102 (2006)

[26] Hansen, N., Kern, S.: Evaluating the CMA evolution strategy on multimodal test functions. In: Yao, X., et al. (eds) Parallel Problem Solving from Nature PPSN VIII, Springer, LNCS, vol. 3242, pp. 282–291 (2004)

[27] Hansen, N., Ostermeier, A.: Completely derandomized self-adaptation in evolution strategies. Evolutionary Computation 9(2):159–195 (2001)

[28] Hoos, H.: Computer-aided design of high-performance algorithms. Tech. Rep. TR-2008-16, University of British Columbia, Department of Computer Science (2008)

[29] Hoos, H., Stützle, T.: Local search algorithms for SAT: An empirical evaluation. Journal of Automated Reasoning 24(4):421–481 (2000)

[30] Hoos, H., Stützle, T.: Stochastic Local Search—Foundations and Applications. Morgan Kaufmann Publishers, USA (2004)

[31] Huang, D., Allen, T. T., Notz, W. I., Zeng, N.: Global optimization of stochastic black-box systems via sequential kriging meta-models. Journal of Global Optimization 34(3):441–466 (2006)

[32] Hutter, F.: Automated configuration of algorithms for solving hard computational problems. Ph.D. thesis, University of British Columbia, Department of Computer Science, Vancouver, BC, Canada (2009)

[33] Hutter, F., Tompkins, D. A., Hoos, H.: Scaling and Probabilistic Smoothing: Efficient Dynamic Local Search for SAT. In: Principles and Practice of Constraint Programming – CP 2002, Springer-Verlag, LNCS, vol. 2470, pp. 233–248 (2002)

[34] Hutter, F., Hamadi, Y., Hoos, H., Leyton-Brown, K.: Performance prediction and automated tuning of randomized and parametric algorithms. In: Principles and Practice of Constraint Programming – CP 2006, Springer-Verlag, LNCS, vol. 4204, pp. 213–228 (2006)

[35] Hutter F., Babić, D., Hoos, H., Hu, A. J.: Boosting verification by automatic tuning of decision procedures. In: Proc. Formal Methods in Computer-Aided Design (FMCAD'07), IEEE Computer Society Press, pp. 27–34 (2007)

[36] Hutter, F., Hoos, H., Stützle, T.: Automatic algorithm configuration based on local search. In: Proceedings of the 22nd National Conference on Artificial Intelligence (AAAI-07), pp. 1152–1157 (2007)

[37] Hutter, F., Hoos, H., Leyton-Brown, K., Murphy, K.: An experimental investigation of model-based parameter optimisation: SPO and beyond. In: Proceedings of the 11th Annual Conference on Genetic and Evolutionary Computation (GECCO'09), ACM, pp. 271–278 (2009)

[38] Hutter, F., Hoos, H., Leyton-Brown, K., Stützle, T.: ParamILS: An automatic algorithm configuration framework. Journal of Artificial Intelligence Research 36:267–306 (2009)

[39] Hutter, F., Hoos, H., Leyton-Brown, K., Stützle, T.: ParamILS: An automatic algorithm configuration framework (extended version). Tech. Rep. TR-2009-01, University of British Columbia, Department of Computer Science (2009)

[40] Hutter, F., Hoos, H., Leyton-Brown K.: Sequential model-based optimization for general algorithm configuration (extended version). Tech. Rep. TR-2010-10, University of British Columbia, Department of Computer Science (2010)

[41] Hutter, F., Hoos, H., Leyton-Brown, K., Murphy, K.: Time-bounded sequential parameter optimization. In: Proceedings of the 4th International Conference on Learning and Intelligent Optimization (LION 4), Springer-Verlag, LNCS, vol. 6073, pp. 281–298 (2010)

[42] Hutter, F., Hoos, H., Leyton-Brown, K.: Automated configuration of mixed integer programming solvers. In: Proceedings of the 7th International Conference on the Integration of AI and OR Techniques in Constraint Programming for Combinatorial Optimization Problems (CPAIOR 2010), Springer-Verlag, LNCS, vol. 6140, pp. 186–202 (2010)

[43] Hutter, F., Hoos, H., Leyton-Brown, K.: Extending sequential model-based optimization to general algorithm configuration. To appear in: *Proceedings of the 5th International Conference on Learning and Intelligent Optimization (LION 5)* (2011)

[44] Jones, D. R., Schonlau, M., Welch, W. J.: Efficient global optimization of expensive black box functions. Journal of Global Optimization 13:455–492 (1998)

[45] KhudaBukhsh, A., Xu, L., Hoos, H., Leyton-Brown, K.: SATenstein: Automatically building local search SAT solvers from components. In: Proceedings of the 21st International Joint Conference on Artificial Intelligence (IJCAI-09), pp 517–524 (2009)

[46] Leyton-Brown, K., Nudelman, E., Andrew, G., McFadden, J., Shoham, Y.: A portfolio approach to algorithm selection. In: Rossi, F. (ed) Principles and Practice of Constraint Programming – CP 2003, Springer Verlag, Berlin, Germany, Lecture Notes in Computer Science, vol. 2833, pp. 899–903 (2003)

[47] Lourenço, H. R., Martin, O., Stützle, T.: Iterated local search. In: Glover, F., Kochenberger, G. (eds) Handbook of Metaheuristics, Kluwer Academic Publishers, Norwell, MA, USA, pp. 321–353 (2002)

[48] Maron, O., Moore, A. W.: Hoeffding races: Accelerating model selection search for classification and function approximation. In: Advances in neural information processing systems 6, Morgan Kaufmann, pp. 59–66 (1994)

[49] Nelder, J. A., Mead, R.: A simplex method for function minimization. The Computer Journal 7(4):308–313 (1965)

[50] Nell, C. W., Fawcett, C., Hoos, H., Leyton-Brown K.: HAL: A framework for the automated design and analysis of high-performance algorithms. To appear in: *Proceedings of the 5th International Conference on Learning and Intelligent Optimization (LION 5)* (2011)

[51] Nocedal, J., Wright, S. J.: Numerical Optimization, 2nd edn. Springer-Verlag (2006)

[52] Nouyan, S., Campo, A., Dorigo, M.: Path formation in a robot swarm: Self-organized strategies to find your way home. Swarm Intelligence 2(1):1–23 (2008)

[53] Pellegrini, P., Birattari, M.: The relevance of tuning the parameters of metaheuristics. a case study: The vehicle routing problem with stochastic demand. Tech. Rep. TR/IRIDIA/2006-008, IRIDIA, Université Libre de Bruxelles, Brussels, Belgium (2006)

[54] Pop, M., Salzberg, S. L., Shumway, M.: Genome sequence assembly: Algorithms and issues. Computer 35(7):47–54 (2002)

[55] Prasad, M. R., Biere, A., Gupta, A.: A survey of recent advances in SAT-based formal verification. International Journal on Software Tools for Technology Transfer 7(2):156–173 (2005)

[56] Rasmussen, C. E., Williams, C. K. I.: Gaussian Processes for Machine Learning. The MIT Press (2006)

[57] Rice, J.: The algorithm selection problem. Advances in Computers 15:65–118 (1976)

[58] Rossi-Doria, O., Sampels, M., Birattari, M., Chiarandini, M., Dorigo, M., Gambardella, L. M., Knowles, J. D., Manfrin, M., Mastrolilli, M., Paechter, B., Paquete, L., Stützle, T.: A comparison of the performance of different metaheuristics on the timetabling problem. In: Burke, E. K., Causmaecker, P. D. (eds) Practice and Theory of Automated Timetabling IV, 4th International Conference, PATAT 2002, Selected Revised Papers, Springer, Lecture Notes in Computer Science, vol. 2740, pp. 329–354 (2003)

[59] Sacks, J., Welch, W., Mitchell, T., Wynn, H.: Design and analysis of computer experiments (with discussion). Statistical Science 4:409–435 (1989)

[60] Santner, T., Williams, B., Notz, W.: The Design and Analysis of Computer Experiments. Springer Verlag, New York (2003)

[61] Schiavinotto, T., Stützle, T.: The linear ordering problem: Instances, search space analysis and algorithms. Journal of Mathematical Modelling and Algorithms 3(4):367–402 (2004)

[62] Schonlau, M., Welch, W. J., Jones, D. R.: Global versus local search in constrained optimization of computer models. In: Flournoy, N., Rosenberger, W., Wong, W. (eds) New Developments and Applications in Experimental Design,

vol. 34, Institute of Mathematical Statistics, Hayward, California, pp. 11–25 (1998)

[63] Spall, J.: Introduction to Stochastic Search and Optimization. John Wiley & Sons, Inc., New York, NY, USA (2003)

[64] Stützle, T., Hoos, H.: MAX-MIN Ant System. Future Generation Computer Systems 16(8):889–914 (2000)

[65] Thachuk, C., Shmygelska, A., Hoos, H.: A replica exchange Monte Carlo algorithm for protein folding in the hp model. BMC Bioinformatics 8(342) (2007)

[66] Tompkins, D., Hoos, H.: Dynamic Scoring Functions with Variable Expressions: New SLS Methods for Solving SAT. In: Proceedings of the 13th International Conference on Theory and Applications of Satisfiability Testing (SAT 2010), Springer-Verlag, LNCS, vol. 6175, pp. 278–292 (2010)

[67] Xu, L., Hutter, F., Hoos, H., Leyton-Brown, K.: SATzilla-07: The design and analysis of an algorithm portfolio for SAT. In: Principles and Practice of Constraint Programming – CP 2007, Springer Berlin / Heidelberg, LNCS, vol. 4741, pp. 712–727 (2007)

[68] Xu, L., Hutter, F., Hoos, H., Leyton-Brown, K.: SATzilla: Portfolio-based algorithm selection for SAT. Journal of Artificial Intelligence Research 32:565–606 (2008)

[69] Xu, L., Hutter, F., Hoos, H., Leyton-Brown, K.: SATzilla2009: An Automatic Algorithm Portfolio for SAT, Solver Description, SAT Competition 2009 (2009)

[70] Xu, L., Hoos, H., Leyton-Brown, K.: Hydra: Automatically configuring algorithms for portfolio-based selection. In: Proceedings of the 24th AAAI Conference on Artificial Intelligence (AAAI-10), pp. 210–216 (2010)

[71] Yuan, Z., Fügenschuh, A., Homfeld, H., Balaprakash, P., Stützle T., Schoch M.: Iterated greedy algorithms for a real-world cyclic train scheduling problem. In: Blesa, M., Blum, C., Cotta, C., Fernández, A., Gallardo, J., Roli, A., Sampels, M. (eds) Hybrid Metaheuristics, Lecture Notes in Computer Science, vol. 5296, Springer Berlin / Heidelberg, pp. 102–116 (2008)

Chapter 4
Case-Based Reasoning for Autonomous Constraint Solving⋆

Derek Bridge, Eoin O'Mahony, and Barry O'Sullivan

4.1 Introduction

Your next conference will take you from your home in Cork, Ireland to Victoria on Vancouver Island, Canada. You recall that you made a similar trip two years previously, when you travelled from Cork, through London Heathrow, New York JFK and Vancouver In: Airport, finishing with a bus connection to Whistler, British Colombia. You can avoid planning your next trip 'from scratch' by reusing your previous itinerary. You make a small adaptation to the final leg: from Vancouver International Airport, you will take a bus and then a ferry to your hotel in Victoria.

Humans often reason from experiences in the way exemplified above. Faced with a new problem, we recall our experiences in solving similar problems in the past, and we modify the past solutions to fit the circumstances of the new problem.

Within Artificial Intelligence (AI), the idea that we can solve problems by recalling and reusing the solutions to similar past problems, rather than reasoning 'from scratch', underlies *Case-Based Reasoning* (CBR), which has been the target of active research and development since the late 1980s [35, 36, 39, 7, 2]. CBR is a problem solving and learning strategy: reasoning is remembered (this is learning); and reasoning is remembering (this is problem-solving) [37]. CBR can be useful in domains where problem types recur, and where similar problems have similar solutions [37]. Its wide range of application areas — from classification and numeric prediction to configuration, design and planning — and domains — from medicine to law to recommender systems — is testimony to its generality [23, 24, 13, 29, 62, 8]. In many cases, it has moved out of the research laboratory and into fielded systems [11].

In this chapter, we review the application of CBR to search and especially to constraint solving. We present our review in three parts. In Section 4.2, using the

---

Cork Constraint Computation Centre, University College Cork, Ireland
e-mail: {d.bridge|e.omahony|b.osullivan}@4c.ucc.ie

⋆ This work was part-funded by Science Foundation Ireland (Grant Number 05/IN/I886).

Y. Hamadi et al. (eds.), *Autonomous Search*,
DOI 10.1007/978-3-642-21434-9_4,
© Springer-Verlag Berlin Heidelberg 2011

application of CBR to route planning as a case study, we explain the CBR process model, i.e., CBR's main subprocesses. This helps establish terminology for later sections. Section 4.3 reviews the roles that CBR can play in search, ranging from its role in general AI search in tasks such as planning, scheduling and game playing to its use in constraint solving. Section 4.4 presents CPHYDRA, a recent successful application of CBR to autonomous constraint solving. In CPHYDRA, CBR is used to inform a portfolio approach to constraint problem solving.

4.2 Case-Based Reasoning

4.2.1 CBR Knowledge Containers and Process Model

There are many ways to characterize Case-Based Reasoning systems. Here, we describe them in terms of their *knowledge containers* and their *process model*.

Their knowledge containers are
- the *case base*;
- the *vocabulary* used to represent the cases in the case base;
- the *similarity measure* used in retrieving cases from the case base; and
- the *solution transformation knowledge* used to reuse the retrieved solution(s) [61].

Their process model typically comprises some or all of the following steps (see Figure 4.1):
- *retrieval* of the most similar case(s);
- *reuse* of the knowledge contained in the retrieved cases to propose a solution to the new problem;
- *revision* of the proposed solution in the light of feedback; and
- *retention* of parts of the newly gained experience for use in future problem solving [1].

The last of these steps is what makes CBR a strategy for learning as well as for problem solving.

4.2.2 The Turas System

By way of a case study that exemplifies CBR's knowledge containers and process model, we present *Turas*, McGinty and Smyth's route planner [45]. A user of *Turas* specifies start and end locations on a city map, and *Turas* finds a route that connects the two.

The *Turas* case base contains routes that the user has travelled in the past and that she has found to be of sufficient quality to store for future reuse. Route quality is subjective: some users may prefer short routes; others may prefer scenic routes,

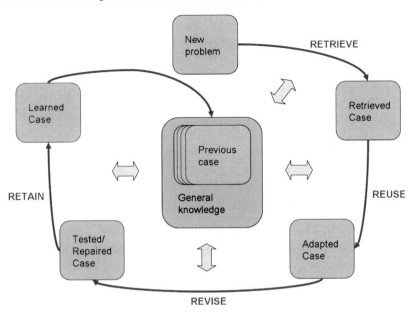

Fig. 4.1: The CBR process model from [1]

or routes that mostly use main roads, and so on. The *Turas* representation of route quality is entirely implicit: it is captured only by the kinds of cases that the user stores in her case base.

In *Turas*, each case or route is represented by an undirected graph, where nodes represent road junctions and edges represent road segments. The set of allowed junction and segment identifiers defines the *Turas* case vocabulary.

Graphs are only one possible case representation. Especially commonplace across other CBR systems are case representations that use feature vectors or sets of attribute-value pairs. In these, the features or attributes and their possible values define the system's vocabulary [7]. CBR systems may even use *feature selection* methods to automatically select from among candidate features [34], or *feature weighting* methods to automatically determine feature importance [76].

Retrieval. In our presentation of *Turas*, we will refer to the pair that comprises the user's start and end locations, *start* and *end*, as a *problem specification*. *Turas* retrieves from the case base the route that is most similar to this specification. Informally, *Turas* retrieves the case that contains a pair of junctions (nodes), n and n', that are close to *start* and *end*. More formally, from case base *CB*, it retrieves:

$$\arg\min_{c \in CB}(\mathrm{dist}(start,c.n) + \mathrm{dist}(end,c.n'))$$

where dist is Euclidean distance.

Turas retrieves only one case. It is not uncommon for CBR systems to retrieve several cases. In particular, many CBR systems score cases by their similarity to the problem specification and retrieve the k most similar cases. In this case, the retrieval component of the system is an implementation of *kNN*, the k nearest neighbours method.

As we shall see in later sections of this chapter, CBR systems that represent cases by feature vectors or sets of attribute-value pairs often define case similarity differently from *Turas*, and according to a so-called *local-global principle* [7]. According to this principle, the system's similarity knowledge container has a *local similarity measure* for each feature or attribute. For example, on a numeric-valued attribute, similarity between two values may be defined as the complement of the normalized absolute difference in the values. Local similarity measures may also be defined in terms of taxonomic knowledge, or by enumeration of all pairs of values. This is one of the ways in which a CBR system may incorporate domain-specific and expert knowledge. A *global similarity measure* aggregates local similarities to measure the overall similarity. Aggregation is often done by averaging, possibly weighted by the feature weights mentioned earlier.

Reuse. A retrieval-only system would barely qualify as a CBR system. It is essential that there be inference, which is the responsibility of the reuse step in the process model. This can take many forms. Where CBR is used for classification, for example, each retrieved case contains a class label, and the system may predict the class of the new problem by taking a vote among the retrieved cases. In other applications, the solutions of the retrieved cases must be *adapted* to suit the circumstances of the new problem [35]. For example, the reuse knowledge container might contain rules, perhaps mined from the case base or elicited from experts, for carrying out simple substitutions to make a past solution apply to new circumstances.

In *Turas*, the user wants a route that goes from *start* to *end*, but the retrieved case c goes from *c.start* to *c.end* and contains junctions $c.n$ and $c.n'$ that are close to *start* and *end*. Adaptation consists of trimming the ends of the retrieved case to produce a new case c' whose start is n and whose end is n'. In other words, *Turas* uses the heuristic that the subsequence of the case that brings us closest to the problem specification locations should be reused unmodified, and leading and trailing parts of the case that take us further away from locations in the problem specification should be discarded.

This, of course, leaves *Turas* with two new subproblems, which we can express using two new problem specifications. The first subproblem is how to get from *start* to $c'.n$; the second subproblem is how to get from $c'.n'$ to *end*. These subproblems are solved by recursive invocation of *Turas*. If the distance between a problem specification's *start* and *end* falls below a threshold, then *Turas* invokes A^* to plan that part of the route.

Revision. By the end of its retrieval and reuse steps, a CBR system has created a solution to the new problem. This solution needs to be tested. In some applications, we might test it by simulating its use. For example, we can execute a case-based plan on a world model to ensure that it leads from start state to end state and that action preconditions are satisfied at every step [28]. In other applications, we test the solution by applying it for real. In *Turas*, for example, the user can drive the recommended route. Subsequently, she might give feedback, e.g., in the form of a rating. She might even edit the route manually to fix any weaknesses that she experiences.

Retention. At this point, we have a new experience: the system proposed a solution to our new problem, and we have tested this solution. The system should learn from this experience. At its simplest, a CBR system such as *Turas* can create a new case from a problem specification and a successful solution and insert the new case into its case base, thus extending its coverage of the problem space. Some CBR systems might allow for the retention of unsuccessful cases too, assuming that retrieval and reuse can handle them intelligently. There is the possibility that new experiences result in updates to the CBR system's other knowledge containers too.

In principle, with greater coverage comes greater likelihood of retrieving cases that are highly similar to a given problem specification, and hence a reduced need for solution adaptation during the reuse step.

In practice, matters are more complicated. New cases may be noisy, to the detriment of solution quality. A CBR system may also have to address the *utility problem*. This was first discussed in the context of explanation-based learning by Minton [49]. Minton's PRODIGY/EBL system learns search control rules to reduce search by eliminating dead-end paths. However, as the number of such rules grows, the cost of rule-matching grows, and may outweigh the savings in search. Similarly, in a CBR system, as the knowledge containers, especially the case base, grow in size, retrieval costs may come to exceed the performance benefits that accrue from greater problem space coverage [49, 17, 18]. Hence, the four-step CBR process model has been extended to include additional steps for analysis and maintenance of the knowledge containers [60], and there is a considerable body of techniques for knowledge container maintenance, especially maintenance of the case base, e.g., [38, 48, 77, 15].

4.2.3 Discussion

We have given a very high-level overview of CBR, illustrated with reference to the *Turas* system. The concepts that we have explained will recur throughout the rest of this chapter, as we review other ways in which search systems can exploit case-based experience.

CBR is compelling because it seems so natural: humans often seem to reason in a similar way. It is compelling too because of the way it integrates problem solving and learning. It is especially applicable in 'weak theory' domains, where deep underlying causal models may not be known or may be difficult and costly to elicit.

Rarely cited as an advantage, but well illustrated by *Turas*, CBR allows for easy personalization of solutions. In *Turas*, each user has her own case base, containing her own preferred routes, implicitly capturing a personalized notion of route quality. The new routes that *Turas* builds from those in a user's case base will reflect the personal biases of that case base. In an extension to the work, McGinty and Smyth present a distributed version of *Turas*: when a user's own case base lacks requisite coverage, cases may be retrieved from the case bases of similar users [46]. This shows how CBR can be used in collaborative problem solving.

From a learning point of view, CBR is usually a form of *instance-based learning* [51]. The training instances are the cases, and these are either presented up front or incrementally as problem solving proceeds. They are simply stored in the case base for later reuse. This is a form of *lazy learning*: generalization takes place at problem-solving time and can therefore take the new problem into account by reasoning from instances (cases) that are close to the new problem. By contrast, in *eager learning*, generalization takes place on receipt of training examples, in advance of presentation of new problems, and therefore eager learning commits to its generalization of the training examples before it sees any new problems.

In the next section, we review the roles that CBR can play in search, especially in constraint solving.

4.3 Case-Based Reasoning and Search

George Luger's textbook on Artificial Intelligence contains the following description of search:

> "Search is a problem-solving technique that systematically explores a space of problem states, i.e., successive and alternative stages in the problem-solving process. Examples of problem states might include the different board configurations in a game or intermediate steps in a reasoning process. This space of alternative solutions is then searched to find an answer. Newell and Simon (1976) have argued that this is the essential basis of human problem solving. Indeed, when a chess player examines the effects of different moves or a doctor considers a number of alternative diagnoses, they are searching among alternatives." [40]

There are many classes of search problem for which experience can help improve the efficiency with which a solution can be found. As we have discussed, the reuse of experience is the core principle upon which Case-Based Reasoning (CBR) is based. Of course, CBR can be applied more generally, and is often combined with other reasoning techniques as the application domain demands. For example, there are many integrations of CBR with rule-based reasoning, model-based reasoning, constraint satisfaction, information retrieval, and planning [44].

The majority of applications of CBR to search have been in game playing, scheduling, and most of all planning. In game playing CBR has been used for plan selection in strategy games [3], plan recognition in games [16], move selection in adversarial games [69], and planning in imperfect information games [68]. Case-based scheduling has also received attention [66, 67].

However, there is a vast body of work on case-based planning algorithms, some of which are concerned with speeding up the planning process by retrieving and adapting a plan from a case base [52]. For example, de la Rosa et al. use CBR in a forward planning algorithm to order nodes for evaluation [63]. In their work, cases capture typical state transitions, which can be learned from past plans; successor nodes are recommended for evaluation if they match these cases. Not only can the cases themselves be learned, their utility in planning can also be monitored to influence their likelihood of being reused [63]. In fact, this planning-specific research can be viewed as an instance of the more general technique of Constructive Adaptation [57]. Constructive Adaptation is a form of best-first search in which cases are used for both hypothesis generation and hypothesis ordering.

The many applications of CBR to planning include planning support in collaborative workflows [19], hierarchical planning [33], hierarchical abstractions for planning [42, 65], adaptation-guided retrieval for planning [73], case adaptation through replanning [74], and constraint-based case-based planning [78].

4.3.1 Integrations of CBR and CSP Techniques

In this chapter we are concerned with integrations of CBR with constraint programming and, in particular, with how such an integration can be used to support autonomous search in constraint solving. Many search problems can be formulated as Constraint Satisfaction Problems [64]. A Constraint Satisfaction Problem (CSP) is defined in terms of a finite set of decision variables where each variable is associated with a finite set of values, called the domain of the variable, and a set of constraints that specify the legal ways of assigning values to the variables. The task in a CSP is to find an assignment of a value to each variable such that the set of constraints is satisfied. The problem is well known to be NP-complete. In addition, in many application domains a natural objective function can be formulated that defines the notion of an optimal solution. CSPs are solved either using systematic search, such as backtracking interleaved with constraint propagation, or non-systematic approaches based on local search [64].

A number of systems have been built that integrate CBR and CSP techniques, e.g., in the fields of configuration and engineering design [30, 71] and forest fire fighting [6]. However, there are a number of integrations through which either CSP techniques are used to enhance CBR, or vice versa. For example, Neagu and Faltings have shown how constraints can be used to support the adaptation phase of the CBR process [54, 55]. Specifically, they show how interchangeable values for variables in solutions can help reason about the effect of change in a solution set so that minimal sets of choices can be found to adapt existing solutions for a new problem. This is particularly interesting since the adaptation phase of CBR has received less attention than other aspects, such as case retrieval.

Conversely, CBR techniques have been used to help find solutions to new sets of constraints by reasoning about solutions to similar sets of constraints [31, 58, 32]. These approaches are similar to CSP techniques that help solve challenging CSPs by caching search states [70]. The majority of the search effort is spent proving the insolubility of unsatisfiable subtrees when using a systematic backtrack-style algorithm. Therefore, by remembering search states that lead to failure, one can improve search efficiency. One can view this as a form of CBR since one stores interesting states, and retrieves these when they are similar to the current search.

A review of applications that integrate CBR and CSP techniques up to the late 1990s is available [41]. In the next subsection we will briefly review the literature on the use of CBR techniques to improve the efficiency of various types of search techniques.

4.3.2 CBR for Constraint-Based Search

CBR has been used to solve a variety of problems that are typically solved using systematic search techniques such as timetabling [9] and route planning [27]. A typical way of using CBR to improve systematic search is through solution reuse, which has been successfully applied in real-time scheduling [12]. However, none of these systems can be regarded as truly supporting autonomous search.

In constraint programming, a major challenge is to choose the appropriate search strategy to use for a given CSP model. Autonomous approaches to addressing this challenge are amongst the most difficult to develop. We briefly survey two specific approaches to autonomous systematic and local search, before discussing an award-winning approach to autonomous constraint solving in Section 4.4.

Autonomous Systematic Search. Gebruers et al. present a simple CBR system for choosing a solution strategy [20]. In their work, the solution strategy is chosen in advance, and not changed dynamically during the search. Each case in the case base contains a description of a problem, along with a good solution strategy for solving that problem. The problems are described by a vector of static, surface features. In their experiment, they work with the Social Golfer Problem, and features include the number of golfers, the number of weeks, and the ratio of these two. Of all the possible features, a subset is used in the CBR; this subset is chosen using the Wrapper method [34], and cases are retrieved using Euclidean distance.

Gebruers et al. compare their CBR approach to a decision tree [59] and several benchmark methods, including a 'use best' approach, which selects the same strategy every time: the one that is a winner most often in the training set. In their experiments, CBR predicts the best strategy about 10% more often than decision trees and 'use best'. Furthermore, in terms of total execution time, regardless of whether a winning strategy is predicted or not, CBR significantly outperforms the other strategies, largely because it predicts far fewer strategies that time out. The approach can therefore be regarded as one of the first that supports autonomous search for solving CSPs.

There is related work that does not use CBR. Minton dynamically constructs constraint programs by performing an incomplete search of the space of possible programs [50]. The contrast between his work and that of Gebruers et al. is that Minton constructs new programs whereas Gebruers et al. seek to reuse existing strategies. Rather different again is the system reported in [10], which executes multiple strategies, gathers information *at runtime* about their relative performance and decides which strategies to continue with. The focus in that system is on domains where optimisation is the primary objective, rather than constraint satisfaction.

Autonomous Local Search. One of the most common local search techniques for solving large-scale constraint optimisation problems is tabu search [21, 22]. In tabu search memory structures are used to remember previous states that should not be revisited, i.e., *tabu* states, in order to help diversify search. Tabu states prevent the local search algorithm from converging at a local minimum.

Grolimund and Ganascia [25, 26] have reported a system called ALOIS which automatically designs the search control operators in tabu search in a problem-independent manner. ALOIS is the acronym for "Analogical Learning for Optimisation to Improve Operator Selection". The design of good tabu search techniques for a specific problem class is a highly technical process which requires that the search algorithm designer has a deep understanding of the problem being considered. ALOIS uses a case base that tries to relate how using a search operator in some state yielded a specific objective function value. The principle that underlies ALOIS is that using an operator in a similar search state will result in a similar reward. The better the reward obtained, the more frequently a particular operator becomes available for retrieval using the CBR system. By rewarding better performing operators in specific search states, ALOIS can learn complex tabu strategies. On hard, (un)capacitated facility location problems ALOIS improves the performance of several generic tabu search strategies.

4.4 CPHYDRA: A Case-Based Portfolio Constraint Solver

None of the systems reviewed above has been applied across a large number and variety of constraint problems. CPHYDRA is an algorithm portfolio for constraint satisfaction that uses Case-Based Reasoning to determine how to solve an unseen problem instance by exploiting a case base of problem-solving experiences [56]. In areas such as satisfiability testing and integer linear programming a carefully chosen combination of solvers can outperform the best individual solver for a given set of problems. This approach is an extension of the concept of algorithm portfolios [43]. In an algorithm portfolio a collection of algorithms is used to solve a problem. The strengths and weaknesses of each of the algorithms are balanced against each other. Algorithm portfolios can be very successful due to the diverse performance of different algorithms on similar problems. Key in the design of algorithm portfolios is a selection process fro choosing an algorithm from the portfolio to run on a given problem. The selection process is usually performed using a machine learning technique based on feature data extracted from the problem. CPHYDRA's algorithm portfolio consists of a set of constraint solvers and its selection process is a CBR system. The system won the 2008 International CSP Solver Competition, demonstrating that a CBR-based approach to autonomous constraint solving could be built and deployed successfully.

4.4.1 The International CSP Solver Competition

The International Constraint Satisfaction Problem (CSP) Solver Competition is an annual competition for Constraint Solvers[1]. The goal of the competition is to solve as many problems as possible given a limit of 30 minutes on each problem. A wide range of solvers participate in the competition. Problem sets from all competitions are freely available online in the standard XCSP format [14]. This is an XML format that is a standardized specification for CSPs.

The competition contains five categories. The categories represent both the type of constraint and the constraint *arity* used to model the problems in the category. The *arity* of a constraint is the number of variables it constrains. The type can be one of the following: an *extensional* constraint is expressed as a table of allowed/disallowed assignments to the variables it constrains; an *intensional* constraint is defined in terms of an expression that defines the relationship that must hold amongst the assignments to the variables it constrains; a *global* constraint is an intensional constraint associated with a dedicated filtering algorithm. The categories of the 2008 competition were as follows:

- 2-ARY-EXT: instances containing only constraints with an *arity* of one (unary) or two (binary) extensional constraints.
- N-ARY-EXT: instances containing extensional constraints where at least one constraint has an *arity* greater than 2.
- 2-ARY-INT: instances containing at least one binary or unary intensional constraint.
- N-ARY-INT: instances containing at least one intensional constraint of *arity* greater than 2.
- GLOBAL: instances involving any kind of constraint but including at least one global constraint.

Entrants could restrict themselves to specific categories within the competition, since a solver dedicated to, say, binary extensional problems might be unable to handle, say, global constraints.

4.4.2 CPHYDRA

The CPHYDRA system was developed with the goal of entering the 2008 International CSP Competition. It is itself unable to solve CSPs, but it has a portfolio of constraint solvers that it can use to solve a given problem. In the version entered into the 2008 competition, this portfolio consisted of three solvers: Mistral, Choco and Abscon. These solvers, although similar in their design (all are finite domain solvers with similar levels of consistency), have diverse strengths and weaknesses.

From a CBR point of view, the CPHYDRA knowledge containers are as follows:

---

[1] http://cpai.ucc.ie

- The *case base*: in CPHYDRA the case base contains all problems from previous CSP competitions. Each case respresents an individual problem. The run times for each of the constituent solvers on this problem are stored as part of the case.
- The *vocabulary* used to represent a case comprises two parts. The first is a vector of real numbers representing features of the problem, discussed below. The second is the time taken for each of CPHYDRA's constituent solvers to solve the problem, or an indication that they timed out.
- The *similarity measure* in CPHYDRA is a simple Euclidean distance calculation between the vector of real numbers representing the problem features and each case.
- The *solution transformation knowledge* of CPHYDRA is an objective function to maximize the chances of solving the problem within the time limit.

CPHYDRA has the following process model (Figure 4.2):

- *Retrieval* in CPHYDRA is achieved through a simple k-nearest neighbour (kNN) approach. In the competition k was set to 10.
- *Reuse* in CPHYDRA involves building a schedule that runs solvers that appear in the cases CPHYDRA retrieves in the previous step for periods that are computed to maximize the chances of solving the problem within the time limit (30 minutes in the competition).
- In its *Revise* step, CPHYDRA tries to solve a new problem instance by running solvers according to the schedule computed in its *Reuse* step. Either a solver solves the instance, and the rest of the schedule need not be executed, or none of the solvers solves the instance, i.e., the schedule is executed to conclusion without success. In the competition there is no opportunity to use this feedback to revise the case base.
- CPHYDRA has no *retention* step, since in the competition setting entrants cannot self-modify incrementally after each problem instance.

We will now explain in detail the components of CPHYDRA.

Case Base. The case base used in CPHYDRA was generated from the problem set of the 2007 International Constraint Solver Competition and problems released before the 2008 competition. The total size of the case base is of the order of 3,000 cases. Each of the cases respresented one of the problems. As mentioned earlier, a problem is represented by a vector of real numbers. This vector contains two separate sets of features.

The first set of *syntactic* features is a representation of the basic properties of the problem. They count things like the number of constraints, ratio of intensional constraints to extensional constraints, average domain size and average constraint *arity*, among others. In CPHYDRA there are 22 of these *syntactic* features. CPHYDRA computes these features by parsing the XCSP file specifying the problem and extracting the information needed.

The second set of *semantic* features captures information on the structure of the problem. These are generated by running a constraint solver on the problem for a short time. In CPHYDRA Mistral is run for two seconds on the problem. Information on search statistics and modelling choices is extracted, such as how many nodes

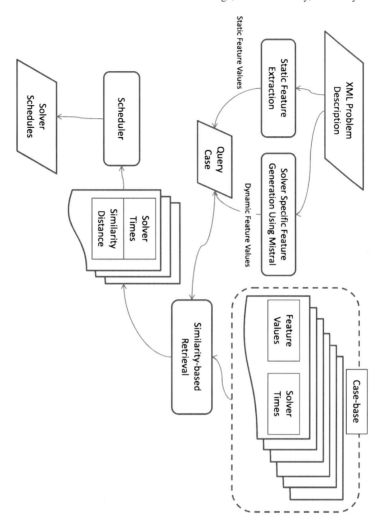

Fig. 4.2: The CPHYDRA system

were searched in the time, how the variables are represented in the solver, and so on. CPHYDRA uses 14 of these *semantic* features.

During the generation of the case base any problems that were solved by Mistral during the extraction of the *semantic* features were not included in the case base. We decided to omit these cases as they are trivially simple. We felt that they would not contribute to the case base. All features, *syntactic* and *semantic*, that represent a quantity are logarithmically scaled whereas the ratios are percentages. This prevents large values from dominating the similarity computation.

Retrieval. As mentioned above, the CPHYDRA case retrieval is done using a simple *k*-nearest neighbour (*k*NN) algorithm. The similarity between cases is calculated as the Euclidean distance between the vectors of real numbers that represent problem features. When the k nearest neighbours are computed, if the kth nearest case has a similarity to the target case equal to that of the $(k+1)^{th}$ case, both are taken. In the competition version of CPHYDRA k was set to 10. This value was chosen following some initial experimentation, but also large values are not suitable given the complexity of the reuse stage, which will be discussed next.

Reuse. CPHYDRA is designed to solve problems within a given time limit. It has a portfolio of solvers to allow it to achieve this task. Earlier versions of CPHYDRA picked one solver to attempt to solve the problem. This was done through a simple majority vote for one of all of the k similar cases. This approach lacked subtlety as all hopes of solving the problem were placed on one solver. Imagine that three cases have been returned, two of which state that Choco has solved a similar problem in 25 minutes and one of which states that Mistral has solved a similar problem in five minutes. A majority vote will choose Choco to be run for the full 30 minutes that are allowed in the competition. Running Choco for 25 minutes and Mistral for five minutes gives a much better cover of the similar cases. This simple schedule reflects much more of the information the similar cases are providing, since all the information returned is being used to make the best decision on how to solve the proposed problem.

To avoid the weakness just mentioned, CPHYDRA generates a schedule of its constituent solvers. The schedule is optimized to maximize the chances of solving the problem within 30 minutes. Given the set of similar cases, C, CPHYDRA computes a schedule such that the number of cases in C that would be solved using the schedule is maximized.

More formally the goal is to compute a solver schedule, that is, a function $f : S \mapsto R$ mapping an amount of CPU time to each solver. Consider a set of similar cases C and a set of solvers S. For a given solver $s \in S$ and a time point $t \in [0..1800]$ we define $C(s,t)$ as the subset of C solved by s if given at least time t. The schedule can be computed by the following constraint program:

$$\text{maximize } \bigcup_{s \in S} |C(s, f(s))|$$
$$\text{subject to } \Sigma_{s \in S} f(s) \leq 1800.$$

This problem is NP-hard as it is a generalisation of the knapsack problem. Fortunately, in practice it is simple because the number of similar cases $k = |C|$ and the number of solvers $|S|$ are small.

In Figures 4.3 and 4.4 CPHYDRA's scheduling is illustrated for two solvers, in this case Mistral and Choco. Each point in the graphs represents a case, whose (x,y) co-ordinates represent the time taken to solve the problem for Choco (x) and Mistral (y). The x and y axes represent time.

In Figure 4.3, 25 minutes are given to Choco and five minutes to Mistral. The cases that would not be solved by this schedule are highlighted. Figure 4.4 has a better solution, with 15 minutes being allocated to each solver, solving one more problem than the first solution. The optimal solution, Figure 4.5, gives 25 minutes to Mistral and five to Choco; only three problems remain unsolved.

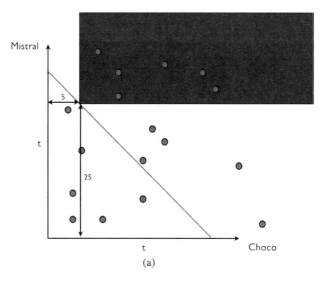

(a)

Fig. 4.3: One solver schedule

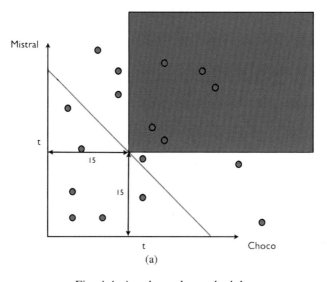

(a)

Fig. 4.4: Another solver schedule

In practice there are some complications associated with the scheduling computation. The first is that some solvers are dominated by others in a schedule. Formally, we define a solver s_1 to be dominated by another solver s_2 if and only if

$$\forall_{t<1800} C(s_1,t) \subseteq C(s_2,t+\varepsilon).$$

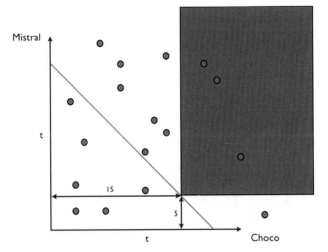

Fig. 4.5: The optimal solver schedule

Informally, s_1 is dominated by s_2 if s_2 solves all the problems s_1 solves in a time no worse than a specified error value (ε). If a solver is dominated by another solver there is nothing to be gained by running them both. When this is the case, during scheduling, we remove the solver that is dominated and allocate its time amongst all other solvers proportionally to the time they received during the computation of the schedule.

The second complication is that during the computation of the schedule it is possible that there will be time unallocated to any solver. Consider the example of two solvers (s_1, s_2) and two cases (c_1, c_2). Solver s_1 solves case c_1 in five minutes and times out on case c_2 whereas solver s_2 solves c_2 in ten minutes but times out on c_1. An optimal schedule for this example is to run s_1 for five minutes and s_2 for ten minutes. This situation is obviously not ideal as not all the time available is used. CPHYDRA allocates the remaining time to solvers proportionally to the time they received during the computation of the schedule.

Once the final schedule has been computed, CPHYDRA begins to execute the schedule. Optimisation to order the schedule to solve the problem quickly is certainly possible but it is ignored in CPHYDRA.

4.4.3 Evaluating CPHYDRA in Practice

CPHYDRA won the 2008 International CSP Solver Competition, winning four in of the five categories as well as winning overall. The results are shown in Figure 4.6. CPHYDRA beat all its constituent solvers in every category, showing that a collection of solvers can be stronger than any of the individual solvers. The single category

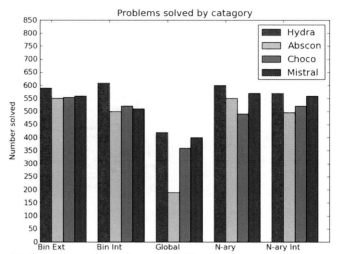

(a) Comparison of the performance of CPHYDRA in each problem category to each of its constituent solvers

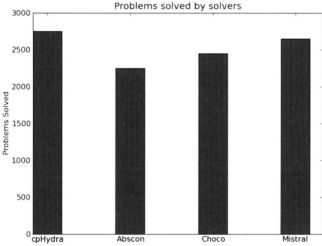

(b) Overall comparison of the number of problem instances solved by CPHYDRA and each of its constituent solvers

Fig. 4.6: Results from all categories in the left and from the whole competition on the right

that CPHYDRA failed to win outright was the *Global* constraint category. This category was won by the solver "Sugar". This is an argument for adding Sugar to the algorithm portfolio as it clearly has something to contribute.

During analysis of competition data, an interesting property of the portfolio approach emerged. In a competition setting, most portfolio approaches are very successful. Figure 4.7 shows the performance of different scheduling approaches com-

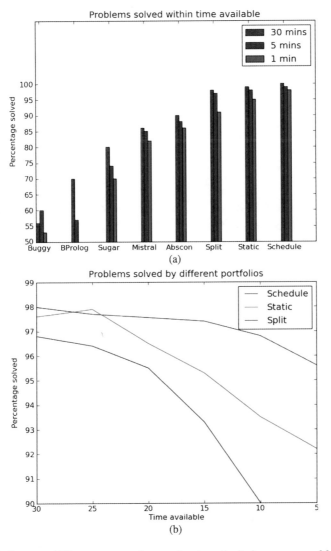

Fig. 4.7: Data on different approaches as the time limit decreases on 2007 data

puted on data from the 2007 competition. The schedules compared in the graph are listed below. The solvers in the graph represent the constituent solvers used for the portfolio on the tests. These schedules vary in complexity, from a simple even split of available time to a scheduling system informed by a CBR system:

- *Schedule* – The CPHYDRA scheduling system.
- *Split* – Each of the constituent solvers receives an equal amount of time.
- *Static* – A schedule generated using the scheduling process described earlier.

It is apparent from these graphs (Figure 4.7) that, as the time available decreases, CPHYDRA begins to clearly outperform the other approaches. We believe this to be a consequence of the majority of the problems in the competition being relatively easy. Most problems, if they are solved, are solved in a short amount of time.

Although CPHYDRA has been very successful, there remains work to be done. CPHYDRA does not utilize many of the more sophisticated CBR techniques such as features selection to remove features that contribute little to the case base, automatic feature weighting to identify features that appear to be critical in case comparison, and case base maintenance to remove cases that do not improve the effectiveness of the system, among many others.

4.5 Concluding Remarks

Within Artificial Intelligence the idea that we can solve problems by recalling and reusing the solutions to similar past problems, rather than reasoning 'from scratch', underlies *Case-Based Reasoning*. In this chapter, we reviewed the application of CBR to search. We presented the application of CBR to route planning as a case study, which was used to explain the case-based reasoning process model. We provided a review of the roles that CBR can play in search in general, and then focused on its use in systematic and incomplete approaches to constraint solving. Finally, we presented CPHYDRA, a CBR-based portfolio approach to constraint problem solving, which won the 2008 CSP Solver Competition, demonstrating that successful autonomous search systems can be built using CBR techniques.

References

[1] Aamodt, A., Plaza, E.: Case-Based Reasoning: Foundational Issues, Methodological Variants, and System Approaches. AI Communications 7(1), 39–59 (1994)
[2] Aha, D. W., Marling, C., Watson, I.: Case-based reasoning commentaries: Introduction. The Knowledge Engineering Review 3(20), 201–202 (2005)
[3] Aha, D. W., Molineaux, M., Ponsen, M. J. V.: Learning to win: Case-based plan selection in a real-time strategy game. In: Muñoz-Avila and Ricci [53], pp. 5–20
[4] Aha, D. W., Watson, I. (eds.): Case-Based Reasoning Research and Development, 4th International Conference on Case-Based Reasoning, ICCBR 2001, Vancouver, BC, Canada, July 30 - August 2, 2001, Proceedings, *Lecture Notes in Computer Science*, vol. 2080. Springer (2001)
[5] Ashley, K. D., Bridge, D. G. (eds.): Case-Based Reasoning Research and Development, 5th International Conference on Case-Based Reasoning, ICCBR

2003, Trondheim, Norway, June 23-26, 2003, Proceedings, *Lecture Notes in Computer Science*, vol. 2689. Springer (2003)

[6]   Avesani, P., Perini, A., Ricci, F.: Interactive case-based planning for forest fire management. Appl. Intell. **13**(1), 41–57 (2000)

[7]   Bergmann, R.: Experience Management: Foundations, Development Methodology, and Internet-Based Applications. LNAI 2432. Springer (2002)

[8]   Bridge, D., Göker, M. H., McGinty, L., Smyth, B.: Case-based recommender systems. The Knowledge Engineering Review **3**(20), 315–320 (2005)

[9]   Burke, E. K., MacCarthy, B. L., Petrovic, S., Qu, R.: Case-based reasoning in course timetabling: An attribute graph approach. In: Aha and Watson [4], pp. 90–104

[10] Carchrae, T., Beck, J. C.: Low-knowledge algorithm control. In: Procs. of the 19th AAAI, pp. 49–54 (2004)

[11] Cheetham, W., Watson, I.: Fielded applications of case-based reasoning. The Knowledge Engineering Review **3**(20), 321–323 (2005)

[12] Coello, J. M. A., dos Santos, R. C.: Integrating CBR and heuristic search for learning and reusing solutions in real-time task scheduling. In: K. D. Althoff, R. Bergmann, K. Branting (eds.) ICCBR, *Lecture Notes in Computer Science*, vol. 1650, pp. 89–103. Springer (1999)

[13] Cox, M. T., Muñoz-Avila, H., Bergmann, R.: Case-based planning. The Knowledge Engineering Review **3**(20), 283–287 (2005)

[14] Organising Committee of the Third International Competition of CSP Solvers, O.C.: XML representation of constraint networks format XCSP 2.1 (2008)

[15] Cummins, L., Bridge, D.: Maintenance by a Committee of Experts: The MACE Approach to Case-Base Maintenance. In: L. McGinty, D. C. Wilson (eds.) Procs. of the 8th International Conference on Case-Based Reasoning, LNAI 5650, pp. 120–134. Springer (2009)

[16] Fagan, M., Cunningham, P.: Case-based plan recognition in computer games. In: Ashley and Bridge [5], pp. 161–170

[17] Francis Jr., A. G., Ram, A.: The Utility Problem in Case-Based Reasoning. In: D. B. Leake (ed.) Procs. of the Workshop on Case-Based Reasoning, p. 160. AAAI Press (1993)

[18] Francis Jr., A. G., Ram, A.: A comparative utility analysis of case-based reasoning and control-rule learning systems. In: Procs. of the AAAI Workshop on Case-Based Reasoning, pp. 36–40. AAAI Press (1994)

[19] Freßmann, A., Maximini, K., Maximini, R., Sauer, T.: CBR-based execution and planning support for collaborative workflows. In: S. Brüninghaus (ed.) ICCBR Workshops, pp. 271–280 (2005)

[20] Gebruers, C., Hnich, B., Bridge, D., Freuder, E.: Using CBR to select solution strategies in constraint programming. In: H. Muñoz-Avila, F. Ricci (eds.) Procs. of the 6th International Conference on Case-Based Reasoning, LNAI 3620, pp. 222–236. Springer (2005)

[21] Glover, F.: Tabu search, part 1. ORSA Journal on Computing **1**, 190–206 (1989)

[22] Glover, F.: Tabu search, part 2. ORSA Journal on Computing **2**, 4–32 (1990)

[23] Goel, A., Craw, S.: Design, innovation and case-based reasoning. The Knowledge Engineering Review 3(20), 271–276 (2005)
[24] Göker, M. H., Howlett, R. J., Price, J. E.: Case-based reasoning for diagnosis applications. The Knowledge Engineering Review 3(20), 277–281 (2005)
[25] Grolimund, S., Ganascia, J. G.: Integrating case based reasoning and tabu search for solving optimisation problems. In: Veloso and Aamodt [75], pp. 451–460
[26] Grolimund, S., Ganascia, J. G.: Driving tabu search with case-based reasoning. European Journal of Operational Research 103(2), 326–338 (1997). URL http://ideas.repec.org/a/eee/ejores/v103y1997i2p326-338.html
[27] Haigh, K. Z., Veloso, M. M.: Route planning by analogy. In: Veloso and Aamodt [75], pp. 169–180
[28] Hammond, K. J.: Explaining and Repairing Plans that Fail. Artificial Intelligence 45, 173–228 (1990)
[29] Holt, A., Bichindaritz, I., Schmidt, R., Perner, P.: Medical applications in case-based reasoning. The Knowledge Engineering Review 3(20), 289–292 (2005)
[30] Hua, K., Smith, I. F. C., Faltings, B.: Integrated case-based building desing. In: S. Wess, K. D. Althoff, M. M. Richter (eds.) EWCBR, Lecture Notes in Computer Science, vol. 837, pp. 436–445. Springer (1993)
[31] Huang, Y.: Using case-based techniques to enhance constraint satisfaction problem solving. Applied Artificial Intelligence 10(4), 307–328 (1996)
[32] Huang, Y., Miles, R.: A case based method for solving relatively stable dynamic constraint satisfaction problems. In: Veloso and Aamodt [75], pp. 481–490
[33] Khemani, D., Prasad, P.: A memory-based hierarchical planner. In: Veloso and Aamodt [75], pp. 501–509
[34] Kohavi, R., John, G.: Wrappers for feature subset selection. Artificial Intelligence 97(1–2), 273–324 (1997)
[35] Kolodner, J.: Case-Based Reasoning. Morgan Kaufmann (1993)
[36] Leake, D. B. (ed.): Case-Based Reasoning: Eperiences, Lessons, & Future Directions. AAAI Press/MIT Press (1996)
[37] Leake, D. B.: CBR in Context: The Present and Future. [36], pp. 3–30
[38] Leake, D. B., Wilson, D. C.: Categorizing case-base maintenance: Dimensions and directions. In: B. Smyth, P. Cunningham (eds.) Procs. of the 4th European Conference on Case-Based Reasoning, pp. 196–207. Springer-Verlag (1998)
[39] Lenz, M., Bartsch-Spörl, B., Burkhard, H. D., Wess, S. (eds.): Case-Based Reasoning Technology: From Foundations to Applications. LNAI 1400. Springer (1998)
[40] Luger, G. F.: Artificial Intelligence: Structures and Strategies for Complex Problem Solving. Addison-Wesley Longman Publishing Co., Inc., Boston, MA, USA (2001)
[41] Squalli, M., Purvis, L., Freuder, E.: Survey of applications integrating constraint satisfaction and case-based reasoning. In: Procs. of the 1st Interna-

tional Conference and Exhibition on The Practical Application of Constraint Technologies and Logic Programming (1999)

[42] Macedo, L., Pereira, F. C., Grilo, C., Cardoso, A.: Plans as structured networks of hierarchically and temporally related case pieces. In: I. F. C. Smith, B. Faltings (eds.) EWCBR, *Lecture Notes in Computer Science*, vol. 1168, pp. 234–248. Springer (1996)

[43] Markowitz, H. M.: Portfolio selection. Journal of Finance **7**(1), 77–91 (1952)

[44] Marling, C., Rissland, E. L., Aamodt, A.: Integrations with case-based reasoning. Knowledge Eng. Review **20**(3), 241–245 (2005)

[45] McGinty, L., Smyth, B.: Personalised Route Planning: A Case-Based Approach. In: Procs. of the 5th European Workshop on Case-Based Reasoning, LNAI 1898, pp. 431–442. Springer (2000)

[46] McGinty, L., Smyth, B.: Collaborative Case-Based Reasoning: Aplications in Personalised Route Planning. In: D. W. Aha, I. Watson (eds.) Procs. of the 4th International Conference on Case-Based Reasoning, LNAI 2080, pp. 362–376. Springer (2001)

[47] McGinty, L., Wilson, D. C. (eds.): Case-Based Reasoning Research and Development, 8th International Conference on Case-Based Reasoning, ICCBR 2009, Seattle, WA, USA, July 20-23, 2009, Proceedings, *Lecture Notes in Computer Science*, vol. 5650. Springer (2009)

[48] McKenna, E., Smyth, B.: Competence-Guided Case-Base Editing Techniques. In: Procs. of the 5th European Workshop on Case-Based Reasoning, LNAI 1898, pp. 186–197. Springer (2000)

[49] Minton, S.: Quantitative results concerning the utility of explanation-based learning. Arificial Intelligence **42**(2–3), 363–392 (1990)

[50] Minton, S.: Automatically configuring constraint satisfaction programs: A case study. Constraints **1**(1), 7–43 (1996)

[51] Mitchell, T. M.: Machine Learning. McGraw-Hill (1997)

[52] Muñoz-Avila, H.: On the role of the cases in case-based planning. In: Ashley and Bridge [5], pp. 2–3

[53] Muñoz-Avila, H., Ricci, F. (eds.): Case-Based Reasoning, Research and Development, 6th International Conference, on Case-Based Reasoning, ICCBR 2005, Chicago, IL, USA, August 23-26, 2005, Proceedings, *Lecture Notes in Computer Science*, vol. 3620. Springer (2005)

[54] Neagu, N., Faltings, B.: Exploiting interchangeabilities for case adaptation. In: Aha and Watson [4], pp. 422–436

[55] Neagu, N., Faltings, B.: Soft interchangeability for case adaptation. In: Ashley and Bridge [5], pp. 347–361

[56] O'Mahony, E., Hebrard, E., Holland, A., Nugent, C., O'Sullivan, B.: Cphydra – an algorithm portfolio for constraint solving. In: Proceedings of AICS(2008)

[57] Plaza, E., Arcos, J. L.: Constructive adaptation. In: S. Craw, A. D. Preece (eds.) Procs. of the 6th European Conference on Case-Based Reasoning, LNCS 2146, pp. 306–320. Springer-Verlag (2002)

[58] Purvis, L., Pu, P.: Adaptation using constraint satisfaction techniques. In: Veloso and Aamodt [75], pp. 289–300

[59] Quinlan, J. R.: C4.5: Programs for Machine Learning. Morgan Kaufmann (1993)

[60] Reinartz, T., Iglezakis, I., Roth-Berghofer, T.: Review and restore for case-based maintenance. Computational Intelligence **17**(2), 214–234 (2001)

[61] Richter, M. M.: Introduction. In: Lenz et al. [39], pp. 1–15

[62] Rissland, E. L., Ashley, K. D., Branting, L. K.: Case-based reasoning and law. The Knowledge Engineering Review **3**(20), 293–298 (2005)

[63] de la Rosa, T., Olaya, A. G., Borrajo, D.: Using cases utility for heuristic planning improvement. In: R. Weber, M. M. Richter (eds.) ICCBR, *Lecture Notes in Computer Science*, vol. 4626, pp. 137–148. Springer (2007)

[64] Rossi, F., Beek, P. v., Walsh, T.: Handbook of Constraint Programming (Foundations of Artificial Intelligence). Elsevier Science Inc., New York, NY, USA (2006)

[65] Sánchez-Ruiz-Granados, A. A., González-Calero, P. A., Díaz-Agudo, B.: Abstraction in knowledge-rich models for case-based planning. In: McGinty and Wilson [47], pp. 313–327

[66] Scott, S., Osborne, H., Simpson, R.: Selecting and comparing multiple cases to maximise result quality after adaptation in case-based adaptive scheduling. In: E. Blanzieri, L. Portinale (eds.) EWCBR, *Lecture Notes in Computer Science*, vol. 1898, pp. 517–528. Springer (2000)

[67] Scott, S., Simpson, R.: Case-bases incorporating scheduling constraint dimensions - experiences in nurse rostering. In: Smyth and Cunningham [72], pp. 392–401

[68] Shih, J.: Sequential instance-based learning for planning in the context of an imperfect information game. In: Aha and Watson [4], pp. 483–501

[69] Sinclair, D.: Using example-based reasoning for selective move generation in two player adversarial games. In: Smyth and Cunningham [72], pp. 126–135

[70] Smith, B. M.: Caching search states in permutation problems. In: P. van Beek (ed.) CP, *Lecture Notes in Computer Science*, vol. 3709, pp. 637–651. Springer (2005)

[71] Smith, I. F. C., Lottaz, C., Faltings, B.: Spatial composition using cases: Idiom. In: Veloso and Aamodt [75], pp. 88–97

[72] Smyth, B., Cunningham, P. (eds.): Advances in Case-Based Reasoning, 4th European Workshop, EWCBR-98, Dublin, Ireland, September 1998, Proceedings, *Lecture Notes in Computer Science*, vol. 1488. Springer (1998)

[73] Tonidandel, F., Rillo, M.: An accurate adaptation-guided similarity metric for case-based planning. In: Aha and Watson [4], pp. 531–545

[74] Tonidandel, F., Rillo, M.: Case adaptation by segment replanning for case-based planning systems. In: Muñoz-Avila and Ricci [53], pp. 579–594

[75] Veloso, M. M., Aamodt, A. (eds.): Case-Based Reasoning Research and Development, First International Conference, ICCBR-95, Sesimbra, Portugal, October 23-26, 1995, Proceedings, *Lecture Notes in Computer Science*, vol. 1010. Springer (1995)

[76] Wettschereck, D., Aha, D. W.: Weighting features. In: M. Veloso, A. Aamodt (eds.) Procs. of the 1st International Conference on Case-Based Reasoning, LNAI 110, pp. 347–358. Springer (1995)
[77] Wilson, D. R., Martinez, T. R.: Reduction Techniques for Instance-Based Learning. Machine Learning **38**, 257–286 (2000)
[78] Zhuo, H., Yang, Q., Li, L.: Constraint-based case-based planning using weighted max-sat. In: McGinty and Wilson [47], pp. 374–388

Chapter 5
Learning a Mixture of Search Heuristics

Susan L. Epstein and Smiljana Petrovic

5.1 Introduction

An important goal of artificial intelligence research is to construct robust autonomous artifacts whose behavior becomes increasingly *expert*, that is, they perform a particular task faster and better than the rest of us [10]. This chapter explores the idea that, if one expert is a good decision maker, the combined recommendations of multiple experts will serve as an even better decision maker. The conjecture that a combination of decision makers will outperform an individual one dates at least from the Marquis de Condorcet (1745–1794) [50]. His Jury Theorem asserted that the judgment of a committee of competent experts, each of whom is correct with probability greater than 0.5, is superior to the judgment of any individual expert.

It is difficult, or even impossible, to specify all the domain knowledge a program requires in a challenging domain. Thus an expert program must *learn*, that is, improve its behavior based on its own problem-solving experience. Machine learning algorithms extract their knowledge from *training examples,* models of desired behavior drawn from experience. Ideally, an oracle labels each training example as correct or incorrect, and the algorithm seeks to learn how to label not only those examples correctly, but also new *testing examples* drawn from the same population. Thus it is important that the learner not *overfit*, that is, not shape decisions so closely to the training examples that it performs less well on the remainder of the population from which they were drawn.

An *autonomous* learner has no oracle or teacher. It must monitor its own performance to direct its own learning, that is, it must create its own training examples

Susan L. Epstein
Department of Computer Science, Hunter College and The Graduate Center of The City University of New York, New York, New York, e-mail: susan.epstein@hunter.cuny.edu

Smiljana Petrovic
Department of Computer Science, Iona College, New Rochelle, New York, e-mail: spetrovic@iona.edu

Y. Hamadi et al. (eds.), *Autonomous Search,*
DOI 10.1007/978-3-642-21434-9_5,
© Springer-Verlag Berlin Heidelberg 2011

and gauge its own performance. In addition, the learner must somehow infer the correct/incorrect labels for those examples from its own performance on the particular problem where the examples arose. Furthermore, the autonomous learner must evaluate its performance more generally: Is the program doing well? Has it learned enough? Should it start over?

Typically, one creates an autonomous learner because there is no oracle at hand. Such is the case in the search for solutions to constraint satisfaction problems (*CSPs*). The constraint literature is rich in heuristics to solve these problems, but the efficacy of an individual heuristic may vary dramatically with the kind of CSP it confronts, and even with the individual problem. A combination of heuristics seems a reasonable approach, but that combination must somehow be learned. This chapter begins, then, with general background on combinations of decision makers and machine learning, followed by specific work on combinations to guide CSP search. Subsequent sections describe ACE, an ambitious project that learns a mixture of heuristics to solve CSPs.

5.2 Machine Learning and Mixtures of Experts

In the face of uncertainty, a *prediction algorithm* draws upon theory and knowledge to forecast a correct decision. Often however, there may be multiple reasonable predictors, and it may be difficult for the system builder to select from among them. Dieterich gives several reasons not to make such a selection, that is, to have a machine learning algorithm employ a mixture of hypotheses [11]. On limited data, there may be different hypotheses that appear equally accurate. In that case, although one could approximate the unknown true hypothesis by the simplest one, averaging or mixing all of them together could produce a better approximation. Moreover, even if the target function cannot be represented by any of the individual hypotheses, their combination could produce an acceptable representation.

An *ensemble method* combines a set of individual hypotheses. There is substantial theoretical and empirical confirmation that the average case performance of an ensemble of hypotheses outperforms the best individual hypothesis, particularly if they are represented as decision trees or neural networks [2, 31, 46]. Indeed, for sufficiently accurate and diverse classifiers, the accuracy of an ensemble of classifiers has been shown to increase with the number of hypotheses it combines [22]. One well-known ensemble method is AdaBoost, which ultimately combines a sequence of learned hypotheses, emphasizing examples misclassified by previously generated hypotheses [14, 42].

More generally, a *mixture of experts algorithm* learns from a sequence of trials how to combine its experts' predictions [26]. In a supervised environment, a trial has three steps: the mixture algorithm receives predictions from each of e experts, makes its own prediction y based on them, and then receives the correct value y'. The objective is to create a mixture algorithm that minimizes the *loss function* (the distance between y and y'). The performance of such an algorithm is often measured by its *relative loss*: the additional loss to that incurred on the same example by the best individual expert. Under the worst-case assumption, mixture of experts algorithms have been proved asymptotically close to the behavior of the best expert [26].

5.3 Constraint Satisfaction and Heuristic Search

A CSP is a set of variables, each with a domain of values, and a set of constraints expressed as relations over subsets of those variables. In a *binary* CSP, each constraint is on at most two variables. A *problem class* is a set of CSPs with the same characterization. For example, binary CSPs in *model B* are characterized by $<n, m, d, t>$, where n is the number of variables, m the maximum domain size, d the density (fraction of constraints out of $n(n-1)/2$ possible constraints) and t the *tightness* (fraction of possible value pairs that each constraint excludes) [19]. A binary CSP can be represented as a *constraint graph*, where vertices correspond to the variables (labeled by their domains), and each edge represents a constraint between its respective variables.

A randomly generated problem class may also mandate a specific structure for its CSPs. For example, each of the *composed problems* used here consists of a subgraph (its *central component*) loosely joined to one or more additional subgraphs (its *satellites*) [1]. Figure 5.1 illustrates a composed problem with two satellites. Geometric CSPs also have non-random structure. A random geometric graph $<n, D>$ has n vertices, each represented by a random point in the unit square [25]. There is an edge between two vertices if and only if their (Euclidean) distance is no larger than D. A class of random geometric CSPs $<n, D, d, t>$ is based on a set of random geometric graphs $<n, D>$. In $<n, D, d, t>$, the variables represent random points, and constraints are on variables corresponding to points close to each other. Additional edges ensure that the graph is connected. Density and tightness are given by the parameters d and t, respectively. Figure 5.2 illustrates a geometric graph with 500 variables. Real-world CSPs typically display some non-random structure in their constraint graphs.

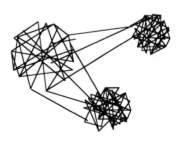

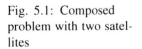

Fig. 5.1: Composed problem with two satellites

Fig. 5.2: Geometric
graph from [25]

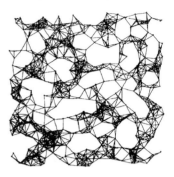

This chapter presents experiments on six CSP classes. *Geo* (the geometric problems <50, 10, 0.4, 0.82>) and *Comp* (the composed problems with central component in model B with <22, 6, 0.6, 0.1>, linked to a single model B satellite with <8, 6, 0.72, 0.45> by edges with density 0.115 and tightness 0.05) are classes of structured CSPs. The others are model B <50, 10, 0.38, 0.2>, which is exceptionally hard for its *size* (*n* and *m*); <50, 10, 0.18, 0.37>, which is the same size but somewhat easier; < 20, 30, 0.444, 0.5 >, whose problems have large domains; and <30, 8, 0.26, 0.34> which are easy compared to the other classes, but difficult for their size. Some of these problems appeared in the First International Constraint Solver Competition at CP-2005.

An instantiation assigns a value to all (*full* instantiation) or some (*partial* instantiation) of the variables. A *solution* to a CSP is a full instantiation that satisfies all the constraints. A *solvable* CSP has at least one solution. A solver *solves* a CSP if it either finds a solution or proves that it is *unsolvable,* that is, that it has no solution. All problems used in the experiments reported here are randomly generated, solvable binary CSPs with at least one solution.

Traditional (*global*) CSP search makes a sequence of decisions that instantiates the variables in a problem one at a time with values from their respective domains. After each value assignment, some form of *inference* detects values in the domains of *future variables* (those not yet instantiated) that are incompatible with the current instantiation. The work reported here uses the MAC-3 inference algorithm to maintain arc consistency during search [40]. MAC-3 temporarily removes currently unsupportable values to calculate *dynamic domains* that reflect the current instantiation. If, after inference, every value in some future variable's domain is *inconsistent* (violates some constraint), a *wipeout* has occurred and the current partial instantiation cannot be extended to a solution. At that point, some *retraction* method is applied. Here we use *chronological backtracking*, which prunes the subtree (*digression*) rooted at an inconsistent *node* (assignment of values to some subset of the variables) and withdraws the most recent value assignment(s).

The efficacy of a constraint solver is gauged by its ability to solve a problem, along with the computational resources (CPU time and search tree size in nodes) required to do so. Search for a CSP solution is NP-complete; the worst-case cost is exponential in the number of variables n for any known algorithm. Often, however, a CSP can be solved with a cost much smaller than that of the worst case. A CSP *search algorithm* specifies heuristics for variable (and possibly value) selection, an inference method, and a backtracking method. It is also possible to *restart* search on a problem, beginning over with no assignments and choosing a new first variable-value pair for assignment.

In global search, there are two kinds of search decisions: select a variable or select a value for a variable. Constraint researchers have devised a broad range of variable-ordering and value-ordering heuristics to speed search. Each heuristic relies on its own *metric*, a measure that the heuristic either maximizes or minimizes when it makes a decision. *Min domain* and *max degree* are classic examples of these heuristics. (A full list of the metrics for the heuristics used in these experiments appears in the Appendix.) A metric may rely upon dynamic and/or learned knowledge. Each such heuristic may be seen as expressing a preference for choices based on the scores returned by its metric. As demonstrated in Section 5.5.1, however, no single heuristic is "best" on all CSP classes. Our research therefore seeks a combination of heuristics.

In the experiments reported here, resources were controlled with a *node limit* that imposed an upper bound on the number of assignments of a value to a variable during search on a given problem. Unless otherwise noted, the node limit per problem was 50,000 for <50, 10, 0.38, 0.2>; 20,000 for <20, 30, 0.444, 0.5>; 10,000 for <50, 10, 0.18, 0.37>; 500 for <30, 8, 0.26, 0.34>; and 5,000 for *Comp* and *Geo* problems. Performance was declared *inadequate* if at least ten out of 50 problems went unsolved under a specified resource limit.

A good mixture of heuristics can outperform even the best individual heuristic, as Table 5.1 demonstrates. The first line shows the best performance achieved by any traditional single heuristic we tested. The second line illustrates the performance of a random selection of heuristics, without any learning. (These experiments are non-deterministic and therefore averaged over a set of ten runs.) On one class, the sets of heuristics proved inadequate on every run, and on the other class, five runs were inadequate and the other five dramatically underperformed every other approach. The third line shows that a good pair of heuristics, one for variable ordering and the other for value ordering, can perform significantly better than an individual heuristic. The last line of Table 5.1 demonstrates that a customized combination of more than two heuristics, discovered with the methods described here, can further improve performance. Of course, the use of more than one heuristic may increase solution time, particularly on easier problems where a single heuristic may suffice. On harder problems, however, increased decision time is justified by the ability to solve more problems. This chapter addresses work on the automatic identification of such particularly effective mixtures.

Table 5.1: Search tree size under individual heuristics and under mixtures of heuristics on two classes of problems. Each class has its own particular combination of more than two heuristics that performs better

Guidance	<20, 30, 0.444, 0.5>			<50, 10, 0.38, 0.2>		
	Nodes	Solved	Time	Nodes	Solved	Time
Best individual heuristic tested	3,403.42	100%	10.70	17,399.06	84%	79.02
Randomly selected combination of more than two heuristics	five inadequate runs			ten inadequate runs		
Best pair of variable-ordering and value-ordering heuristic identified	1,988.10	100%	17.73	10,889.00	96%	76.16
Best learned weighted combination of more than 2 heuristics found by ACE	1,956.62	100%	29.22	8,559.66	98%	111.20

5.4 Search with More than One Heuristic

Many well-respected machine-learning methods have been applied to combine algorithms to solve constraint satisfaction problems. This work can be characterized along several dimensions: whether more than one algorithm is used on a single problem, whether the algorithms and heuristics are known or discovered, and whether they address a single problem or a class of problems.

5.4.1 Approaches that Begin with Known Algorithms and Heuristics

Given a set of available algorithms, one approach is to choose a single algorithm to solve an entire problem based upon experience with other problems (not necessarily in the same class). Case-based reasoning has been used this way for CSP search. A feature vector characterizes a solved CSP and points to a set of strategies (a model, a search algorithm, a variable-ordering heuristic, and a value-ordering heuristic) appropriate for that instance [17]. Given a new CSP, majority voting by strategies associated with similar CSPs chooses a strategy for the current one. In a more elaborate single-selection method, a Support Vector Machine (*SVM*) with a Gaussian kernel learns to select the best heuristic during search at checkpoints parameterized by the user [3]. Training instances are described by static features of a CSP, dynamic features of the current partial instantiation, and labels on each checkpoint that indicate whether each heuristic has a better runtime than the default heuristic there. The solver applies the default heuristic initially but, after restart, replaces the default with a randomly selected heuristic chosen from those that the SVM preferred.

A slightly more complex approach permits a different individual heuristic to make the decisions at each step, again based on experience solving other problems not necessarily in the same class. A *hyperheuristic* decides which heuristic to apply at each decision step during search [7]. A hyperheuristic is a set of problem states, each labeled by a condition-action rule of the form "if the state has these properties, then apply this heuristic." A genetic algorithm evolves a hyperheuristic for a given set of states [45]. To solve a CSP, the best among the evolved hyperheuristics is chosen, and then the heuristic associated with the problem state most similar to the current partial instantiation is applied.

Alternatively, a solver can be built for a single CSP from a set of known search algorithms that take turns searching. For example, *REBA* (Reduced Exceptional Behavior Algorithm) applies more complex algorithms only to harder problems [5]. It begins search with a simple algorithm, and when there is no indication of progress, switches to a more complex algorithm. If necessary, this process can continue through a prespecified sequence of complex algorithms. The complexity ranking can be tailored to a given class of CSPs, and is usually based on the median cost of solution and an algorithm's sensitivity to exceptionally hard problems from the class. In general the most complex of these algorithms have better worst-case performance but a higher average cost when applied to classes with many easy problems that could be quickly solved by simpler algorithms. Another approach that alternates among solvers is CPHydra. It maintains a database of cases on not necessarily similar CSPs, indexed both by static problem features and by modeling selections [30]. Each case includes the time each solver available to CPHydra took to solve the problem. CPHydra retrieves the most similar cases and uses them to generate a schedule that interleaves the fastest solvers on those cases.

It is also possible to race algorithms against one another to solve a single problem. An *algorithm portfolio* selects a subset from among its available algorithms according to some schedule. Each of these algorithms is run in parallel to the others (or interleaved on a single processor with the same priority), until the fastest one solves the problem [21]. With a *dynamic algorithm portfolio*, schedule selection changes the proportion of CPU time allocated to each heuristic during search. Dynamic algorithm portfolios favor more promising algorithms [16] or improve average-case running time relative to the fastest individual solver [44]. A particularly successful example is SATzilla-07. It builds a portfolio for each *SAT* (propositional satisfiability) problem instance *online* as it searches [49]. Based on features of the instance and each algorithm's past performance, SATzilla-07 uses linear regression to build a computationally inexpensive model of empirical hardness that predicts each algorithm's runtime on a given SAT problem.

5.4.2 Approaches that Discover Their Own Algorithms

Other autonomous learners seek to discover their own heuristics, given inference and backtracking mechanisms. One approach is to discover a combination of existing algorithms appropriate for a class of problems. For example, Multi-TAC learns an ordered list of variable-ordering heuristics for a given class of CSPs [28]. Beginning with an initially empty list of heuristics, each remaining heuristic is attached to the parent to create a different child. The utility of a child is the number of instances it solves within a given time limit, with total time as a tiebreaker. The child with the highest utility becomes the new parent. This process repeats recursively until it does not produce any child that improves upon its parent. The resulting list is consulted in order. Whenever one heuristic cannot discriminate among the variables, those it ranks at the top are forwarded to the next heuristic on the list. Another way to link a class of problems to a combination of algorithms is to construct a class description based on the algorithms' performance. For example, one effort constructs and stores portfolios for quantified Boolean formulae that are generalizations of SAT problems [41].

Local search is an alternative paradigm to global search. *Local search* only considers full instantiations, and moves to another full instantiation that changes the value of some metric. Heuristics specify the metric and how to select among instantiations that qualify. Multiple search heuristics have been integrated with value-biased stochastic sampling [9]. On a given problem instance, multiple restarts sample the performance of different base heuristics. Then the program applies extreme value theory to construct solution quality distributions for each heuristic, and uses this information to bias the selection of a heuristic on subsequent iterations. Extensive research has also been conducted on optimization problems to learn which low-level local search heuristics should be applied in a region of the solution space [8]. One approach combined local search heuristics for SAT problems [29]. Each step chose a constraint to be adjusted based on some measure of its inconsistency in the current instantiation. Then a heuristic was selected probabilistically, based on its expected *utility value* (ability to ameliorate the violation of that constraint). All utility values were initially equal, and then positively or negatively reinforced based on the difference between the current total cost and the total cost the last time that constraint was selected.

The building blocks for a new algorithm need not be standalone algorithms themselves. For example, CLASS discovers local search variable-ordering heuristics for a class of SAT problems with a genetic algorithm on prespecified heuristic primitives [15]. Its primitives describe the currently satisfied clauses and search experience. CLASS uses them to construct LISP-like s-expressions that represent versions of standard local search variable-selection heuristics. Its initial population is a set of randomly generated expressions. Each population is scored by its performance on a set of problems. Then, to construct the next generation, instead of using traditional crossover and mutation, CLASS creates ten new children from its primitives and makes the ten lowest-scoring expressions less likely to survive. The best heuristic found during the course of the search is returned.

5.5 A Plan for Autonomous Search

As constraint satisfaction is increasingly applied to real-world problems, some work on mixtures of heuristics is more applicable than others. Local search, for example, will not halt on an unsolvable problem, and many real-world problems are unsolvable. Classes of problems recur, particularly in real-world environments (e.g., scheduling the same factory every week), and so it seems reasonable to learn to solve a class of CSPs rather than address each new problem in isolation. The success of a case-based method depends in large part on its index and ability to detect similarity, but real-world CSPs vary broadly, and that breadth challenges both the index and the similarity matching. Methods that learn heuristics are designed to replace the traditional heuristics, which perform far better on some problem classes than on others. Nonetheless, the research supporting these heuristics is extensive, and when they work, they work very well; discarding them seems premature.

The remainder of this chapter therefore uses global search on a class of similar CSPs with a large set of known heuristics. Although there are a great many well-tested heuristics, the successful methods that combine or interleave them have dealt with relatively few *candidates* (possible heuristics), despite the burgeoning literature. There are three challenges here: which candidates to consider for a mixture, how to extract training examples, and how autonomous search should gauge its own performance.

5.5.1 Candidate Heuristics

Learning is necessary for CSP solutions because even well-trusted individual heuristics vary dramatically in their performance on different classes. Consider, for example, the performance of five popular variable-ordering heuristics on the three classes in Table 5.2. The *Min domain/dynamic degree* heuristic is the most successful on $< 20, 30, 0.444, 0.5 >$ problems, but it is inadequate on *Comp* problems.

Table 5.2: Average number of nodes explored by traditional variable-ordering heuristics (with lexical value ordering) on 50 problems from each of three classes. The best (in bold) and the worst (in italics) performance by a single heuristic vary with the problem class. Problem classes were defined in Section 5.3

	Geo		*Comp*		*<20, 30, 0.444, 0.5>*	
Heuristics	*Nodes*	*Solved*	*Nodes*	*Solved*	*Nodes*	*Solved*
Min domain/dynamic degree	258.1	98%	*inadequate*		**3403.4**	100%
Min dynamic domain/weighted-degree	**246.4**	100%	57.67	100%	3534.3	100%
Min domain/static degree	254.6	98%	*inadequate*		3561.3	100%
Max static degree	*397.7*	98%	*inadequate*		4742.1	96%
Max weighted degree	343.3	98%	**50.44**	100%	*5827.9*	98%

On real-world problems and on problems with non-random structure, the opposite of a traditional heuristic may provide better guidance during search [33, 27, 32]. The *dual* of a heuristic reverses the import of its metric (e.g., *max domain* is the dual of *min domain*). Table 5.3 demonstrates the superior performance of some duals on *Comp* problems. Recall that a *Comp* problem has a central component that is substantially larger, looser (has lower tightness), and sparser (has lower density) than its satellite. Once a solution to the subproblem defined by the satellite is found, it is relatively easy to extend that solution to the looser and sparser central component. In contrast, if one extends a partial solution for the subproblem defined by the central component to the satellite variables, inconsistencies eventually arise deep within the search tree. Despite the low density of the central component in such a problem, its variables' degrees are often larger than those in the significantly smaller satellite. The central component proves particularly attractive to two of the traditional heuristics in Table 5.2, which then flounder there. We emphasize again that the characteristics of such composed problems are often found in real-world problems. Our approach, therefore, is to take as candidates many popular heuristics, along with their duals.

Table 5.3: Average number of nodes explored by three traditional heuristics (in italics) and their duals on *Comp* problems (described in Section 5.3). Note the better performance of two of the duals here

Heuristics	*Nodes*	*Solved*
Min static degree	33.15	100%
Max static degree	inadequate	—
Max domain/dynamic degree	532.22	95%
Min domain/dynamic degree	inadequate	—
Max domain	1168.71	90%
Min domain	373.22	97%

5.5.2 *Training Examples*

From a successful search, the solver extracts both positive and negative training examples. Each training example is the current instantiation and the decision made there. Here, too, difficulties arise. The search trace is not necessarily the best way to solve that CSP. There may have been a better (smaller search tree or less elapsed time) way to solve it. Thus the training examples selected may not be those an oracle would have provided. Nonetheless, decisions not subsequently retracted on the path to the solution can be considered better than those that were retracted, so we take them as positive examples. We also know that the roots of retracted subtrees (*disgressions*) were errors, and therefore take them as negative examples. (Note that

Fig. 5.3: The extraction
of positive and negative
training instances from
the trace of a successful
CSP search

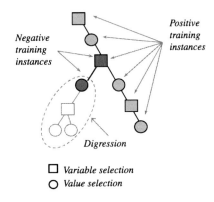

☐ *Variable selection*
○ *Value selection*

assigning value v to variable X is not an error if in some solution of the problem $X = v$; what we address here is whether assigning v to X given the current partial instantiation is an error.)

As in Figure 5.3, *positive training instances* are those made along an error-free path extracted from a solution trace. *Negative training instances* are value selections that led to a digression, as well as variable selections whose subsequent value assignment failed. (Given correct value selections, any variable ordering can produce a backtrack-free solution; ACE deems a variable selection inadequate if the subsequent value assignment to that variable failed.) Decisions below the root of a digression do not become training instances.

Although machine learning assumes that training and testing examples come from the same population, a solver's experience in a class of CSPs will not be uniform, and thus the solver is likely to be misled. The difficulty a solver has on problems in the same class, despite their common characterization, has been shown to have a heavy tail, that is, to have a Pareto-normal distribution. This means that a portion of problems in the class will be very difficult for the solver's algorithm, and that portion will not decrease exponentially. Furthermore, although CSPs in the same class are ostensibly similar, there is evidence that their difficulty may vary substantially for a given search algorithm [23]. Thus our learner will inevitably be confronted with a non-uniform, heuristically selected set of training examples.

5.5.3 Performance Assessment

CSP search performance is traditionally gauged by the size of the search tree expanded (number of partial instantiations) and elapsed CPU time. Here, a solver executes in a *run* of two phases: a *learning phase* during which it attempts to solve a sequence of CSPs from a given class, and a *testing phase* during which it attempts to solve a sequence of hitherto unseen problems drawn from the same class. Because the class is of uneven difficulty, the problems in the learning phase may not be indicative of the class as a whole. Thus we average the solver's performance over an *experiment*, here a set of ten runs.

Ideally, a solver should also consider its performance more broadly. It should be aware of its general progress on the class. If it believes it can learn no more, it should terminate learning itself and proceed to testing. And if it believes that learning is not going well, it should elect to discard what it has learned and start over. The solver constructed to meet these expectations is called ACE.

5.6 ACE

When *ACE* (the Adaptive Constraint Engine) learns to solve a class of binary CSPs, it customizes a weighted mixture of heuristics for the class [13]. ACE is based on *FORR*, an architecture for the development of expertise from multiple heuristics [12]. ACE's search algorithm (in Figure 5.4) alternately selects a variable and then selects a value for it from its domain. The size of the resultant search tree depends upon the order in which values and variables are selected.

5.6.1 Decision Hierarchy

Decision-making procedures in ACE are called *Advisors*. They are organized into three tiers, and presented with the current set of choices (variables or values). Tier 1 Advisors are correct and quick, and preordered by the user. If any of them approves a choice, it is executed. (For example, *Victory* recommends any value from the domain of the final unassigned variable. Since inference has already removed inconsistent values, any remaining value produces a solution.) Disapproval from any tier 1 Advisor eliminates some subset of choices; the remaining choices are passed to the next Advisor. (The set of choices is not permitted to go empty.) Tier 2 Advisors address subgoals; they are outside the scope of this chapter and not used in the experiments reported here.

The work described here focuses on the Advisors in tier 3. Each tier 3 Advisor *comments* upon (produces a strength for) some of its favored choices, those whose metric scores are among the f most favored. Because a metric can return identical values for different choices, an Advisor usually makes many more than f comments. (Here $f = 5$, unless otherwise stated.) The *strength* $s(A, c)$ is the degree of support from Advisor A for choice c. Each tier 3 Advisor's view is based on a descriptive metric. All tier 3 Advisors are consulted together. As in Figure 5.4, a decision in tier 3 is made by *weighted voting*, where the strength $s(A, c)$ given to choice c by Advisor A is multiplied by the *weight* $w(A)$ of Advisor A. All weights are initialized to 0.05, and then learned for a class of problems by the processes described below. The discount factor $q(A)$ in $(0,1]$ modulates the influence of Advisor A until it has commented often enough during learning. As data is observed on A, $q(A)$ moves

Search (p, Avar, Aval)
Until problem p is solved or the allocated resources are exhausted
  Select unvalued variable v

$$v = \arg\max_{c_{\text{var}} \in V} \sum_{A \in A_{var}} q(A) \cdot w(A) \cdot s(A, c_{\text{var}})$$

  Select value d for variable v from v's domain D_v

$$d = \arg\max_{c_{\text{val}} \in D_v} \sum_{A \in A_{val}} q(A) \cdot w(A) \cdot s(A, c_{\text{val}})$$

  Update domains of all unvalued variables   *inference*
  Unless domains of all unvalued variables are nonempty
        return to a previous alternative value   *retraction*

Fig. 5.4: Search in ACE with a weighted mixture of variable-ordering Advisors from A_{var}, and value-ordering Advisors from A_{val}. $q(A)$ is the discount factor. $w(A)$ is the weight of Advisor A. $s(A, c)$ is the strength of Advisor A for choice c

toward 1, effectively increasing the impact of A on a given class as its learned weight becomes more trustworthy.

Weighted voting selects the choice with the greatest sum of weighted strengths from all Advisors. (Ties are broken randomly.) Each tier 3 Advisor's heuristic view is based on a descriptive metric. For each metric, there is a dual pair of Advisors, one that favors smaller values for the metric and one that favors larger values. Typically, only one of the pair has been reported in the literature as a heuristic. Weights are learned from problem-solving experience.

5.6.2 Weight Learning

Given a class of binary, solvable problems, ACE's goal is to formulate a mixture of Advisors whose joint decisions lead to effective search on a class of CSPs. Its learning scenario specifies that the learner seeks only one solution to one problem at a time, and learns only from problems that it solves. There is no information about whether a single different decision might have produced a far smaller search tree. This is therefore a form of incremental, self-supervised reinforcement learning based only on limited search experience and incomplete information. As a result, any weight-learning algorithm for ACE must select training examples from which to learn, determine what constitutes a heuristic's support for a decision, and specify a way to assign credits and penalties.

ACE learns weights only after it solves a problem. ACE's two most successful approaches to weight learning are Digression-based Weight Learning (*DWL*) [13] and Relative Support Weight Learning (*RSWL*) [35]. It uses them to update the weights of its tier 3 Advisors. Both weight-learning algorithms glean training instances from their own (likely imperfect) successful searches (as described in Section 5.5.2).

Learn Weights
Initialize all weights to 0.05
Until termination of the learning phase
    Identify learning problem p
    Search (p, A_{var}, A_{val})
    If p is solved
       then for each training instance t from p
          for each Advisor A that supports t
             when t is a positive training instance, increase $w(A)$ *credit*
             when t is a negative training instance, decrease $w(A)$ *penalize*
       else when full restart criteria are satisfied
          initialize all weights to 0.05

Fig. 5.5: Learning weights for Advisors. The *Search* algorithm is defined in Figure 5.4

A weight-learning algorithm defines what it means to support a decision. Under DWL, an Advisor is said to support only those decisions to which it assigned the highest strength. In contrast, RSWL considers all strengths. The *relative support* of an Advisor for a choice is the normalized difference between the strength the Advisor assigned to that choice and the average strength it assigned to all available choices at that decision point. For RSWL, an Advisor supports a choice if its relative support for that choice is positive, and opposes that choice if its relative support is negative. As in Figure 5.5, heuristics that support positive training instances receive credits, and heuristics that support negative training instances receive penalties. For both DWL and RSWL, an Advisor's weight is the averaged sum of the credits and penalties it receives, but the two weight-learning algorithms determine credits and penalties differently.

DWL reinforces Advisors' weights based on the size of the search tree and the size of each digression. An Advisor that supports a positive training instance is rewarded with a weight increment that depends upon the size of the search tree, relative to the minimal size of the search tree in all previous problems. An Advisor that supports a negative training instance is penalized in proportion to the number of search nodes in the resultant digression. Small search trees indicate a good variable order, so the variable-ordering Advisors that support positive training instances from a successful small tree are highly rewarded. For value ordering, however, a small search tree is interpreted as an indication that the problem was relatively easy (i.e., any value selection would likely have led to a solution), and therefore results in only small weight increments. In contrast, a successful but large search tree suggests that a problem was relatively difficult, so value-ordering Advisors that support positive training instances from it receive substantial weight increments [13].

RSWL is more local in nature. With each training instance RSWL reinforces weights based upon the distribution of each heuristic's preferences across all the available choices. RSWL reinforces weights based both upon relative support and upon an estimate of how difficult it is to make the correct decision. For example, an Advisor that strongly singles out the correct decision in a positive training instance receives more credit than a less discriminating Advisor, and the penalty for a wrong choice from among a few is harsher than for a wrong choice from among many.

In addition to an input set of Advisors, ACE has one benchmark for variable ordering and another for value ordering. Each *benchmark Advisor* models random advice; it makes random comments with random strengths. Although the benchmarks' comments never participate in decision making, the benchmarks themselves earn weights. That weight serves as a filter for the benchmark's associated Advisors; an Advisor must have a learned weight higher than its benchmark's (that is, provide better than random advice) to be constructive.

5.7 Techniques that Improve Learning

This section describes four techniques that use both search performance and problem difficulty to adapt learning: full restart, random subsets of heuristics, consideration of decision difficulty, and the nuances of preferences. Section 5.8 provides empirical demonstrations of their efficacy.

5.7.1 Full Restart

From a large initial list of heuristics that contains minimizing and maximizing versions of many metrics, some perform poorly on a particular class of problems (*class-inappropriate heuristics*) while others perform well (*class-appropriate heuristics*). In some cases, class-inappropriate heuristics occasionally acquire high weights on an initial problem and then control subsequent decisions. As a result, subsequent problems may have extremely large search trees.

Given unlimited resources, DWL will recover from class-inappropriate heuristics with high weights, because they typically generate large search trees and large digressions. In response, DWL will impose large penalties and provide small credits to the variable-ordering Advisors that lead decisions. With their significantly reduced weights, class-inappropriate Advisors will no longer dominate the class-appropriate Advisors. Nonetheless, solving a hard problem without good heuristics is computationally expensive. If adequate resources are unavailable under a given node limit and a problem goes unsolved, no weight changes occur at all.

Under *full restart*, however, ACE monitors the frequency and the order of unsolved problems in the problem sequence. If it deems the current learning attempt not promising, ACE abandons the learning process (and any learned weights) and begins learning on new problems with freshly initialized weights [34]. Note that full restart is different from restart on an individual problem, discussed in Section 5.3, which diversifies the search for a solution to that problem alone [20]. We focus here only on the impact of full restart of the entire learning process.

The node limit is a critical parameter for full restart. Because ACE abandons a problem if it does not find a solution within the node limit, the node limit is the criterion for unsuccessful search. Since the full restart threshold directly depends upon the number of failures, the node limit is the performance standard for full restart. The node limit also controls resources; lengthy searches permitted under high node limits are expensive.

Resource limits and full restart impact the cost of learning in complex ways. With higher node limits, weights can eventually recover without the use of full restart, but recovery is more expensive. With lower node limits, the cost of learning (total number of nodes across all learning problems) with full restart is slightly higher than without it. The learner fails on all the difficult problems, and even on some of medium difficulty, repeatedly triggering full restart until the *weight profile* (the set of tier 3 weights) is good enough to solve almost all the problems. Full restart abandons some problems and uses additional problems, which increases the cost of learning. The difference in cost is small, however, since each problem's cost is subject to a relatively low node limit. As the node limit increases, full restart produces fewer inadequate runs, but at a higher cost. It takes longer to trigger full restart because the learned weight profile is deemed good enough and failures are less frequent. Moreover, with a high node limit, every failure is expensive. When full restart eventually triggers, the prospect of relatively extensive effort on further problems is gone. Because it detects and eliminates unpromising learning runs early, full restart avoids many costly searches and drastically reduces overall learning cost. Experimental results and further discussion of full restart appear in Section 5.8.1.

5.7.2 *Learning with Random Subsets*

The interaction among heuristics can also serve as a filter during learning. Given a large and inconsistent initial set of heuristics, many class-inappropriate ones may combine to make bad decisions, and thereby make it difficult to solve any problem within a given node limit. Because only solved problems provide training instances for weight learning, no learning can take place until some problem is solved. Rather than consult all its Advisors at once, ACE can randomly select a new subset of Advisors for each problem, consult them, make decisions based on their comments, and update only their weights [37]. This method, *learning with random subsets*, eventually uses a subset in which class-appropriate heuristics predominate and agree on choices that solve a problem.

It is possible to preprocess individual heuristics on representatives of a problem class (possibly by racing [4]), and then eliminate from the candidate heuristics those with poor individual performance. That approach, however, requires multiple solution attempts on some set of problems. Multiple individual runs would consume more computational resources, because many different Advisors reference (and share the values of) the same metrics in our approach. Moreover, elimination of poorly performing heuristics after preprocessing might prevent the discovery of important synergies (e.g., tiebreakers) between eliminated heuristics and retained ones.

For a fixed node limit and set of heuristics, an underlying assumption here is that the ratio of class-appropriate to class-inappropriate heuristics determines whether a problem is likely to be solved. When class-inappropriate heuristics predominate in a set of heuristics, the problem is unlikely to be solved and no learning occurs. The selection of a new random subset of heuristics for each new problem, however, should eventually produce some subset S with a majority of class-appropriate heuristics that solves its problem within a reasonable resource limit. As a result, the Advisors in S will have their weights adjusted. On the next problem instance, the new random subset S' is likely to contain some low-weight Advisors outside of S, and some reselected from S. Any previously successful Advisors from S that are selected for S' will have larger positive weights than the other Advisors in S', and will therefore heavily influence search decisions. If S succeeded because it contained more class-appropriate than class-inappropriate heuristics, $S \cap S'$ is also likely to have more class-appropriate heuristics and therefore solve the new problem, so again those that participate in correct decisions will be rewarded. On the other hand, in the less likely case that the majority of $S \cap S'$ consists of reinforced, class-inappropriate heuristics, the problem will likely go unsolved, and the class-inappropriate heuristics will not be rewarded further.

Learning with random subsets manages a substantial set of heuristics, most of which may be class-inappropriate and contradictory. It results in fewer *early failures* (problems that go unsolved with the initial weights, before any learning occurs) within the given node limit, and thereby makes training instances available for learning sooner. Learning with random subsets is also expedited by faster decisions during learning because it often solicits advice from fewer Advisors.

When there are roughly as many class-appropriate as class-inappropriate Advisors, the subset sizes are less important than when class-inappropriate Advisors outnumber class-appropriate ones. Intuitively, if there are few class-appropriate heuristics available, the probability that they are selected as a majority in a larger subset is small (indeed, 0 if the subset size is more than twice the number of class-appropriate Advisors). For example, given a class-appropriate Advisors and b class-inappropriate Advisors, the probability that the majority of a subset of r randomly-selected Advisors is class-appropriate is

$$ p = \sum_{k=\lfloor \frac{r}{2}+1 \rfloor}^{r} \frac{\binom{a}{k}\binom{b}{r-k}}{\binom{a+b}{r}} \tag{5.1} $$

and the expected number of trials until the subset has a majority of class-appropriate Advisors is

$$ p = \sum_{i=1}^{\infty} i(1-p)^{i-1} p = \frac{1}{p}. \tag{5.2} $$

When there are more class-inappropriate Advisors ($a < b$), a smaller set is more likely to have a majority of class-appropriate Advisors. For example, if $a = 6$, $b = 9$, and $r = 4$, equation (1) evaluates to 0.14 and equation (2) to 7. For $a = 6$, $b = 9$, and $r = 10$, however, the probability of a randomly selected subset with a majority of class-appropriate heuristics is only 0.04 and the expected number of trials until the subset has a majority of class-appropriate Advisors is 23.8.

Weight convergence is linked to subset size. Weights converge faster when subsets are larger. When the random subsets are smaller, subsequent random subsets are less likely to overlap with those that preceded them, and therefore less likely to include Advisors whose weights have been revised. As a result, failures occur often, even after some class-appropriate heuristics receive high weights. ACE monitors its learning progress, and can adapt the size of random subsets. In this scenario, random subsets are initially small, but as learning progresses and more Advisors participate and obtain weights, the size of the random subsets increase. This makes overlap more likely, and thereby speeds learning. Experimental results and further discussion of learning with random subsets appear in Section 5.8.2.

5.7.3 Learning Based on Decision Difficulty

Correct easy decisions are less significant for learning; it is correct difficult decisions that are noteworthy. Thus it may be constructive to estimate the difficulty of each decision the solver faces as if it were a fresh problem, and adjust Advisors' weights accordingly. Our rationale for this is that, on easy problems, any decision leads to a solution. Credit for an easy decision effectively increases the weight of Advisors that support it, but if the decision was made during search, those Advisors probably already had high weights. ACE addresses this issue with two algorithms, each dependent upon a single parameter: RSWL-κ and RSWL-d.

Constrainedness, as measured by κ, has traditionally been used to identify hard classes of CSPs [18]. κ depends upon n, d, m, and t, as defined in Section 5.3:

$$\kappa = \frac{n-1}{2} \cdot d \cdot \log_m \frac{1}{1-t}. \qquad (5.3)$$

For every search algorithm, and for fixed n and m, hard problem classes have κ close to 1. *RSWL-κ* uses equation (5.3) to measure the difficulty κ_P of subproblem P at each decision point. For a given parameter k, RSWL-κ gives credit to an Advisor only when it supports a positive training instance derived from a search state where $|\kappa_P - 1| < k$. RSWL-κ penalizes an Advisor only when it supports a negative training instance derived from a search state where $|\kappa_P - 1| > k$. The calculation of κ_P on every training instance is computationally expensive.

RSWL-d uses the number of unassigned variables at the current search node as a rough, quick estimate of problem hardness. Decisions at the top of the search tree

Fig. 5.6: A constraint
graph for a CSP problem
on 12 variables

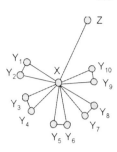

are known to be more difficult [38]. For a given parameter h, no penalty is given at all for any decision in the top h percent of the nodes in the search tree, and no credit is given for any decision below them. Experimental results and further discussion of learning with decision difficulty appear in Section 5.8.3.

5.7.4 Combining Heuristics' Preferences

The preferences expressed by heuristics can be used to make decisions during search. The intuition here is that comparative nuances, as expressed by preferences, contain more information than just what is "best." Recall that each heuristic reflects an underlying metric that returns a *score* for each possible choice. Comparative opinions (here, *heuristics' preferences*) can be exploited in a variety of ways that consider both the scores returned by the metrics on which these heuristics rely and the distributions of those scores across a set of possible choices.

The simplest way to combine heuristics' preferences is to scale them into some common range. Mere ranking of these scores, however, reflects only the preferences for one choice over another, not the extent to which one choice is preferred over another. For example in Figure 5.6, the degrees of variables X and Y_1 differ by 9, while the degrees of Y_1 and Z differ by only 1. Nonetheless, ranking by degree assigns equally spaced strengths (3, 2 and 1, respectively) to X, Y_1, and Z. Ranking also ignores how many choices share the same score. For example, in Table 5.4, the ranks of choices Y_1 and Z differ by only 1, although the heuristic prefers only one choice over Y_1 and 11 choices over Z. We have explored several methods that express Advisors' preferences and address those shortcomings [36].

Linear interpolation not only considers the relative position of scores, but also the actual differences between them. Under linear interpolation, strength differences are proportional to score differences. For example, in Table 5.4, strengths can be determined by the value of the linear function through the points (11, 3) and (1, 1). Instead of applying strength 2 for all the Y variables, linear interpolation gives them strength 1.2, which is closer to the strength 1 given to variable Z, because the degrees of the Y variables are closer to the degree of Z. The significantly higher degree of

Table 5.4: The impact of different preference expression methods on a single metric

Variables	X	$Y_1, Y_2, ..., Y_{10}$	Z
Degree metric scores	11	2	1
Rank strength	3	2	1
Linear strength	3.0	1.2	1.0
Borda-w strength	11.0	1.0	0.0
Borda-wt strength	12.0	11.0	1.0

variable X is reflected in the distance between its strength and those given to the other variables.

The *Borda methods* were inspired by an election method devised by Jean-Charles de Borda in the late eighteenth century [39]. Borda methods consider the total number of available choices, the number of choices with a smaller score and the number of choices with an equal score. Thus the strength for a choice is based on its position relative to the other choices.

The first Borda method, *Borda-w*, awards a point for each *win* (metric score higher than the score of some other commented choice). Examples for *Borda-w* strengths are also shown in Table 5.4. The set of lowest-scoring variables (here, only Z) always has strength 0. Because every Y variable outscored only Z, the strength of any Y variable is 1. The highest-scoring choice X outscored 11 choices, so X's strength is 11.

The second Borda method, *Borda-wt*, awards one point for each win and one point for each *tie* (score equal to the score of some other choice). It can be interpreted as emphasizing losses. The highest-scoring set of variables (here, only X) always has strength equal to the number of variables to be scored. For example, in Table 5.4, no choice outscored the highest-scoring choice X, so its strength is 12, one choice (X) outscored the Y variables, so their strengths are reduced by one point ($12 - 1 = 11$), and 11 choices outscored Z, resulting in strength 1.

The difference between the two Borda methods is evident when many choices share the same score. *Borda-w* considers only how many choices score lower, so that a large subset results in a big gap in strength between that subset and the previous (more preferred) one. Under *Borda-wt*, a large subset results in a big gap in strength between that subset and the next (less preferred) one. In Table 5.4, for example, the 10 Y variables share the same score. Under *Borda-w*, the difference between the strength of every Y variable and X is 10, while the difference between the strength of any Y variable and Z is only 1. Under *Borda-wt*, however, the difference between the strength of any Y and X is only 1, while the difference between the strength of any Y and Z is 10. The Borda approach can be further emphasized by making a point inversely proportional to the number of subsets of tied values. Experimental results and further discussion of learning with preferences using this emphasis appear in Section 5.8.4.

5.8 Results

The methods in the previous section are investigated here with ACE. As described in Section 5.5.3, each run under ACE is a learning phase followed by a testing phase. During a learning phase ACE refines its Advisor weights; the testing phase provides a sequence of fresh problems with learning turned off. All of the following experiments average results across ten runs, and differences cited are statistically significant at the 95% confidence level.

In the experiments that follow, learning terminated after 30 problems, counting from the first solved problem, or terminated if no problem in the first 30 was solved at all. Under full restart, more learning problems can be used, but the upper bound for the total number of problems in a learning phase was always 80. For each problem class, every testing phase used the same 50 problems. When any ten of the 50 testing problems went unsolved within the node limit, learning in that run was declared *inadequate* and further testing was halted. In every learning phase, ACE had access to 40 tier 3 Advisors, 28 for variable selection and 12 for value selection (described in the Appendix). During a testing phase, ACE used only those Advisors whose learned weights exceeded those of their respective benchmarks.

5.8.1 Full Restart Improves Performance

The benefits of full restart are illustrated here on $<30, 8, 0.26, 0.34>$ problems with DWL, where the node limit during learning is treated as a parameter and the node limit during testing is 10,000 nodes. A run was declared *successful* if testing was not halted due to repeated failures. The *learning cost* is the total number of nodes during the learning phase of a run, calculated as the product of the average number of nodes per problem and the average number of problems per run. The *restart strategy* is defined by a *full restart threshold* (k, l), which performs a full restart after failure on k problems out of the last l. (Here, $k = 3$ and $l = 4$.) This seeks to avoid full restarts when multiple but sporadic failures are actually due to uneven problem difficulty rather than to an inadequate weight profile. Problems that went unsolved under initial weights before any learning occurred (*early failures*) were not counted toward full restart. If the first 30 problems went unsolved under the initial weights, learning was terminated and the run judged unsuccessful. The learner's performance here is measured by the number of successful runs (out of ten) and the learning cost across a range of node limits.

Under every node limit tested, full restart produced more runs that were successful, as Figure 5.7 illustrates. At lower node limits, Figure 5.8 shows that better testing performance came with a learning cost similar to or slightly higher than the cost without full restart. At higher node limits, the learning cost was considerably lower with full restart. With very low node limits (200 or 300 nodes), even with full restart DWL could not solve all the problems. During learning, many problems went unsolved under a low node limit and therefore did not provide training in-

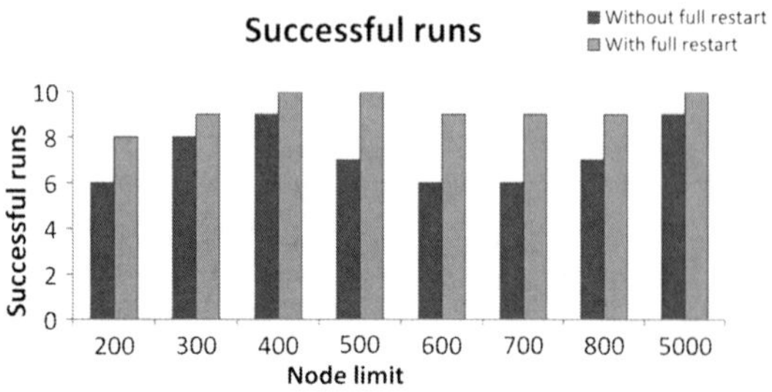

Fig. 5.7: Number of suc-
cessful runs (out of ten)
on <30, 8, 0.26, 0.34>
problems under different
node limits

Successful runs

■ Without full restart
■ With full restart

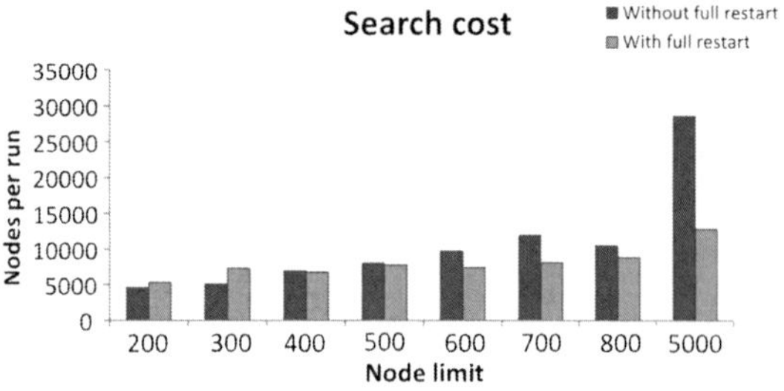

Fig. 5.8: Learning cost,
measured by the average
number of nodes per run,
on <30, 8, 0.26, 0.34>
problems under different
node limits

Search cost

■ Without full restart
■ With full restart

stances. On some (inadequate) runs, no solution was found to any of the first 30
problems, so learning was terminated without any weight changes. When the node
limit was somewhat higher (400 nodes), more problems were solved, more training
instances became available and more runs were successful. These reasonably low
node limits set a high standard for the learner; only weight profiles well tuned to the
class will solve problems within them and thereby provide good training instances.
Further increases in the node limit (500, 600, 700 and 800 nodes), however, did
not further increase the number of successful runs. Under higher node limits, problems were solved even with weight profiles that were not particularly good for the
class, and may have produced training instances that were not appropriate. Under
extremely high node limits (5,000 nodes), problems were solved even under inadequate weight profiles, but the weight-learning mechanism was able to recover a
good weight profile, and again the number of successful runs increased.

Similar performance was observed under RSWL and on *Geo* and the other model
B classes identified in Section 5.3, but not on the *Comp* problems. Some *Comp*
problems go unsolved under any node limit, while many others are solved. Because
failures there are sporadic, they do not trigger full restart. The use of full restart on
them does not improve learning, but it does not harm it either. (Data omitted.) Full
restart is therefore used throughout the remainder of these experiments.

5.8.2 Random Subsets Improve Performance

When the full complement of Advisors was present but resource limits were strict,
even with full restart ACE sometimes failed to solve problems in some difficult

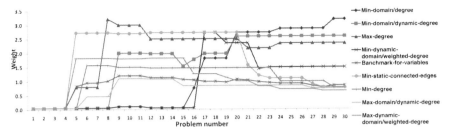

Fig. 5.9: Weights of eight variable-ordering Advisors and their common benchmark during learning after each of 30 problems in <50, 10, 0.38, 0.2> on a single run

classes. Random subsets corrected this. Figure 5.9 illustrates weight convergence during learning with random subsets. On this run some Advisors (e.g., *Min domain/-dynamic degree* and *Min static degree*) recovered from initially inadequate weights. The figure tracks the weights of eight variable-ordering heuristics and their common benchmark (described in Section 5.6.2) after each of 30 problems. Here the problems were drawn from <50, 10, 0.38, 0.2>, and 30% of the variable-ordering Advisors and 30% of the value-ordering Advisors were randomly selected for each problem. Plateaus in weights correspond to problems where the particular heuristic was not selected for the current random subset, or the problem went unsolved, so that no learning or weight changes occurred. The first four problems were *early failures* (problems that went unsolved under initial weights, before any learning occurred). When the fifth problem was solved, some class-inappropriate Advisors received high weights from its training instances. On the next few problems, either heavily weighted but class-inappropriate heuristics were reselected and the problem went unsolved and no weights changed, or some class-appropriate Advisors were selected and gained high weights. Eventually the latter began to dominate decisions, so that the disagreeing class-inappropriate Advisors had their weights reduced. After the 21st problem, when the weight of *Min static connected edges* had significantly decreased, the weights clearly separated the class-appropriate Advisors from the class-inappropriate ones. Afterwards, as learning progressed, the weights stabilized.

Experiments that illustrate the benefits of random subsets tested four ways to choose the Advisors from each problem:

1. *All* used all the Advisors on every problem.
2. *Fixed* chose a fixed percentage q (30% or 70%), and then chose q variable-ordering Advisors and q value-ordering Advisors, without replacement.
3. *Varying* chose a random value r in [30, 70] for each problem, and then chose r percent of the variable-ordering Advisors and r percent of the value-ordering Advisors, without replacement.
4. *Incremental* initially selected q of the variable-ordering Advisors and q of the value-ordering Advisors. Then, for each subsequent problem, it increased the sizes of the random subsets in proportion to the number of Advisors whose weight was greater than their initially assigned weight.

On problems in $<50, 10, 0.18, 0.37>$, Table 5.5 compares these approaches for learning with random subsets of Advisors to learning with all the Advisors at once. When all 40 Advisors were consulted, the predominance of class-inappropriate Advisors sometimes prevented the solution of any problem under the given node limit, so that some learning phases were terminated after 30 unsolved problems. In those runs no learning occurred. With random subsets, however, adequate weights were learned on every run, and there were fewer early failures.

Table 5.5: Early failures, successful runs, decision time and percentage of computation time during learning with random subsets of Advisors, compared to computation time with all the Advisors on problems in $<50, 10, 0.18, 0.37>$

Advisors	Early failures per run	Successful runs	Time per learning decision	Learning time per run
All	27.0	4	100.00%	100.00%
Fixed $q = 70\%$	5.2	10	74.30%	24.39%
Varying $r \in [30\%, 70\%]$	1.9	10	65.84%	20.74%
Incremental $q = 30\%$	1.8	10	75.35%	19.76%
Fixed $q = 30\%$	3.1	10	55.39%	21.85%

Table 5.5 also demonstrates that using random subsets significantly reduces learning time. The time to select a variable or a value is not necessarily directly proportional to the number of selected Advisors. This is primarily because dual pairs of Advisors share the same fundamental computational cost: calculating their common metric. For example, the bulk of the work for *Min product domain value* lies in the one-step lookahead that calculates (and stores) the products of the domain sizes of the neighbors after each potential value assignment. Consulting only *Min product domain value* and not *Max product domain value* will therefore not significantly reduce computation time. Moreover, the metrics for some Advisors are based upon metrics already calculated for others that are not their duals. The reduction in total computation time per run also reflects any reduction in the number of learning problems.

The robustness of learning with random subsets is demonstrated with experiments documented in Table 5.6 that begin with fewer Advisors, a majority of which are class-inappropriate. Based on weights from successful runs with all Advisors, Advisors were first identified as class-appropriate or class-inappropriate for problems in $<50, 10, 0.18, 0.37>$. ACE was then provided with two different sets A_{var} of variable-ordering Advisors in which class-inappropriate Advisors outnumbered class-appropriate ones (nine to six or nine to four). When all the provided Advisors were consulted, the predominance of class-inappropriate Advisors effectively prevented the solution of any problem under the given node limit and no learning took place. When learning with random subsets, as the size of the random subsets decreased, the number of successful runs increased. As a result, ran-

Table 5.6: Learning with more class-inappropriate than class-appropriate Advisors on problems in $<50, 10, 0.18, 0.37>$. Smaller, fixed-size random subsets appear to perform best

Advisors	6 class-appropriate 9 class-inappropriate		4 class-appropriate 9 class-inappropriate	
	Early failures	*Successful runs*	*Early failures*	*Successful runs*
All	30.0	0	30.0	0
Fixed $q = 70\%$	21.2	7	30.0	0
Varying $r \in [30\%, 70\%]$	8.4	10	17.6	6
Incremental $q = 30\%$	3.8	10	12.0	9
Fixed $q = 30\%$	5.1	10	13.3	10

dom subsets with fixed $q = 30\%$ is used throughout the remainder of these experiments.

5.8.3 The Impact of Decision Difficulty

Consideration of relative support and some assessment of problem difficulty can improve testing performance, as shown in Table 5.7. RSWL solved more problems in $<50, 10, 0.38, 0.2>$ during both learning and testing than did DWL, and required fewer full restarts. Moreover, both RSWL-κ and RSWL-d solved more problems with fewer nodes during testing.

5.8.4 Performance with Preferences

We tested both linear interpolation and the Borda methods on the CSP classes identified in Section 5.3. On the unstructured model B problems, preference expression made no significant difference. On *Comp*, however, across a broad range of node limits, preference expression had an effect, as shown in Table 5.8.

Initially, most variables in the central component score similarly on most metrics, and most variables in the satellite score similarly to one another but differently from those in the central component. Under *Borda–wt*, if only a few choices score higher, the strength of the choices from the next lower-scoring subset is close enough to influence the decision. If there are many high-scoring choices, in the enhanced version the next lower subset will have a much lower strength, which decreases its influence. Moreover, when many choices share the same score, they are penalized for failure to discriminate, and their strength is lowered. When *Borda–w* assigns lower strengths

Table 5.7: Learning and testing performance with different weight learning methods on <50, 10, 0.38, 0.2> problems. Bold figures indicate statistically significant reductions, at the 95% confidence level, in the number of nodes and in the percentage of solved problems compared to search under DWL

| | Learning | | | Testing | |
| | | Unsolved | Full | | |
Weight-learning algorithm	Problems	problems	restarts	Nodes	Solved
DWL	36.8	14.1	0.8	13,708.58	91.8%
RSWL	31.5	7.5	0.2	13,111.44	**95.2%**
RSWL-*d*, *h*=30%	30.9	8.4	0.1	**11,849.00**	94.6%
RSWL-*κ*, *k*=0.2	32.9	8.9	0.3	**11,231.60**	95.0%

Table 5.8: Testing performance with RSWL on *Comp* problems with reduced node limits and a variety of preference expression methods

| Node | Ranking | | Borda-w | | Borda-wt | | Linear | |
Limit	Nodes	Solved	Nodes	Solved	Nodes	Solved	Nodes	Solved
5000	161.1	97.8%	134.1	98.0%	638.5	88.6%	164.2	97.8%
1000	161.1	97.8%	134.1	98.0%	564.7	89.8%	164.2	97.8%
500	161.5	97.8%	121.1	98.2%	728.5	86.6%	164.2	97.8%
100	161.4	97.8%	111.6	98.4%	642.1	88.2%	163.7	97.8%
35	160.4	97.8%	111.7	98.4%	660.0	89.0%	**33.5**	100.0%

to large subsets from the central component, it makes them less attractive. That encourages search to explore variables in the satellites to be selected first; this is often the right way to solve such problems.

Linear interpolation performed similarly to RSWL with ranking on *Comp* problems, except under the lowest node limit tested. Given only 35 nodes, RSWL with linear preference expression was able to solve every problem during testing. The 35-node limit imposes a very high learning standard; it allows learning only from small searches. (A backtrack-free solution would expand exactly 30 nodes for a *Comp* problem.) Only with the nuances of information provided by linear preference expression did ACE develop a weight profile that solved all the testing problems in every run.

5.9 Conclusions and Future Work

Our fundamental underlying assumption is that the right way to make search decisions in a class of CSPs has some uniformity, that is, that it is possible to learn from one problem how to solve another. ACE tries to solve a CSP from a class, and, if it finds a solution, it extracts training examples from the search trace. If it fails to solve the CSP, however, given the size of the search space, the solver will learn nothing from the effort it expended.

Given a training example, the learning algorithms described here reinforce heuristics that prove successful on a set of problems and discard those that do not. Our program represents its learned knowledge about how to solve problems as a weighted sum of the output from some subset of its heuristics. Thus the learner's task is both to choose the best heuristics and to weight them appropriately.

ACE is a successful, adaptive solver. It learns to select a weighted mixture of heuristics for a given problem class, one that produces search trees smaller than those from outstanding individual heuristics in the CSP literature. ACE learns from its own search performance, based upon the accuracy, intensity, frequency and distribution of its heuristics' preferences. ACE adapts its decision making, its reinforcement policy, and its heuristic selection mechanisms effectively.

Our current work extends these ideas on several fronts. Under an option called *Pusher*, ACE consults the single highest-weighted tier 3 variable-ordering heuristic below the maximum search depth at which it has experienced backtracking on other problems in the same class [13]. Current work includes learning different weight profiles for different stages in solving a problem, where stages are determined by search tree depth or the constrainedness of the subproblem at the decision point. A generalization of that approach would associate weight profile(s) with an entire benchmark family of problems, and begin with the weights of the most similar benchmark family for each new problem instance.

Rather than rely on an endless set of fresh problems, we plan to reuse unsolved problems and implement boosting with little additional effort during learning [42]. A major focus is the automated selection of good parameter settings for an individual class (including the node limit and full-restart parameters), given the results in [24]. We also intend to extend our research to classes containing both solvable and unsolvable problems, and to optimization problems. Finally, we plan to study this approach further in light of the factor analysis evidence of strong correlations between CSP ordering heuristics [47]. Meanwhile, ACE proficiently tailors a mixture of search heuristics for each new problem class it encounters.

Appendix

Two vertices with an edge between them are *neighbors*. Here, the *degree of an edge* is the sum of the degrees of its endpoints, and the *edge degree of a variable* is the sum of the edge degrees of the edges on which it is incident. Each of the following metrics produces two Advisors.

Metrics for variable selection were static degree, dynamic domain size, FF2 [43], dynamic degree, number of valued neighbors, ratio of dynamic domain size to dynamic degree, ratio of dynamic domain size to degree, number of acceptable constraint pairs, static and dynamic edge degree with preference for the higher or lower degree endpoint, weighted degree [6], and ratio of dynamic domain size to weighted degree.

Metrics for value selection were number of value pairs for the selected variable that include this value, and, for each potential value assignment: minimum resulting domain size among neighbors, number of value pairs from neighbors to their neighbors, number of values among neighbors of neighbors, neighbors' domain size, a weighted function of neighbors' domain size, and the product of the neighbors' domain sizes.

Acknowledgements ACE is a joint project with Eugene Freuder and Richard Wallace of The Cork Constraint Computation Centre. Dr. Wallace made important contributions to the DWL algorithm. This work was supported in part by the National Science Foundation under IIS-0328743, IIS-0739122, and IIS-0811437.

References

[1] Aardal, K. I., Hoesel, S. P. M. v., Koster, A. M. C. A., Mannino, C., Sassano, A.: Models and solution techniques for frequency assignment problems. A Quarterly Journal of Operations Research, 1(4), pp. 261–317 (2003)

[2] Ali, K., Pazzani, M.: Error reduction through learning multiple descriptions. Machine Learning, 24, pp. 173–202 (1996)

[3] Arbelaez, A., Hamadi, Y., Sebag, M.: Online heuristic selection in constraint programming. International Symposium on Combinatorial Search (SoCS-2009), Lake Arrowhead, CA

[4] Birattari, M., Stützle, T., Paquete, L., Varrentrapp, K.: (2002). A Racing Algorithm for configuring metaheuristics. Genetic and Evolutionary Computation Conference (GECCO-2002), New York, NY, pp. 11–18 (2009)

[5] Borrett, J. E., Tsang, E. P. K., Walsh, N. R.: Adaptive constraint satisfaction: The quickest first principle. 12th European Conference on AI (ECAI-1996), Budapest, Hungary, pp. 160–164 (1996)

[6] Boussemart, F., Hemery, F., Lecoutre, C., Sais, L.: Boosting systematic search by weighting constraints. 16th European Conference on AI (ECAI-2004), Valencia, Spain, pp. 146–150 (2004)

[7] Burke, E., Hart, E., Kendall, G., Newall, J., Ross, P., Schulenburg, S.: Hyperheuristics: an emerging direction in modern research technology. Handbook of Metaheuristics, Dordrecht: Kluwer Academic Publishers, pp. 457–474 (2003)

[8] Chakhlevitch, K., Cowling, P.: Hyperheuristics: Recent developments. C. Cotta, M. Sevaux, K. Sorensen (Eds.), Adaptive and Multilevel Metaheuristics, Studies in Computational Intelligence, Springer, pp. 3–29 (2008)

[9] Cicirello, V. A., Smith, S. F.: Heuristic selection for stochastic search optimization: Modeling solution quality by extreme value theory. Principles and Practice of Constraint Programming (CP-2004), Toronto, Canada, LNCS 3258, pp. 197–211 (2004)

[10] D'Andrade, R. G.: Some propositions about the relations between culture and human cognition. J. W. Stigler, R. A. Shweder, G. Herdt (Eds.), Cultural Psychology: Essays on Comparative Human Development, Cambridge University Press, Cambridge 1990, pp. 65–129 (1990)

[11] Dietterich, T. G.: Ensemble methods in machine learning. J. Kittler, F. Roli (Eds.) First International Workshop on Multiple Classifier Systems, MCS-2000, Cagliari, Italy, LNCS 1857, pp. 1–15 (2000)

[12] Epstein, S. L.: For the Right Reasons: The FORR architecture for learning in a skill domain. Cognitive Science, 18, 479–511 (1994)

[13] Epstein, S. L., Freuder, E. C., Wallace, R.: Learning to support constraint programmers. Computational Intelligence, 21(4), 337–371 (2005)

[14] Freund, Y., Schapire, R.: Experiments with a new boosting algorithm. Thirteenth International Conference on Machine Learning (ICML-96), Bari, Italy, pp. 148–156 (1996)

[15] Fukunaga, A. S.: Automated discovery of composite SAT variable-selection heuristics. Eighteenth National Conference on Artificial Intelligence (AAAI-2002), Edmonton, Canada, pp. 641–648 (2002)

[16] Gagliolo, M., Schmidhuber, J.: Dynamic algorithm portfolios. Ninth International Symposium on Artificial Intelligence and Mathematics, Special Issue of the Annals of Mathematics and Artificial Intelligence, 47(3-4), pp. 295–328 (2006)

[17] Gebruers, C., Hnich, B., Bridge, D., Freuder, E.: Using CBR to select solution strategies in constraint programming. Case-Based Reasoning Research and Development. LNCS 3620, pp. 222–236 (2005)

[18] Gent, I. P., Prosser, P., Walsh, T.: The constrainedness of search. The Thirteenth National Conference on Artificial Intelligence (AAAI-1996), Portland, Oregon, pp. 246–252 (1996)

[19] Gomes, C., Fernandez, C., Selman, B., Bessière, C.: Statistical regimes across constrainedness regions. Principles and Practice of Constraint Programming (CP-2004), Toronto, Canada, LNCS 3258, pp. 32–46 (2004)

[20] Gomes, C. P., Selman, B., Kautz, H. A.: Boosting combinatorial search through randomization, The Fifteenth National Conference on Artificial Intelligence (AAAI-1998), Madison, Wisconsin, pp. 431–437 (1998)

[21] Gomes, C. P., Selman, B.: Algorithm portfolios. Artificial Intelligence, 126(1-2), pp. 43–62 (2001)

[22] Hansen, L., Salamon, P.: Neural network ensembles. IEEE Transactions on Pattern Analysis and Machine Intelligence, 12, pp. 993–1001 (1990)

[23] Hulubei, T., O'Sullivan, B.: Search heuristics and heavy-tailed behavior. Principles and Practice of Constraint Programming (CP-2005), Barcelona, Spain, pp. 328–342 (2005)

[24] Hutter, F., Hamadi, Y., Hoos, H. H., Leyton-Brown, K.: Performance prediction and automated tuning of randomized and parametric algorithms. Principles and Practice of Constraint Programming (CP-2006), Nantes, France, pp. 213–228 (2006)

[25] Johnson, D. S., Aragon, C. R., McGeoch, L. A., Schevon, C.: Optimization by simulated annealing: An experimental evaluation; Part I, Graph Partitioning. Operations Research, 37(6), 865–892 (1989)

[26] Kivinen, J., Warmuth, M. K.: Averaging expert predictions. Computational Learning Theory: 4th European Conference (EuroCOLT-1999), Nordkirchen, Germany, pp. 153–167 (1999)

[27] Lecoutre, C., Boussemart, F., Hemery, F.: Backjump-based techniques versus conflict directed heuristics. 16th IEEE International Conference on Tools with Artificial Intelligence (ICTAI-2004), Boca Raton, Florida, USA, pp. 549–557 (2004)

[28] Minton, S., Allen, J. A., Wolfe, S., Philpot, A.: An overview of learning in the Multi-TAC system. First International Joint Workshop on Artificial Intelligence and Operations Research, Timberline, Oregon, USA (1995)

[29] Nareyek, A.: Choosing search heuristics by non-stationary reinforcement learning. Metaheuristics: Computer Decision-Making, Kluwer Academic Publishers, Norwell, MA, USA, pp. 523–544 (2004)

[30] O'Mahony, E., Hebrard, E., Holland, A., Nugent, C., O'Sullivan, B.: Using case-based reasoning in an algorithm portfolio for constraint solving. 19th Irish Conference on Artificial Intelligence and Cognitive Science (AICS-2008), Cork, Ireland (2008)

[31] Opitz, D., Shavlik, J.: Generating accurate and diverse members of a neural-network ensemble. Advances in Neural Information Processing Systems, 8, pp. 535–541 (1996)

[32] Otten, L., Gronkvist, M., Dubhashi, D. P.: Randomization in constraint programming for airline planning. Principles and Practice of Constraint Programming (CP-2006), Nantes, France, pp. 406–420 (2006)

[33] Petrie, K. E., Smith, B. M.: Symmetry breaking in graceful graphs. Principles and Practice of Constraint Programming (CP-2003), Kinsale, Ireland, LNCS 2833, pp. 930–934 (2003)

[34] Petrovic, S., Epstein, S. L.: Full restart speeds learning. 19th International FLAIRS Conference (FLAIRS-06), Melbourne Beach, Florida (2006)

[35] Petrovic, S., Epstein, S. L.: Relative support weight learning for constraint solving. Workshop on Learning for Search, AAAI-06, Boston, Massachusetts, USA, pp. 115–122 (2006)

[36] Petrovic, S., Epstein, S. L.: Preferences improve learning to solve constraint problems. Workshop on Preference Handling for Artificial Intelligence (AAAI-07), Vancouver, Canada, pp. 71–78 (2007)

[37] Petrovic, S., Epstein, S. L.: Random subsets support learning a mixture of heuristics. International Journal on Artificial Intelligence Tools (IJAIT), 17(3), pp. 501–520 (2008)

[38] Ruan, Y., Kautz, H., Horvitz, E.: The backdoor key: A path to understanding problem hardness. Nineteenth National Conference on Artificial Intelligence (AAAI-2004), San Jose, CA, USA, pp. 124–130 (2004)

[39] Saari, D. G.: Geometry of voting. Studies in Economic Theory, Vol. 3, Springer (1994)

[40] Sabin, D., Freuder, E. C.: Understanding and improving the MAC algorithm. Principles and Practice of Constraint Programming (CP-1997), Linz, Austria, LNCS 1330, pp. 167–181 (1997)

[41] Samulowitz, H., Memisevic, R.: Learning to solve QBF. 22nd National Conference on Artificial intelligence (AAAI-2007), Vancouver, Canada, pp. 255–260 (2007)

[42] Schapire, R. E.: The strength of weak learnability. Machine Learning, 5(2): 197–227 (1990)

[43] Smith, B., Grant, S.: Trying harder to fail first. European Conference on Artificial Intelligence (ECAI-1998), pp. 249–253 (1998)

[44] Streeter, M., Golovin, D., Smith, S. F.: Combining multiple heuristics online. 22nd National Conference on Artificial Intelligence (AAAI-07), Vancouver, Canada, pp. 1197–1203 (2007)

[45] Terashima-Marin, H., Ortiz-Bayliss, J., Ross, P., Valenzuela-Rendón, M.: Using hyper-heuristics for the dynamic variable ordering in binary Constraint Satisfaction Problems, MICAI 2008, Advances in Artificial Intelligence, LNCS 5317, pp. 407–417 (2008)

[46] Valentini, G., Masulli, F.: Ensembles of learning machines. Neural Nets (WIRN-2002), Vietri sul Mare, Italy, LNCS 2486, pp. 3–22 (2002)

[47] Wallace, R. J.: Factor analytic studies of CSP heuristics. Principles and Practice of Constraint Programming (CP-2005), Barcelona, Spain, LNCS 3709, pp. 712–726 (2005)

[48] Whitley, D.: The genitor algorithm and selection pressure: Why rank-based allocation of reproductive trials is best. International Conference on Genetic Algorithms (ICGA-1989), pp. 116–121 (1989)

[49] Xu, L., Hutter, F., Hoos, H., Leyton-Brown, K.: SATzilla-07: The design and analysis of an algorithm portfolio for SAT. Principles and Practice of Constraint Programming (CP-2007), Providence, RI, USA, pp. 712–727 (2007)

[50] Young, H. P.: Condorcet's theory of voting. The American Political Science Review, 82(4), 1231–1244 (1988)

Part II
On-line Control

Chapter 6
An Investigation of Reinforcement Learning for Reactive Search Optimization

Roberto Battiti and Paolo Campigotto

6.1 Introduction

Reactive Search Optimization (RSO) [7, 6, 3] proposes the integration of subsymbolic machine learning techniques into search heuristics for solving complex optimization problems. The word *reactive* hints at a ready response to events during the search through an internal online feedback loop for the self-tuning of critical parameters. When RSO is applied to local search, its objective is to maximize a given function $f(x)$ by analyzing the past local search history (the trajectory of the tentative solution in the search space) and by learning the appropriate balance of intensification and diversification. In this manner the knowledge about the task and about the local properties of the *fitness surface* surrounding the current tentative solution can influence the future search steps to render them more effective.

Reinforcement learning (RL) arises in the different context of machine learning, where there is no guiding teacher, but *feedback signals from the environment* are used by the learner to modify its future actions. In RL one has to make a sequence of decisions. The outcome of each decision is not fully predictable. In fact, in addition to an immediate *reward*, each action causes a change in the system state and therefore a different context for the next decisions. To complicate matters, the reward is often delayed and one aims at maximizing not the immediate reward, but some form of *cumulative reward* over a sequence of decisions. This means that greedy policies do not always work. In fact, it can be better to go for a smaller immediate reward if this action leads to a state of the system where bigger rewards

Roberto Battiti
DISI – Dipartimento di Ingegneria e Scienza dell'Informazione, Università di Trento, Italy, e-mail: battiti@disi.unitn.it

Paolo Campigotto
DISI – Dipartimento di Ingegneria e Scienza dell'Informazione, Università di Trento, Italy, e-mail: campigotto@disi.unitn.it

Y. Hamadi et al. (eds.), *Autonomous Search*,   131
DOI 10.1007/978-3-642-21434-9_6,
© Springer-Verlag Berlin Heidelberg 2011

can be obtained in the future. Goal-directed learning from interaction with an (un-known) environment with trial-and-error search and delayed reward is the main feature of RL.

As was suggested, for example, in [2], the issue of learning from an initially unknown environment is shared by RSO and RL. A basic difference is that RSO optimizes a function and the environment is provided by a fitness surface to be explored, while RL optimizes the long-term reward obtained by selecting actions at the different states. The sequential decision problem and therefore the non-greedy nature of choices is also common. For example, in Reactive Tabu Search, the application of RSO in the context of Tabu Search, steps leading to worse configurations need in some cases to be performed to escape from a basin of attraction around a local optimizer. It is therefore of interest to investigate the relationship in more detail, to see whether specific techniques of reinforcement learning can be profitably used in RSO.

In this paper, we discuss the application of reinforcement learning methods for tuning the parameters of stochastic local search (SLS) algorithms. In particular, we select the Least Squares Policy Iteration (LSPI) algorithm [22] to implement the reinforcement learning mechanism. This investigation is done in the context of the Maximum Satisfiability (MAX-SAT) problem. The reactive versions of the SLS algorithms for the SAT/MAX-SAT problem, like RSAPS, Adaptive Walksat and AdaptNovelty$^+$, perform better than their original non-reactive versions. While solving a single instance, the level of diversification of the reactive SLS algorithms is adaptively increased or decreased if search stagnation is or is not detected.

Reactive approaches in the above methods consist of ad hoc. ways to detect the search stagnation and to adapt the value of one or more parameters determining the diversification of the algorithm.

In this investigation, a generic RL-based reactive scheme is designed, which can be customized to replace the ad hoc. method of the algorithm considered.

To test its performance, we select three SAT/MAX-SAT SLS solvers: the (adaptive) Walksat, the Hamming-Reactive Tabu Search (H-RTS) and the Reactive Scaling and Probabilistic Smoothing (RSAPS) algorithms. Their ad hoc. reactive methods are replaced with our RL-based strategy and the results obtained over a random MAX-3-SAT benchmark and a structured MAX-SAT benchmark are discussed.

This paper is organized as follows. Previous approaches of reinforcement learning applied to optimization are discussed in Section 6.2, while the basics of reinforcement learning and dynamic programming are given in Section 6.3. In particular, the LSPI algorithm is detailed.

The considered Reactive SLS algorithms for the SAT/MAX-SAT problem are introduced in Section 6.4. Section 6.5 explains our integration of RSO with the RL strategy. Finally, the experimental comparison of the RL-based reactive approach with the original ad hoc. reactive schemes (Section 6.6) concludes the work.

6.2 Reinforcement Learning for Optimization

Many are the intersections between optimization, dynamic programming and reinforcement learning. Approximated versions of DP/RL comprise challenging optimization tasks. Consider, for example, the maximization operations in determining the best action when an action value function is available, the optimal choice of approximation architectures and parameters in dynamic programming, or the optimal choice of algorithm details and parameters for a specific RL instance.

A recent paper about the interplay of optimization and machine learning (ML) research is [8], which mainly shows how recent advances in optimization can be profitably used in ML.

This work, however, goes in the opposite direction: which techniques of RL can be used to improve heuristic algorithms for a standard optimization task such as minimizing a function? Interesting summaries of statistical machine learning methods applied for large-scale optimization are presented in [1]. Autonomous search methods that adapt the search performance to the problem at hand are described in [15], which defines a metric to classify problem solvers based on their computation characteristics. The authors first identify a set of rules describing search processes, and then they classify existing search paradigms from different but related areas, such as the Satisfiability and the Constraint Satisfaction Problems.

In [10] RL is applied in the area of local search for solving $\max_x f(x)$: the rewards from a local search method π starting from an initial configuration x are given by the size of improvements of the best-so-far value f_{best}. In detail, the value function $V^\pi(x)$ of configuration x is given by the expected best value of f seen on a trajectory starting from state x and following the local search method π. The curse of dimensionality discourages directly using x for state description: informative *features* extracted from x are used to compress the state description to a shorter vector $s(x)$, so that the value function becomes $V^\pi(s(x))$. The introduced algorithm STAGE is a smart version of iterated local search which alternates between two phases. In one phase, new training data for $V^\pi(F(x))$ are obtained by running local search from the given starting configuration. Assuming that local search is memoryless, the estimates of $V^\pi(F(x'))$ for all the points s' along the local search trajectory can be obtained from a single run. In the other phase, one optimizes the value function $V^\pi(F(x))$ (instead of the original f), so that a hopefully new and promising starting point is obtained. A suitable approximation architecture $V^\pi(F(x); w)$ and a supervised machine learning method to train $V^\pi(F(x); w)$ from the examples are needed. The memory-based version of the RASH (Reactive Affine Shaker) optimizer introduced in [11] adopts a similar strategy. An open issue is the proper definition of the features by the researcher, which are chosen by insight in [10]. Preliminary results are also presented about *transfer*, i.e., the possibility to use a value function learnt on one task for a different task with similar characteristics.

A second application of RL to local search is that of supplementing f with a "scoring function" to help in determining the appropriate search option at every step. For example, different basic moves or entire different neighborhoods can be applied. RL can in principle make more systematic some of the heuristic approaches involved in designing appropriate "objective functions" to guide the search process. An example is the RL approach to job-shop scheduling in [31, 32], where a neural network-based $TD(\lambda)$ scheduler is demonstrated to outperform a standard iterative repair (local search) algorithm. The RL problem is designed to pick the better among two possible scheduling actions (moves transforming the current solution). The reward scheme is designed to prefer actions which *quickly* find a *good* schedule. For this purpose, a fixed negative reinforcement is given for each action leading to a schedule still containing constraint violations. In addition, a scale-independent measure of the final admissible schedule length gives feedback about the schedule quality. Features are extracted from the state, and are either handcrafted or determined in an automated manner [32]; the δ values along the trajectory are used to gradually update a parametric model of the state value function of the optimal policy.

ML can be profitably applied also in tree search techniques. Variable and value ordering heuristics (choosing the right order of variables or values) can noticeably improve the efficiency of complete search techniques, e.g., for Constraint Satisfaction Problems. For example, RLSAT [21] is a DPLL solver for the SAT problem which uses experience from previous executions to learn how to select appropriate branching heuristics from a library of predefined possibilities, with the goal of minimizing the total size of the search tree, and therefore the CPU time. Lagoudakis and Littman [20] extend algorithm selection for recursive computation, which is formulated as a sequential decision problem: the selection of an algorithm at a given stage requires an immediate cost – in the form of CPU time – and leads to a different state with a new decision to be executed. For SAT, features are extracted from each sub-instance to be solved, and TD learning with Least Squares is used to learn an appropriate action value function, approximated as a linear function of the extracted features. Some modifications of the basic methods are needed. First, two new sub-problems are obtained instead of one; second, an appropriate re-weighting of the samples is needed to prevent the huge number of samples close to the leaves of the search tree from practically hiding the importance of samples close to the root. According to the authors, their work demonstrates that "some degree of reasoning, learning, and decision making on top of traditional search algorithms can improve performance beyond that possible with a fixed set of hand-built branching rules."

In [9] RL is applied in the context of the constructive search technique, which builds a complete solution by selecting the value for one solution component at a time. One starts from the task data d and repeatedly picks a new index m_n to fix a value u_{m_n}, until values for all N components are fixed. The decision to be made from a partial solution $x_p = (d, m_1, u_{m_1}, \ldots, m_n, u_{m_n})$ is about which index to consider next and which value to assign. Let us assume that K fixed construction algorithms are available for the problem. The application consists of combining in the most appropriate manner the information obtained by the set of construction algorithms in order to fix the next index and value. The approximation architecture suggested is

$$V(d, x_p) = \psi_0(x_p, w_0) + \sum_{k=1}^{K} \psi_k(x_p, w_k) H_{k,d}(x_p),$$

where $H_{k,d}(x_p)$ is the value obtained by completing the partial solution with the kth method, and ψ are tunable coefficients depending on the parameters w which are learnt from many runs on different instances of the problem. While in the previously mentioned application one evaluates starting points for local search, here one evaluates the promise of partial solutions leading to good, complete solutions. The performance of this technique in terms of CPU time is poor: in addition to being run for the separate training phase, the K construction algorithms must be run for each partial solution (for each variable-fixing step) in order to allow for the picking of the next step in an optimal manner.

A different parametric combination of costs of solutions obtained by two fixed heuristics is considered for the *stochastic programming problem* of maintenance and repair [9]. In the problem, one has to decide whether to immediately repair breakdowns by consuming the limited amount of spare parts, or to keep spare parts for breakdowns at a later phase.

In the context of continuous function optimization, [24] uses RL for replacing a priori-defined adaptation rules for the step size in Evolution Strategies with a reactive scheme which adapts step sizes automatically during the optimization process. The states are characterized only by the success rate after a fixed number of mutations; the three possible actions consist of increasing (by a fixed multiplicative amount), decreasing or keeping the current step size. SARSA learning with various reward functions is considered, including combinations of the difference between the current function value and the one evaluated at the last reward computation and the movement in parameter space (the distance traveled in the last phase). On-the-fly parameter tuning, or on-line calibration of parameters for evolutionary algorithms by reinforcement learning (crossover, mutation, selection operators, population size), is suggested in [12]. The EA process is divided into episodes; the state describes the main properties of the current population (mean fitness, or f values, standard deviation, etc.); the reward reflects the progress between two episodes (improvement of the best fitness value); the action consists of determining the vector of control parameters for the next episode.

The trade-off between exploration and exploitation is another issue shared by RL and stochastic local search techniques. In RL an optimal policy is learnt by generating samples; in some cases samples are generated by a policy which is being evaluated and then improved. If the initial policy is generating states only in a limited portion of the entire state space, there will be no way to learn the true optimal policy. A paradigmatic example is the n-armed bandit problem [14], so named because of its similarity with a slot machine. In the problem, one is faced repeatedly with a choice among different options, or actions. After each choice a numerical reward is generated from a stationary probability distribution that depends on the selected action. The objective is to maximize the expected total reward over some time period. In this case the action value function depends only on the action, not on the state (which is unique in this simplified problem). In the more general case of many states, no theoretically sound ways exist to combine exploration and exploitation in an optimal manner. Heuristic mechanisms come to the rescue, for example, by selecting actions not in a greedy manner given an action value function, but based on a simulated annealing-like probability distribution. Alternatively, actions which are within ε of the optimal value can be chosen with a nonzero probability.

A recent application of RL in the area of stochastic local search is [25]. The noise parameter of the Walksat/SKC algorithm [27] is self-tuned while solving a single problem instance by an average reward reinforcement learning method: R-learning [26]. The R-learning algorithm learns a state-action value function $Q(s,a)$ estimating the reward for following an action a from a state s. To improve the trade-off between the exploitation of the current state of learning with exploration for further learning, the state-action value function selects with probability $1 - \varepsilon$ the action a with the best estimated reward; otherwise it chooses a random action. Once the selected action is executed, a new state s' and a reward r are observed and the average reward estimate and the state-action value function $Q(s,a)$ are updated. In [25], the state of the MDP is represented by the current number of unsatisfied clauses and an action is the selection of a new value for the noise parameter from the set $(0.05, 0.1, 0.15, \ldots, 0.95, 1.00)$ followed by a local move. The reward is the reduction in the number of unsatisfied clauses since the last local minimum. In this context the average reward measures the average progress towards the solution of the problem, i.e., the average difference in the objective function before and after the local move. The modification performed by the approach in [25] to the Walksat/SKC algorithm consists of (re)setting its noise parameter at each iteration. The result of this approach is compared over a benchmark of SAT instances with the results obtained when the noise parameter is hand-tuned. Even if the approach obtains uniformly good results on the selected benchmark, the hand-tuned version of the Walksat/SKC algorithm performs better. Furthermore, the proposed approach is not robust with the setting of the parameters of the R-learning algorithm.

The work in [30] applies a reinforcement learning technique, Q-learning [28], in the context of the Constraint Satisfaction Problem (CSP). A CSP instance is usually solved by iteratively selecting one variable and assigning a value from its domain. The value assigned to the selected variable is then propagated, updating the domains of the remaining variables and checking the consistency of the constraints. If a conflict is detected, a backtracking procedure is executed. A complete assignment generating no conflicts is a solution to the CSP instance. The algorithm proposed in [30] learns which variable ordering heuristic should be used at each iteration, formulating the search process executed by the CSP solver as a reinforcement learning task. In particular, each state of the MDP process consists of a partial or total assignment of values to the variables, while the set of the actions is given by the set of the possible variable ordering heuristic functions. A reward is assigned when the CSP instance is solved. In this way, the search space of the RL algorithm corresponds to the search space of the input CSP instance and the transition function of the MDP process corresponds to the decision made when solving a CSP instance. However, several open questions related to the proposed approach remain to be solved, as pointed out by the authors.

Another example of the application of machine learning to the CSP is the Adaptive Constraint Engine (ACE) introduced in [13], which learns to solve Constraint Satisfaction Problems. The solving process of a CSP is modeled as a sequence of decisions, where each decision either selects a variable or assigns a value to a variable from its associated value set. After each assignment, propagation rules infer its effect on the domains of the unassigned variables. Each decision is obtained by the weighted combination of a set of prespecified search heuristics. The learning task consists of searching the appropriate set of weights for the heuristics. The Adaptive Constraint Engine can learn the optimal set of weights for *classes* of CSP problems, rather than for a single problem.

6.3 Reinforcement Learning and Dynamic Programming Basics

In this section, *Markov decision processes* are formally defined and the standard dynamic programming technique is summarized in Section 6.3.2, while the Policy Iteration technique to determine the optimal policy is defined in Section 6.3.3. In many practical cases exact solutions must be abandoned in favor of approximation strategies, which are the focus of Section 6.3.4.

6.3.1 Markov Decision Processes

A standard Markov process is given by a set of states $\mathscr{S}$ with transitions between them described by probabilities $p(i,j)$. Let us note the fundamental property of Markov models: earlier states do not influence the transition probabilities to later states. The process evolution cannot be controlled, because it lacks the notion of decisions, *actions* taken depending on the current state and leading to a different state and to an immediate *reward*.

A Markov decision process (MDP) is an extension of the classical Markov process designed to capture the problem of *sequential decision making under uncertainty*, with states, decisions, unexpected results, and "long-term" goals to be reached. An MDP can be defined as a quintuple $(\mathscr{S}, \mathscr{A}, P, R, \gamma)$, where $\mathscr{S}$ is a set of states, $\mathscr{A}$ is a finite set of actions, $P(s,a,s')$ is the probability of transition from state $s \in \mathscr{S}$ to state $s' \in \mathscr{S}$ if action $a \in \mathscr{A}$ is taken, $R(s,a,s')$ is the corresponding reward, and γ is the discount factor, in order to exponentially decrease future rewards. This last parameter is fundamental in order to evaluate the overall value of a choice when considering its consequences on an infinitely long chain. In particular, given the following evolution of an MDP

$$s(0) \xrightarrow{a(0)} s(1) \xrightarrow{a(1)} s(2) \xrightarrow{a(2)} s(3) \xrightarrow{a(3)} \dots \tag{6.1}$$

the cumulative reward obtained by the system is given by

$$\sum_{t=0}^{\infty} \gamma^t R(s(t), a(t), s(t+1)).$$

Note that state transitions are not deterministic; nevertheless their distribution can be controlled by the action a. The goal is to control the system in order to maximize the expected cumulative reward.

Given an MDP $(\mathscr{S}, \mathscr{A}, P, R, \gamma)$, we define a *policy* as a probability distribution $\pi(\cdot|s) : \mathscr{A} \to [0,1]$, where $\pi(a|s)$ is the probability of choosing action a when the system is in state s. In other words, π maps states onto probability distributions over $\mathscr{A}$. Note that we are only considering stationary policies. If a policy is deterministic, then we shall resort to the more compact notation $a = \pi(s)$.

6.3.2 The Dynamic Programming Approach

The learning task consists of the selection of a policy that maximizes a measure of the total reward accumulated during an infinite chain of decisions (infinite horizon). To achieve this goal, let us define the *state-action value function* $Q^{\pi}(s,a)$ of the policy π as the expected overall future reward for applying a specified action a when the system is in status s, under the hypothesis that the ensuing actions are taken according to policy π. A straightforward implementation of the Bellman principle leads to the following definition:

$$Q^{\pi}(s,a) = \sum_{s' \in \mathscr{S}} P(s,a,s') \left(R(s,a,s') + \gamma \sum_{a' \in \mathscr{A}} \pi(a'|s') Q^{\pi}(s',a') \right) \qquad (6.2)$$

where the sum over $\mathscr{S}$ can be interpreted as an integral in the case of a continuous state set. The interpretation is that the value of selecting action a in state s is given by the expected value of the immediate reward plus the value of the future rewards which one expects by following policy π from the new state. These rewards have to be discounted by γ (they are a step in the future w.r.t. the new state) and properly weighted by transition probabilities and action-selection probabilities given the stochasticity in the process.

The expected reward of a state/action pair $(s,a) \in \mathscr{S} \times \mathscr{A}$ is

$$R(s,a) = \sum_{s' \in \mathscr{S}} P(s,a,s') R(s,a,s'),$$

so that (6.2) can be rewritten as

$$Q^{\pi}(s,a) = R(s,a) + \gamma \sum_{s' \in \mathscr{S}} \left(P(s,a,s') \sum_{a' \in \mathscr{A}} \pi(a'|s') Q^{\pi}(s',a') \right)$$

or, in a more compact linear form,

$$Q^{\pi} = R + \gamma P \Pi_{\pi} Q^{\pi} \qquad (6.3)$$

where R is the $|\mathscr{S}||\mathscr{A}|$-entry column vector corresponding to $R(s,a)$, P is the $|\mathscr{S}||\mathscr{A}| \times |\mathscr{S}|$ matrix of $P(s,a,s')$ values having (s,a) as row index and s' as column, and Π_π is a $|\mathscr{S}| \times |\mathscr{S}||\mathscr{A}|$ matrix whose entry $(s,(s,a))$ is $\pi(a|s)$.

Equation (6.3) can be seen as a non-homogeneous linear problem with unknown Q^π,

$$(I - \gamma P\Pi_\pi)Q^\pi = R, \tag{6.4}$$

or, alternatively, as a fixed-point problem,

$$Q^\pi = T_\pi Q^\pi, \tag{6.5}$$

where $T_\pi : x \mapsto R + \gamma P\Pi_\pi x$ is an affine functional.

If the state set $\mathscr{S}$ is finite, then (6.3-6.5) are matrix equations and the unknown Q^π is a vector of size $|\mathscr{S}||\mathscr{A}|$.

In order to solve these equations explicitly, a model of the system is required, i.e., full knowledge of functions $P(s,a,s')$ and $R(s,a)$. When the system is too large, or the model is not completely available, approximations in the form of *reinforcement learning* come to the rescue. As an example, if a *generative model* is available, i.e., a black box that takes state and action in input and produces the reward and next state as output, one can estimate $Q^\pi(s,a)$ through *rollouts*. In each rollout, the generator is used to simulate action a followed by a sufficiently long chain of actions dictated by policy π. The process is repeated several times because of the inherent stochasticity, and averages are calculated.

The above described state-action value function Q, or its approximation, is instrumental in the basic methods of dynamic programming and reinforcement learning.

6.3.3 Policy Iteration

A method to obtain the optimal policy π^* is to generate an improving sequence (π_i) of policies by building a policy π_{i+1} upon the value function associated with policy π_i:

$$\pi_{i+1}(s) = \arg\max_{a \in \mathscr{A}} Q^{\pi_i}(s,a). \tag{6.6}$$

Policy π_{i+1} is never worse than π_i, in the sense that $Q^{\pi_{i+1}} \geq Q^{\pi_i}$ over all state/action pairs.

In the following, we assume that the optimal policy π^* exists in the sense that for all states it attains the minimum of the right-hand side of Bellman's equation; see [9] for more details.

The Policy Iteration (PI) method consists of the alternate computation shown in Fig. 6.1: given a policy π_i, the *policy evaluation* procedure (also known as the "Critic") generates its state-action value function Q^{π_i}, or a suitable approximation. The second step is the *policy improvement* procedure (the "Actor"), which computes a new policy by applying (6.6).

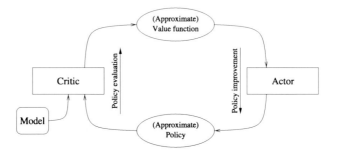

Fig. 6.1: the Policy Iteration (PI) mechanism

The two steps are repeated until the value function does not change after iterating, or the change between consecutive iterations is less than a given threshold.

6.3.4 Approximations: Reinforcement Learning and LSPI

To carry out the above discussion by means of exact methods, in particular using (6.4) as the Critic component, the system model has to be known in terms of its transition probability $P(s,a,s')$ and reward $R(s,a)$ functions. In many cases this detailed information is not available but we have access to the system itself or to a *simulator*. In both cases, we have a black box which given the current state and the performed action determines the next state and reward. In both cases, and more conveniently with a simulator, several sample trajectories can be generated, providing the information about the system behavior needed to learn the optimal control.

A brute force approach can be that of estimating the system model functions $P(\cdot,\cdot,\cdot)$ and $R(\cdot,\cdot)$ by executing a very large series of simulations. The *reinforcement learning* methodology bypasses the system model and directly learns the value function.

Assume that the system simulator (the "Model" box in Fig. 6.1) generates quadruples in the form

$$(s,a,r,s')$$

where s is the state of the system at a given step, a is the action taken by the simulator, s' is the state in which the system falls after the application of a, and r is the reward received. In the setting described by this paper, the (s,a) pair is generated by the simulator.

A viable method to obtain an approximation of the state-action value function is to approximate it with respect to a functional linear subspace having basis $\Phi = (\phi_1,\ldots,\phi_k)$. The approximation $\hat{Q}^\pi \approx Q^\pi$ is in the form

$$\hat{Q}^\pi = \Phi^T w^\pi.$$

The weights vector w^π is the solution of the linear system $Aw^\pi = b$, where

$$A = \Phi^T(\Phi - \gamma P\Pi_\pi\Phi), \qquad b = \Phi^T R. \tag{6.7}$$

An approximate version of (6.7) can be obtained if we assume that a finite set of samples is provided by the "Model" box of Fig. 6.1:

$$\mathscr{D} = \{(s_1, a_1, r_1, s_1'), \ldots, (s_l, a_l, r_l, s_l')\}.$$

In this case, matrix $\mathscr{A}$ and vector b are "learnt" as sums of rank-one elements, each obtained by a sample tuple:

$$A = \sum_{(s,a,r,s')\in\mathscr{D}} \Phi(s,a)\Big(\Phi(s,a) - \gamma\Phi(s',\pi(s'))\Big)^T, \qquad b = \sum_{(s,a,r,s')\in\mathscr{D}} r\Phi(s,a).$$

Variable	Scope	Description
$\mathscr{D}$	In	Set of sample vectors $\{(s,a,r,s')\}$
k	In	Number of basis functions
Φ	In	Vector of k basis functions
γ	In	Discount factor
π	In	Policy
A	Local	$k \times k$ matrix
b	Local	k-entry column vector
w^π	Out	k-entry weight vector

1. **function** LSTDQ ($\mathscr{D}$, k, Φ, γ, π)
2. $\quad A \leftarrow 0$;
3. $\quad b \leftarrow 0$;
4. $\quad$ **for each** $(s,a,r,s') \in D$
5. $\quad\quad A \leftarrow A + \Phi(s,a)\big(\Phi(s,a) - \gamma\Phi(s',\pi(s'))\big)^T$
6. $\quad\quad b \leftarrow b + r\Phi(s,a)$
7. $\quad w^\pi \leftarrow A^{-1}b$

Fig. 6.2: the LSTDQ algorithm

Such approximations lead to the Least Squares Temporal Difference for Q (LSTDQ) algorithm proposed in [22], and shown in Figure 6.2, where the functions $R(s,a)$ and $P(s,a,s')$ are assumed to be unknown and are replaced by a finite sample set $\mathscr{D}$.

Note that the LSTDQ algorithm returns the weight vector that best approximates the value function of a given policy π within the spanned subspace and according to the sample data. It therefore acts as the "Critic" component of the Policy Iteration algorithm. The "Actor" component is straightforward, because it is an application of (6.6). The policy does not need to be explicit: if the system is in state s and the current value function is defined by weight vector w, the best action to take is

Variable	Scope	Description
$\mathcal{D}$	In	Set of sample vectors $\{(s,a,r,s')\}$
k	In	Number of basis functions
Φ	In	Vector of k basis functions
γ	In	Discount factor
ε	In	Weight vector tolerance
w_0	In	Initial value function weight vector
w'	Local	Weight vectors in subsequent iterations
w	Out	Optimal weight vector

1. **function** LSPI $(D, k, \Phi, \gamma, \varepsilon, w_0)$
2. $w' \leftarrow w_0$;
3.   **do**
4. $w \leftarrow w'$;
5. $w' \leftarrow$ LSTDQ (D, k, Φ, γ, w);
6.   **while** $\|w - w'\| > \varepsilon$

Fig. 6.3: the LSPI algorithm

$$a = \arg\max_{a \in \mathcal{A}} \Phi(s,a)^T w. \qquad (6.8)$$

The complete LSPI algorithm is given in Fig. 6.3. Note that, because of the identification between the weight vector w and the ensuing policy π, the code assumes that the previously declared function LSTDQ() accepts its last parameter, i.e., the policy π, in the form of a weight vector w.

6.4 Reactive SAT/MAX-SAT Solvers

This investigation considers as starting point the following methods: the Walksat/SKC algorithm [27] and its reactive version, the Adaptive Walksat [16], the Hamming-Reactive Tabu Search (H-RTS) [5] and the Reactive Scaling and Probabilistic Smoothing (RSAPS) algorithm [18].

The Walksat/SKC algorithm adopts a two-stage variable selection scheme. First, one of the clauses violated by the current assignment is selected uniformly at random. If the selected clause contains variables that can be flipped without violating any other clause, one of these is randomly chosen. When such an improving step does not exist, a random move is executed with a certain probability p: a variable appearing in the selected unsatisfied clause is selected uniformly at random and flipped. Otherwise with probability $1 - p$ the least worsening move is greedily selected. The parameter p is often referred to as noise parameter. The Adaptive Walksat algorithm automatically adjusts the noise parameter, increasing or decreasing it when search stagnation is or is not detected. Search stagnation is measured in terms of the time elapsed since the last reduction in the number of unsatisfied clauses achieved.

RSAPS is a reactive version of the Scaling and Probabilistic Smoothing (SAPS) algorithm [18]. In the SAPS algorithm, once the search process becomes trapped in a local minimum, the weights of the current unsatisfied clauses are multiplied by a factor larger than 1 in order to encourage diversification. After updating, with a certain probability P_{smooth} the clause weights are smoothed back towards uniform values. The smoothing phase comprises forgetting the collateral effects of the earlier weighting decisions that affect the behaviour of the algorithm when visiting future local minima. The RSAPS algorithm dynamically adapts the smoothing probability parameter P_{smooth} during the search. In particular, RSAPS adopts the same stagnation criterion as Adaptive Walksat to trigger a diversification phase. If no reduction in the number of the unsatisfied clauses is observed in the last search iterations, in the case of RSAPS the smoothing probability parameter is reduced while in the case of Adaptive Walksat the noise parameter is increased.

H_RTS is a prohibition-based algorithm that dynamically adjusts the prohibition parameter during the search. For the purpose of this investigation, we consider a "simplified" version of H_RTS, where the non-oblivious search phase[1] is omitted. As matter of fact, we are interested in evaluating the performance of different reactive strategies for adjusting the prohibition parameter. To the best of our knowledge, non-oblivious functions are defined only in the case of k-SAT instances. Furthermore, the benchmark used in the experimental part of this work is not limited to k-SAT instances. Finally, there is some evidence that the performance of H_RTS is determined by the Reactive Tabu Search phase rather than the non-oblivious search phase [17]. For the rest of this work, the term H-RTS refers to the "simplified" version of H-RTS. Its pseudo-code is in Fig. 6.4. Once the initial truth assignment is generated in a random way, the search proceeds by repeating phases of local search followed by phases of Tabu Search (TS) (lines 6–12 in Figure 6.4), until $10\,n$ iterations are completed. The variable t, initialized to zero, contains the current iteration and increases after a local move is applied, while t_r contains the iteration in which the last random assignment was generated and f represents the score function counting the number of unsatisfied clauses. During each combined phase, first the local optimum X_l of f is reached, then $2(T + 1)$ moves of Tabu Search are executed, where T is the prohibition parameter. The design principle underlying this choice is that prohibitions are necessary for diversifying the search only after local search (LS) reaches a local optimum. Finally, an "incremental ratio" test is executed to see whether in the last $T + 1$ iterations the trajectory tends to move away or come closer to the starting point X_l. The test measures the Hamming distance from X_l covered in the last $T + 1$ iterations and a possible reactive modification of T_f is executed depending on the test's results; see [5]. The fractional prohibition T_f (the prohibition T is obtained as $T_f\, n$)

---

[1] The non-oblivious search phase is a local search procedure guided by a non-oblivious function, i.e., a function that weights in different ways the satisfied clauses according to the number of matched literals. For example, the non-oblivious function for MAX-2-SAT problems is the weighted linear combination of the number of clauses with one and two matched literals. See [5] for details.

is therefore changed during the run to obtain a proper balance of diversification and bias.

The random restart executed after $10\,n$ moves guarantees that the search trajectory is not confined to a localized portion of the search space.

procedure H_RTS
1. **repeat**
2. $\quad\lceil t_r \leftarrow t$
3. $\quad\mid X \leftarrow$ random truth assignment
4. $\quad\mid T \leftarrow \lfloor T_f\, n \rfloor$
5. $\quad\mid$ **repeat**
6. $\quad\mid\quad\lceil$ **repeat**   { *local search* }
7. $\quad\mid\quad\mid\quad [X \leftarrow$ BEST-MOVE (LS, f)
8. $\quad\mid\quad\mid$ **until** largest $\Delta f = 0$
9. $\quad\mid\quad\mid X_I \leftarrow X$
10. $\quad\mid\quad\mid$ **for** $2(T+1)$ *iterations* { *reactive tabu search* }
11. $\quad\mid\quad\mid\quad [X \leftarrow$ BEST-MOVE (TS, f)
12. $\quad\mid\quad\mid X_F \leftarrow X$

13. $\quad\mid\quad\lfloor T \leftarrow$ REACT(T_f, X_F, X_I)
14. $\quad\lfloor$ **until** $(t - t_r) > 10\,n$
15. **until**   solution is acceptable or maximum number
    of iterations reached

Fig. 6.4: The simplified version of the H_RTS algorithm considered in this work

6.5 The RL-Based Approach for Reactive SAT/MAX-SAT Solvers

A local search algorithm operates through a sequence of elementary actions (*local moves*, e.g., bit flips). The choice of the local move is driven by many different factors; in particular, most algorithms are *parametric*: their behavior, and their efficiency, depends on the values attributed to some free parameters, so that different instances of the same problem and different search states within the same instance may require different parameter values.

In this work we introduce a generic RL-based approach for adapting during the search the parameter determining the trade-off between intensification and diversification in the three reactive SLS algorithms considered in Section 6.4. From now on, the specific parameter modified by the reactive scheme is referred to as the *target* parameter. The target parameter is the noise parameter in the case of the Walksat algorithm, the prohibition parameter in the case of Tabu Search and the smoothing probability parameter in the case of the SAPS algorithm.

Furthermore, in this paper the term *offline* defines an action executed before the search phase of the local search algorithm, while the term *online* describes an action executed during the search phase. This work proposes an RSO approach for the online tuning of the target parameter based on the LSPI algorithm, which is trained offline.

In order to apply the LSPI method introduced in Section 6.3.4 for tuning the target parameter, first the search process of the SLS algorithms is modelled as a Markov decision process. Each state of the MDP is created by observing the behavior of the considered algorithm over an epoch of *epoch_length* consecutive variable flips. In fact, the effect of changing the target parameter on the algorithm's behavior can only be evaluated after a reasonable number of local moves. Therefore the algorithm's traces are divided into *epochs* $(E_1, E_2, \ldots)$ composed of a suitable number of local moves, and changes of the target parameter are allowed only between epochs. Given the subdivision of the reactive local search algorithm's trace into a sequence of epochs $(E_1, E_2, \ldots)$, the state at the end of epoch E_i is a collection of features extracted from the algorithm's execution up to that moment in the form of a tuple: $s(E_1, \ldots, E_i) \in \mathbb{R}^d$, where d is the number of features that form the state.

In the case of the Reactive Tabu Search algorithm, $epoch\_length = 2 * T_{max}$, where T_{max} is the maximum allowed value for the prohibition parameter. Because in a prohibition mechanism with prohibition parameter T, during the first T steps the Hamming distance keeps increasing and only in the subsequent steps may it decrease, an epoch is long enough to monitor the behavior of the algorithm in the case of the largest allowed T value. A preliminary application of RL to Reactive Tabu Search for SAT has been presented in [4].

In the case of the target parameter in the Walksat and the RSAPS algorithms, each epoch lasts 100 and 200 variable flips, respectively, including null flips (i.e., search steps where no variable is flipped) in the case of the RSAPS algorithm. These values for the epoch length, ranging from 10 to 400, are the optimal values selected during the experiments over a set of candidates.

The scope of this work is the design of a reactive reinforcement-based method that is independent of the specific SAT/MAX-SAT reactive algorithm considered. Therefore, the features selected for the definition of the states of the MDP underlying the reinforcement learning approach do not depend on the target parameter considered, allowing for generalization.

As specified above, the state of the system at the end of an epoch describes the algorithm's behavior during the epoch, and an "action" is the modification of the target parameter before the algorithm enters the next epoch. The target parameter is responsible for the diversification of the search process once stagnation is detected. The state features therefore describe the intensification-diversification trade-off observed in the epoch.

In particular, let us define the following:

- n and m are the number of variables and clauses of the input SAT instance, respectively;

- $f(x)$ is the score function counting the number of unsatisfied clauses in the truth assignment x;
- x_{bsf} is the "best-so-far" (BSF) configuration *before* the current epoch;
- $\overline{f}_{epoch}$ is the average value of f during the current epoch;
- $\overline{H}_{epoch}$ is the average Hamming distance during the current epoch from the configuration at the beginning of the current epoch.

The compact state representation chosen to describe an epoch is the following couple:

$$s \equiv \left(\Delta f, \frac{\overline{H}_{epoch}}{n} \right), \text{ where } \Delta f = \frac{\overline{f}_{epoch} - f(x_{bsf})}{m}.$$

The first component of the state is the mean change of f in the current epoch with respect to the best value. It represents the preference for configurations with low objective function values. The second component describes the ability to explore new configurations in the search space by moving away from local minima. In the literature, this behaviour is often referred to as the diversification bias trade-off. For the purpose of addressing uniformly SAT instances with different numbers of variables, the first and the second state components have been normalized.

The reward signal is given by the normalized change of the best value achieved in the observed epoch with respect to the "*best-so-far*" value *before* the epoch: $(f(x_{bsf}) - f(x_{localBest}))/m$.

The state of the system at the end of an epoch describes the algorithm's behavior during the last epoch, while an action is the modification of the algorithm's parameters before it enters the next epoch. In particular, the action consists of setting the target parameter from scratch at the end of the epoch. The noise and the smoothing probability parameters are set to one of 20 uniformly distributed values in the range $[0.01, 0.2]$. In the Tabu Search context, we consider the fractional prohibition parameter (the prohibition parameter is $T = \lfloor nT_f \rfloor$), equal to one of 25 uniformly distributed values in the range $[0.01, 0.25]$. Therefore the actions set $A = \{a_i, i = 1..n\}$ is composed of $n = 20$ choices in the case of the noise and the smoothing probability parameters and $n = 25$ choices in the case of the prohibition parameter. The effect of the action a_i consists of setting target parameter to $0.01 * i$, $i \in [1, 20]$ in the case of the noise and the smoothing probability parameters and $i \in [1, 25]$ in the case of the prohibition parameter.

Once a reactive local search algorithm is modeled as a Markov decision process, a reinforcement learning method such as LSPI can be used to control the evolution of its parameters. To tackle large state and action spaces, the LSPI algorithm approximates the value function as a linear weighted combination of basis functions (see Eq. 6.8). In this work, we consider the value function space spanned by the following finite set of basis functions:

$$\Phi(s,a) = \begin{pmatrix} I_{a==a_1}(a) \\ I_{a==a_1}(a) \cdot \Delta f \\ I_{a==a_1}(a) \cdot \overline{H}_{epoch} \\ I_{a==a_1}(a) \cdot \overline{H}_{epoch} \cdot \Delta f \\ I_{a==a_1}(a) \cdot (\Delta f)^2 \\ I_{a==a_1}(a) \cdot \overline{H}^2_{epoch} \\ \vdots \\ I_{a==a_n}(a) \\ I_{a==a_n}(a) \cdot \Delta f \\ I_{a==a_n}(a) \cdot \overline{H}_{epoch} \\ I_{a==a_n}(a) \cdot \overline{H}_{epoch} \cdot \Delta f \\ I_{a==a_n}(a) \cdot (\Delta f)^2 \\ I_{a==a_n}(a) \cdot \overline{H}^2_{epoch} \end{pmatrix}. \tag{6.9}$$

The set is composed of $6 * n$ elements, and $I_{a==a_i}$, with $i = 1..n$, are the indicator functions for the actions, evaluating to 1 if the action is the indicated one, and 0 otherwise. The indicator function is used to discern the "state-action" features for the different actions considered: the learning for an action is entirely decoupled from the learning for the remaining actions. For example, consider the case of $n = 20$ actions. The vector $\Phi(s,a)$ has 120 elements and for each of the 20 possible actions only six different elements are not 0.

The adoption of LSPI for Reactive Search Optimization requires a training phase that identifies the optimal policy for the tuning of the parameter(s) of the considered SLS algorithm. During the training phase, a given number of runs of the SLS algorithm are executed over a subset of instances from the considered benchmark. The initial value for the target parameter is selected uniformly at random over the set of the allowed values. Each run is divided into epochs and the target parameter value is modified only at the end of each epoch using a random policy. An example (s,a,r,s') is composed of the states s and s' of the Markov process observed at the end of two consecutive epochs, the action a modifying the target parameter at the end of the first epoch and the reward r observed for the action. The collected examples are used by the LSPI algorithm to identify the optimal policy (see Fig. 6.3). In this work, the training phase is executed off-line.

During the testing phase, the state observed at the end of the current epoch is given as input to the LSPI algorithm. The learnt policy determines the appropriate modification of the target parameter (see Eq. 6.8). For the testing phase, the difference between the RL-based and the original version of the considered SLS algorithm consists only of the modification of the target parameter at the end of each epoch.

6.6 Experimental Results

To measure the performance of our Reinforcement-based approach, we implemented C++ functions for the Walksat/SKC, RSAPS and Reactive Tabu Search methods described in Section 6.4 and interfaced them to the Matlab LSPI implementation by Lagoudakis and Parr, available at http://www.cs.duke.edu/research/AI/LSPI/ (as of December 15, 2009).

The experiments performed in this work were carried out using a 2 GHz Intel Xeon processor computer, with 6 GB RAM. The efficient UBCSAT [29] implementation of the Walksat and (R)SAPS algorithms is considered, while for the H_RTS algorithm the original code of its authors is executed.

While in this paper we base our comparisons on the solution quality after a given number of iterations, the CPU time required by RTS_RL, Walksat_RL and RSAPS_RL is analogous to that of the basic H_RTS, Walksat and SAPS algorithms, respectively. The only negligible CPU time overhead is due to the computation of 20 floating-point 120-element vector products in order to compute $\hat{Q}(s,a)$ (see Eq. 6.8) for the 20 actions at the end of each epoch.

For brevity, the applications of our Reinforcement-based approach to the tuning of the noise parameter, the prohibition value and the smoothing probability parameter are termed "Walksat_RL", "RTS_RL" and "RSAPS_RL", respectively.

We consider two kinds of instances: random MAX-3-SAT instances and instances derived from relevant industrial problems.

For the training phase, we selected one instance from the considered benchmark and performed four runs of the algorithm with different randomly chosen starting truth assignments. We created 8,000 examples for the Walksat_RL and RSAPS_RL algorithms and 4,000 examples for the RTS_RL method. The instance selected for the training is not included in the set of instances used for the testing phase.

6.6.1 Experiments on Difficult Random Instances

The experiments performed are over selected MAX-3-SAT random instances defined in [23]. Each instance has $2,000$ variables and $8,400$ clauses, thus lying near the satisfiability threshold of 3-SAT instances.

The comparison of our approach based on reinforcement learning w.r.t. the original Walksat/SKC, RTS and RSAPS algorithms is in Table 6.1. Each entry in the table contains the mean and the standard error of the mean over ten runs of the best-so-far number of unsatisfied clauses at iteration $210,000$.

The Walksat/SKC algorithm has been executed with the default 0.5 value for the noise parameter. The setting for the SAPS parameters is the default one described in [18], as empirical results in [18] show it is the most robust configuration.

Table 6.1 shows the superior performance of the RSAPS algorithm to that of the SAPS algorithm, motivating the need for reactive parameter tuning over this benchmark. Over all the benchmark instances, the RSAPS algorithm exhibits a lower best-so-far number of unsatisfied clauses at iteration $210,000$ than the SAPS algorithm. The Kolmogorov-Smirnov test for two independent samples indicates that there is statistical evidence (with a confidence level of 95%) for the superior performance of RSAPS to SAPS (except in the case of the instance sel_04.cnf). In the case of

Instance	H_RTS	RTS_RL	Walksat	Walksat_RL	SAPS	RSAPS	SAPS_RL
sel_01.cnf	17.7 (.349)	8.6 (.142)	4.8 (.308)	7.3 (.163)	21.6 (.347)	16.2 (.329)	3.9 (.268)
sel_02.cnf	12.2 (.454)	6.1 (.179)	3.1 (.087)	4.5 (.190)	21.7 (.343)	12.3 (.235)	1.1 (.112)
sel_03.cnf	16.5 (.427)	7.3 (.194)	7.5 (.347)	8.6 (.320)	26.0 (.294)	18.2 (.364)	4.5 (.195)
sel_04.cnf	12.9 (.351)	7.6 (.183)	5.6 (.134)	8.0 (.298)	20.7 (.388)	16.2 (.304)	3.1 (.172)
sel_05.cnf	17.5 (.279)	8.0 (.169)	7.4 (.206)	8.6 (.271)	28.4 (.313)	20.2 (.244)	6.9 (.317)
sel_06.cnf	22.4 (.347)	8.3 (.262)	7.0 (.418)	10.2 (.274)	29.4 (.275)	21.4 (.291)	6.7 (.249)
sel_07.cnf	16.8 (.507)	7.7 (.200)	6.0 (.182)	8.2 (.345)	24.5 (.535)	18.2 (.198)	4.4 (.236)
sel_08.cnf	14.6 (.283)	6.7 (.216)	4.5 (.177)	8.6 (.356)	21.9 (.260)	16.6 (.231)	3.9 (.166)
sel_09.cnf	15.9 (.462)	9.0 (.274)	6.1 (.237)	8.6 (.291)	23.9 (.341)	17.2 (.322)	6.3 (.249)
sel_10.cnf	15.5 (.330)	8.5 (.263)	6.2 (.385)	6.4 (.189)	23.0 (.298)	16.2 (.161)	3.8 (.139)

Table 6.1: A comparison of the mean BSF values of selected SAT/MAX-SAT solvers. The values in the table are the mean and the standard error of the mean over ten runs of the best-so-far number of unsatisfied clauses at iteration 210,000

the sel_04.cnf instance the Kolmogorov-Smirnov test accepts the null hypothesis, while, e.g., in the case of instance sel_03.cnf it rejects the null hypothesis.

Over the considered benchmark, our RL-based approach improves the performance of the underlying algorithm when considering RSAPS and H_RTS.

In the case of the smoothing probability parameter, for each benchmark instance the RL-based algorithm shows a lower number of unsatisfied clauses in the best-so-far number as of iteration 210,000 than the original RSAPS algorithm. For example, over the instance sel_07.cnf, RSAPS_RL on average reaches 4.4 unsatisfied clauses, while the original RSAPS algorithm reaches 18.2 unsatisfied clauses. The Kolmogorov-Smirnov test for two independent samples confirms this observation. Figure 6.5 shows the evolution of the value $\|\mathbf{w} - \mathbf{w}'\|$ during the training phase of the LSPI algorithm in the case of RSAPS_RL. The LSPI algorithm converges in seven iterations, when the value $\|\mathbf{w} - \mathbf{w}'\|$ becomes smaller than 10^{-6}.

In the case of the RL-based tuning of the prohibition parameter, the RTS_RL algorithm outperforms H_RTS over all the benchmark instances. For example, consider the 7.3. unsatisfied clauses reached on average by the RTS_RL algorithm compared to the 16.5 clauses of the H_RTS algorithm over the instance sel_03.cnf. Again, the Kolmogorov-Smirnov test for two independent samples rejects the null hypothesis at the standard significance level of 0.05 for all the instances of the considered benchmark.

However, over the MAX-3-SAT benchmark considered, no improvement is observed for the RL-based reactive tuning of the noise parameter of the Walksat/SKC algorithm. The Kolmogorov-Smirnov test does not show statistical evidence (with a confidence level of 95%) for the different performance of Walksat_RL compared to Walksat/SKC, except in the case of instances 1 and 4, where the latter algorithm reaches on average a lower number of unsatisfied clauses than the former one. In the case of instances 1 and 4, the Kolmogorov-Smirnov test rejects the null hypothesis.

Note that during the offline learning phase of the RL-based version of Walksat, the LSPI algorithm does not converge even after 20 iterations over the set of examples extracted from the considered benchmark (even if 64,000 instead of 8,000 examples are generated).

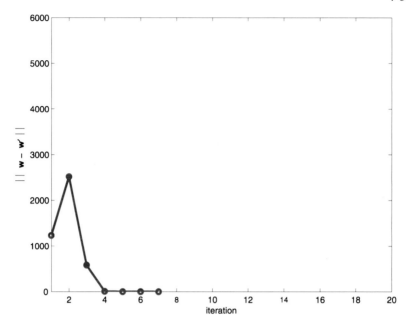

Fig. 6.5: Training of the LSPI algorithm over the MAX-3-SAT random instances considered in the case of RSAPS_RL

6.6.2 Experiments on Structured Industrial Instances

We repeat the experiment over a benchmark of MAX-SAT industrial instances (Table 6.2). Again, note first the superior performance of the RSAPS algorithm to the SAPS algorithm (the Kolmogorov-Smirnov test indicates that there is statistical evidence of different performance, except in the case of instance en_6_nd.cnf). This observation motivates the selection of this benchmark to test the smarter reactive approach based on RL.

Over this benchmark, the RL-based reactive approach does not show appreciable improvement.

Instance	H_RTS	RTS_RL	Walksat	Walksat_RL	SAPS	RSAPS	SAPS_RL
en_6.cnf	1073.7 (.006)	1034.3 (.542)	67.0 (.875)	81.9 (.369)	26.3 (.563)	11.8 (.742)	33.5 (.256)
en_6_case1.cnf	1091.3 (.776)	1059.5 (.112)	65.6 (.228)	92.9 (.299)	25.5 (.343)	6.0 (.222)	25.8 (.997)
en_6_nd.cnf	1087.9 (.242)	1050.7 (.117)	75.6 (.358)	99.0 (.709)	27.8 (.597)	21.7 (.964)	28.3 (.562)
en_6_nd_case1.cnf	1075.9 (.469)	1063.9 (.343)	68.9 (.563)	83.1 (.785)	20.0 (.878)	5.0 (.728)	55.9 (.930)

Table 6.2: A comparison of the mean BSF values of selected SAT/MAX-SAT solvers over a benchmark of big industrial unsat instances. The values in the table are the mean and the standard error of the mean over ten runs of the best-so-far number of unsatisfied clauses at iteration 210,000

In the case of the prohibition parameter, the Kolmogorov-Smirnov test for two independent samples indicates that the difference between the results of the RTS_RL and H_RTS approaches is not statistically significant (at the standard significance level of 0.05) in the case of instances en_6_nd.cnf and en_6_nd_case1.cnf. However, over the instances en_6.cnf and en_6_case1.cnf, RTS_RL shows superior performance to H_RTS. E.g., in the case of instance en_6_nd.cnf the Kolmogorov-Smirnov test accepts the null hypothesis since the p-value is 0.11 while in the case of instance en_6.cnf it rejects the null hypothesis since the p-value is 0.031.

In the case of the smoothing probability and the noise parameters, the RL-based approach shows poor performance even compared to the non-reactive versions of the original methods considered (the Walksat and the SAPS algorithms). The Kolmogorov-Smirnov test does not reject the null hypothesis comparing the results of SAPS to RSAPS_RL (except in the case of instance en_6_nd_case1.cnf, where SAPS outperforms RSAPS_RL). Furthermore, the statistical test confirms the worse results of Walksat_RL to the Walksat algorithm (except in the case of instance en_6_nd_case1.cnf, where there is no statistical evidence of different performance).

Note that during the offline learning phase of the RL-based version of Walksat and RSAPS, the LSPI algorithm does not converge even after 20 iterations over the set of examples extracted from the considered benchmark (even if 64,000 instead of 8,000 examples are generated). Figure 6.6 shows the evolution of the value $\|\mathbf{w} - \mathbf{w}'\|$ during the training phase of the LSPI algorithm in the case of Walksat_RL. The algorithm does not converge: from iteration 12, the L2-norm of the vector $\mathbf{w} - \mathbf{w}'$ oscillates from the value $4, 150$ to $5, 700$.

Furthermore, qualitatively different results w.r.t. Table 6.2 are not observed if a different instance is used for training, or if the training set includes two instances rather than one.

Table 6.2 shows the poor performance of prohibition-based reactive algorithms. To understand possible causes for this lack of performance we analyzed the evolution of the prohibition parameter during the run. Figure 6.7 depicts the evolution of the fractional prohibition parameter T_f during a run of the H_RTS algorithm over two instances of the Engine benchmark.

The fact that the maximum prohibition value is reached suggests that the prohibition mechanism is not sufficient to diversify the search in these more structured instances. The performance of H_RTS is not significantly improved even when smaller values for the maximum prohibition are considered.

6.6.3 A Comparison with Offline Parameter Tuning

The experimental results show that over the MAX-3-SAT benchmark the RL-based approach outperforms the ad hoc. reactive approaches in the case of the RSAPS and H_RTS algorithms. The values in Tables 6.1 and 6.2 are obtained by using the

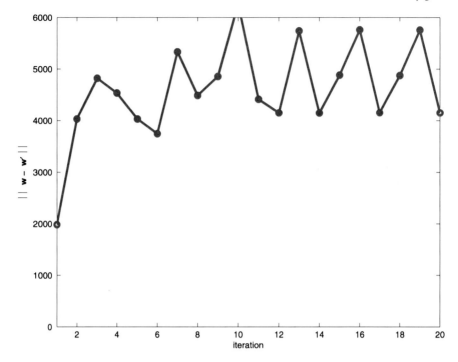

Fig. 6.6: Training of the LSPI algorithm over the industrial instances considered in the case of Walksat_RL

original default settings of the algorithm's parameters. In particular, the default configuration for SAPS parameters is the following:
- the scaling parameter is set to 1.3;
- the smoothing parameter is set to 0.8;
- the smoothing probability parameter is set to 0.05;
- the noise parameter is set to 0.01.

This configuration is also the original default setting of RSAPS, where the reactive mechanism adapts the smoothing probability parameter during the search history. The initial default value for the fractional prohibition parameter in H_RTS is 0.1. The reactive mechanism implemented in H_RTS is responsible for the adaptation of the prohibition parameter during the algorithm execution.

For many domains, a significant improvement in the performance of the SAT solvers can be achieved by setting their parameters differently from the original default configuration. The tuning is executed offline, and the determined parameters are then fixed while the algorithm runs on new instances.

In order to measure the beneficial effect of the RL-based reactive mechanism observed for the smoothing probability parameter and the prohibition over the MAX-3-SAT benchmark, we compare RTS_RL and RSAPS_RL with the best fixed parameter tuning of SAPS and H_RTS algorithms for those instances. The best fixed

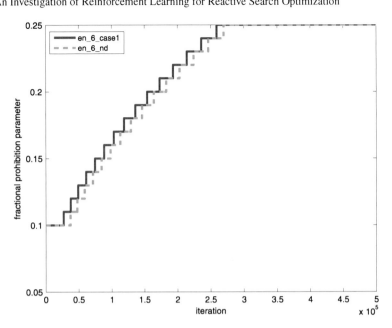

Fig. 6.7: Evolution of the T_f parameter during a run of the H_RTS algorithm over two selected instances of the Engine benchmark

configuration for SAPS and H_RTS parameters are obtained by ParamILS [19], an automatic tool for the parameter tuning problem. In particular, the best fixed setting for SAPS parameters:

- the scaling parameter is set to 1.256;
- the smoothing parameter is set to 0.5;
- the smoothing probability parameter is set to 0.1;
- the noise parameter is set to 0.05.

The best fixed value for the fractional prohibition parameter over the MAX-3-SAT benchmark is 0.13. Note that, by fixing the prohibition parameter in H_RTS, one obtains the standard prohibition-based approach known as GSAT/tabu. Table 6.3 shows the results of the best fixed parameter setting approach.

The Kolmogorov-Smirnov test shows that there is no statistical evidence of different performance between SAPS_RL and SAPS$_{best}$, except in the case of instance sel_5.cnf where the latter algorithm outperforms the former one (the p-value is 0.030). Therefore, over the MAX-3-SAT benchmark SAPS_RL is competitive with *any* fixed setting of the smoothing probability parameter.

In the case of the prohibition parameter, a better performance of the GSAT/tabu$_{best}$ algorithm compared to RTS_RL is observed over three instances (sel_1.cnf, sel_9.cnf and sel_10.cnf) where the Kolmogorov-Smirnov test for two independent samples rejects the null hypothesis at the standard significance level of 0.05. Over the remaining instances, the performance of RTS_RL is competitive with the performance of GSAT/tabu$_{best}$. E.g., in the case of instance sel_8.cnf the Kolmogorov-

Instance	SAPS$_{best}$	SAPS_RL	GSAT/tabu$_{best}$	RTS_RL
sel_01.cnf	2.9 (.099)	3.9 (.268)	5.8 (.187)	8.6 (.142)
sel_02.cnf	2.1 (.099)	1.1 (.112)	4.0 (.163)	6.1 (.179)
sel_03.cnf	2.8 (.103)	4.5 (.195)	5.9 (.119)	7.3 (.194)
sel_04.cnf	2.7 (.094)	3.1 (.172)	6.2 (.161)	7.6 (.183)
sel_05.cnf	3.6 (.084)	6.9 (.317)	7.3 (.094)	8.0 (.169)
sel_06.cnf	4.8 (.261)	6.7 (.249)	7.2 (.122)	8.3 (.262)
sel_07.cnf	2.9 (.110)	4.4 (.236)	6.5 (.108)	7.7 (.200)
sel_08.cnf	3.5 (.117)	3.9 (.166)	5.9 (.179)	6.7 (.216)
sel_09.cnf	4.2 (.220)	6.3 (.249)	6.7 (.115)	9.0 (.274)
sel_10.cnf	2.5 (.135)	3.8 (.139)	5.1 (.137)	8.5 (.263)

Table 6.3: Best fixed tuning of SAPS and and GSAT/tabu algorithms. The values in the table are the mean and the standard error of the mean over ten runs of the BSF value observed at iteration 210,000

Smirnov test accepts the null hypothesis since the p-value is 0.974, while in the case of instance sel_1.cnf it rejects the null hypothesis since the p-value is 0.006.

6.6.4 Interpretation of the Learnt Policy

For RTS_RL, Walksat_RL and SAPS_RL, the LSPI algorithm has been applied to the training sample set, and with (6.9) as approximate space basis. Figures 6.8 and 6.9 and 6.10 and 6.11 show the policy learnt when executing RTS_RL and Walksat_RL over the MAX-3-SAT benchmark, respectively.

Each point in the graph represents the action determined by Eq. 6.9 for the state $s \equiv \left(\Delta f, \frac{\overline{H}_{epoch}}{n} \right)$ of the Markov decision process. On the x- and y-axes, the features of the state are represented. In particular, the x-axis shows the first component on the MDP state $\frac{\overline{H}_{epoch}}{n}$, which represents the normalized mean Hamming distance during the current epoch from the configuration at its beginning (see Section 6.5). The y-axis refers to Δf, the second component of the state, representing the mean change of f in the current epoch with respect to the best value. The range for the x- and y-axes are loose bounds on the maximum and minimum values for the $\frac{\overline{H}_{epoch}}{n}$ and Δf features observed during the offline training phase. The z-axis contains the values for the action.

In the case of RTS_RL, the LSPI algorithm converges and the distribution of the actions over the state space is consistent with the intuition. A high value for the prohibition parameter is suggested in cases where the mean Hamming distance between the configurations explored in the current epoch and the configuration at the beginning of the current epoch does not exceed a certain value, provided that the current portion of the landscape is worse than the previously explored regions. This policy is consistent with intuition: a higher value of T causes a larger differentia-

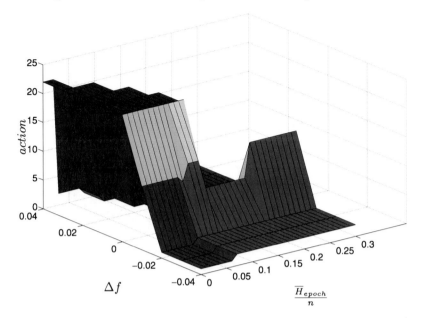

Fig. 6.8: Distribution of the actions in the significant portion of the state space for the RTS_RL algorithm over the Selman benchmark

tion of visited configurations (more different variables need to be flipped), and this is desired when the algorithm needs to escape the neighborhood of a local minimum; in this case, in fact, movement is limited because the configuration is trapped at the "bottom of a valley". On the other hand, when the trajectory is not within the attraction basin of a minimum, a lower value of T enables a better exploitation of the neighborhood. Furthermore, small values for the prohibition parameter are suggested when the current portion of landscape is better than the previously explored regions: the algorithm is currently investigating a promising region of the search space. These observations, consistent with the intuition, confirm the positive behavior of RTS_RL over the Selman benchmark, improving the performance of the H_RTS algorithm.

For Walksat_RL over the MAX-3-SAT instances, the LSPI algorithm does not converge. The distribution of the best action obtained after 20 iterations of the LSPI algorithm is in Figures 6.10 and 6.11. A very similar diversification level is suggested when the current portion of the landscape is both better and worse than the previously explored regions. In particular, a counterintuitive diversification action is suggested when the Δf feature assumes negative values while high values are observed for the $\frac{H_{epoch}}{n}$ feature. These observations confirm the poor results of the Walksat_RL algorithm, performing worse even than the non-reactive Walksat algorithm.

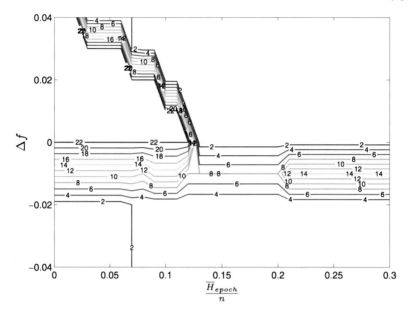

Fig. 6.9: Distribution of the actions in the significant portion of the state space for the RTS_RL algorithm over the Selman benchmark (contour lines)

6.7 Conclusion

This paper describes an application of reinforcement learning for Reactive Search Optimization. In particular, the LSPI algorithm is applied for the online tuning of the prohibition value in H_RTS, the noise parameter in Walksat and the smoothing probability in the SAPS algorithm.

On the one side, the experimental results are promising: over the MAX-3-SAT benchmark considered, RTS_RL and SAPS_RL outperform the H_RTS and RSAPS algorithms, respectively. On the other hand, this appreciable improvement is not observed over the structured instances benchmark. Apparently, the wider differences among the structured instances render the adaptation scheme obtained on some instances inappropriate and inefficient on the instances used for testing.

Some more weaknesses of the proposed reinforcement learning approach were observed during the experiments: the LSPI algorithm does not converge over both the benchmarks considered in the case of Walksat_RL and over the structured instances benchmark in the case of SAPS_RL. Furthermore, the interval size and the discretization interval of the target parameter in the case of Walksat_RL and SAPS_RL have to be hand-tuned.

Finally, the diversification-bias trade-off determining the performance of SLS algorithms is encountered also during the exploration of the states of the Markov decision process performed by any RL schema.

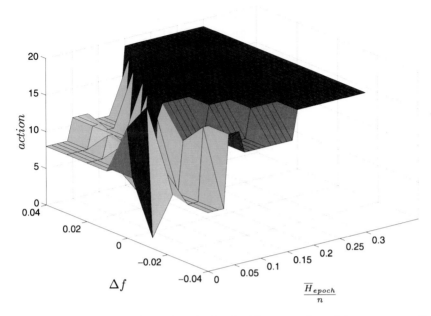

Fig. 6.10: Distribution of the actions in the significant portion of the state space for the Walksat_RL algorithm over the Selman benchmark

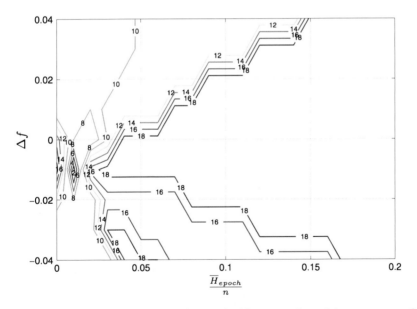

Fig. 6.11: Distribution of the actions in the significant portion of the state space for the Walksat_RL algorithm over the Selman benchmark (contour lines)

The application of RL in the context of Reactive Search Optimization is far from
trivial and the number of research problems it generates appears in fact to be larger
than the number of problems it solves. This is positive if considered from the point
of view of the future of this RSO and of Autonomous Search fields: the less one
can adapt techniques from other areas, the greater the motivation to develop novel
and independent methods. Our preliminary results seem to imply that more direct
techniques specifically designed for the new RSO context will show much greater
effectiveness than techniques inherited from the RL context.

References

[1] Baluja S., Barto A., Boese K., Boyan J., Buntine W., Carson T., Caruana R.,
Cook D., Davies S., Dean T., et al.: Statistical machine learning for large-scale
optimization. Neural Computing Surveys 3:1–58 (2000)
[2] Battiti R.: Machine learning methods for parameter tuning in heuristics. In:
5th DIMACS Challenge Workshop: Experimental Methodology Day, Rutgers
University (1996)
[3] Battiti R., Brunato M.: Reactive search: Machine learning for memory-based
heuristics. In: Gonzalez T.F. (ed.) Approximation Algorithms and Metaheuristics, Taylor and Francis Books (CRC Press), Washington, DC, chap. 21, pp.
21-1–21-17 (2007)
[4] Battiti R., Campigotto P.: Reinforcement learning and reactive search: An
adaptive MAX-SAT solver. In: Proceedings of the 2008 Conference on ECAI
2008: 18th European Conference on Artificial Intelligence, IOS Press, pp.
909–910 (2008)
[5] Battiti R., Protasi M.: Reactive search, a history-sensitive heuristic for
MAX-SAT. ACM Journal of Experimental Algorithmics 2 (article 2),
http://www.jea.acm.org/ (1997)
[6] Battiti R., Tecchiolli G.: The Reactive Tabu Search. ORSA Journal on Computing 6(2):126–140 (1994)
[7] Battiti R., Brunato M., Mascia F.: Reactive Search and Intelligent Optimization, Operations research/Computer Science Interfaces, vol. 45. Springer Verlag (2008)
[8] Bennett K., Parrado-Hernández E.: The interplay of optimization and machine
learning research. The Journal of Machine Learning Research 7:1265–1281
(2006)
[9] Bertsekas D., Tsitsiklis J.: Neuro-dynamic programming. Athena Scientific
(1996)
[10] Boyan J.A., Moore A.W.: Learning evaluation functions for global optimization and Boolean satisfiability. In: Press A. (ed.) Proc. of 15th National Conf.
on Artificial Intelligence (AAAI), pp. 3–10 (1998)
[11] Brunato M., Battiti R., Pasupuleti S.: A memory-based rash optimizer. In:
Geffner A.F.R.H.H. (ed.) Proceedings of AAAI-06 Workshop on Heuristic

Search, Memory Based Heuristics and Their applications, Boston, Mass., pp. 45–51, ISBN 978-1-57735-290-7 (2006)

[12] Eiben A., Horvath M., Kowalczyk W., Schut M.: Reinforcement learning for online control of evolutionary algorithms. In: Brueckner S.A., Hassas S., Jelasity M., Yamins D. (eds.) Proceedings of the 4th International Workshop on Engineering Self-Organizing Applications (ESOA'06), Springer Verlag, LNAI 4335, pp. 151–160 (2006)

[13] Epstein S.L., Freuder E.C., Wallace R.J.: Learning to support constraint programmers. Computational Intelligence 21(4):336–371 (2005)

[14] Fong P.W.L.: A quantitative study of hypothesis selection. In: International Conference on Machine Learning, pp. 226–234 (1995) URL citeseer. ist.psu.edu/fong95quantitative.html

[15] Hamadi Y., Monfroy E., Saubion F.: What is Autonomous Search? Tech. Rep. MSR-TR-2008-80, Microsoft Research (2008)

[16] Hoos H.: An adaptive noise mechanism for WalkSAT. In: Proceedings of the National Conference on Artificial Intelligence, AAAI Press; MIT Press, vol. 18, pp. 655–660 (1999)

[17] Hoos H., Stuetzle T.: Stochastic Local Search: Foundations and applications. Morgan Kaufmann (2005)

[18] Hutter F., Tompkins D., Hoos H.: Scaling and probabilistic smoothing: Efficient dynamic local search for sat. In: Proc. Principles and Practice of Constraint Programming - CP 2002, Ithaca, NY, Sept. 2002, Springer LNCS, pp. 233–248 (2002)

[19] Hutter F., Hoos H.H., Stützle T.: Automatic algorithm configuration based on local search. In: Proc. of the Twenty-Second Conference on Artifical Intelligence (AAAI '07), pp. 1152–1157 (2007)

[20] Lagoudakis M., Littman M.: Algorithm selection using reinforcement learning. Proceedings of the Seventeenth International Conference on Machine Learning, pp. 511–518 (2000)

[21] Lagoudakis M., Littman M.: Learning to select branching rules in the DPLL procedure for satisfiability. LICS 2001 Workshop on Theory and Applications of Satisfiability Testing (SAT 2001) (2001)

[22] Lagoudakis M., Parr R.: Least-Squares Policy Iteration. Journal of Machine Learning Research 4(6):1107–1149 (2004)

[23] Mitchell D., Selman B., Levesque H.: Hard and easy distributions of SAT problems. In: Proceedings of the Tenth National Conference on Artificial Intelligence (AAAI-92), San Jose, CA, pp. 459–465 (1992)

[24] Muller S., Schraudolph N., Koumoutsakos P.: Step size adaptation in Evolution Strategies using reinforcement learning. Proceedings of the 2002 Congress on Evolutionary Computation, 2002 CEC'02 1, pp. 151–156 (2002)

[25] Prestwich S.: Tuning local search by average-reward reinforcement learning. In: Proceedings of the 2nd Learning and Intelligent OptimizatioN Conference (LION II), Trento, Italy, Dec. 10–12, 2007, Springer, Lecture Notes in Computer Science (2008)

[26] Schwartz A.: A reinforcement learning method for maximizing undiscounted rewards. In: ICML, pp. 298–305 (1993)

[27] Selman B., Kautz H., Cohen B.: Noise strategies for improving local search. In: Proceedings of the national conference on artificial intelligence, John Wiley & sons Ltd, USA, vol. 12 (1994)

[28] Sutton R. S., Barto A. G.: Reinforcement Learning: An introduction. MIT Press (1998)

[29] Tompkins D.: UBCSAT. http://www.satlib.org/ubcsat/#introduction (as of Oct. 1, 2008)

[30] Xu Y., Stern D., Samulowitz H.: Learning adaptation to solve Constraint Satisfaction Problems. In: Proceedings of the 3rd Learning and Intelligent OptimizatioN Conference (LION III), Trento, Italy, Jan. 14–18, 2009, Springer, Lecture Notes in Computer Science (2009)

[31] Zhang W., Dietterich T.: A reinforcement learning approach to job-shop scheduling. Proceedings of the Fourteenth International Joint Conference on Artificial Intelligence, 1114–1120 (1995)

[32] Zhang W., Dietterich T.: High-performance job-shop scheduling with a time-delay TD (λ) network. Advances in Neural Information Processing Systems 8:1024–1030 (1996)

Chapter 7
Adaptive Operator Selection and Management in Evolutionary Algorithms

Jorge Maturana, Álvaro Fialho, Frédéric Saubion, Marc Schoenauer, Frédéric Lardeux, and Michèle Sebag

7.1 Introduction

Evolutionary Algorithms (EAs) constitute efficient solving methods for general optimization problems. From an operational point of view, they can be considered as basic computational processes that select and apply variation operators over a set of possible configurations of the problem to be solved, guided by the Darwinian "survival of the fittest" paradigm.

In order to efficiently apply an EA to an optimization problem, there are many choices that need to be made. Firstly, the design of the general skeleton of the algorithm must include the selection of a suitable encoding for the search space at hand, the management of the population (i.e., size setting, selection and replacement processes, and so on), and the definitions of the variation operators that will be used, namely the mutation and recombination operators. These components can be

Jorge Maturana
Instituto de Informática, Universidad Austral de Chile, Valdivia, Chile.
e-mail: jorge.maturana@inf.uach.cl

Álvaro Fialho
Microsoft Research – INRIA Joint Centre, Orsay, France. e-mail: alvaro.fialho@inria.fr

Frédéric Saubion
LERIA, Université d'Angers, Angers, France. e-mail: saubion@info.univ-angers.fr

Marc Schoenauer
Project-Team TAO, INRIA Saclay – Île-de-France and LRI (UMR CNRS 8623), Orsay, France and Microsoft Research – INRIA Joint Centre, Orsay, France.
e-mail: marc.schoenauer@inria.fr

Frédéric Lardeux
LERIA, Université d'Angers, Angers, France. e-mail: lardeux@info.univ-angers.fr

Michèle Sebag
Project-Team TAO, LRI (UMR CNRS 8623) and INRIA Saclay – Île-de-France, Orsay, France and Microsoft Research – INRIA Joint Centre, Orsay, France. e-mail: michele.sebag@inria.fr

Y. Hamadi et al. (eds.), *Autonomous Search*,   161
DOI 10.1007/978-3-642-21434-9_7,
© Springer-Verlag Berlin Heidelberg 2011

considered as the *structural parameters* of the algorithms that define its operational architecture. However, once the structure is defined, a difficult and crucial task remains: how to control the general computation process? This control is usually embedded in a set of *behavioral parameters* that can be related to the data structures or to the computation steps, e.g., the application rate of each variation operator.

All these parameters should then be tuned, depending on the problem at hand. In the early days of Evolutionary Computation, these numerous possible choices were in fact considered as richness, providing very useful flexibility to EAs, that could indeed be applied to a very wide range of scientific fields. Nowadays, the contemporary view of EAs acknowledges that specific problems require specific setups in order to obtain satisfactory performance [13]. In other words, when it comes to solving a given problem, all practitioners know that parameter setting is in fact the Achilles' heel of EAs (together with their high computational cost).

As the efficiency of Evolutionary Algorithms has already been experimentally demonstrated on many difficult problems out of the reach of other optimization methods, more and more scientists (researchers, engineers) are trying them out on their specific problems. However, they very often fail in getting interesting results precisely because there is a lack of general methods for tuning at least some of the involved parameters (and also because they are not, and do not want to become, "Evolutionary Engineers"). Of course, the specialization of an EA to a given problem has an impact on its performance on other problems (according to the No Free Lunch Theorems for Optimization [44]). In fact, through parameter setting, the main challenge is to set (or select) the right algorithm for the given problem, which is an old problem in computer science [39]. Therefore, it is time that we *Cross the Chasm* [33], bringing the benefits of EAs to the whole scientific community without its main burden, that of parameter setting.

A current trend in EA is to focus on the definition of more autonomous solving processes, which aim at allowing the basic user to benefit from a more efficient and easy-to-use algorithmic framework. Parameter setting in EAs is a major issue that has received much attention in recent years [14], and its research is still very active nowadays, as a book recently published [27], and from the numerous recent references cited in this document. This subject is not limited to EAs, being also investigated in operations research and constraint programming communities, where the current solving technologies that are included in efficient solvers require huge expertise to be fully used (see, for instance, [4]).

Parameter setting in EAs may be considered at two complementary levels as follows [36]:

Design: In many application domains that directly pertain to standard representations, users who are not EA experts can simply use off-the-shelf EAs with classic (and thus non-specialized) variation operators to solve their problems. However, the same users will encounter great difficulties when faced with problems outside the basic frameworks. Even if standard variation operators exist in the literature (such as the uniform crossover [40]), the achievement of acceptable results depends necessarily on the specialization of the algorithmic scheme, which usually requires the definition of appropriate operators. The design of problem-specific operators requires much expertise, though some advanced tools are now available [11]. In any case, the impact on the computation process of problem-specific operators is even more difficult to forecast than of well-known operators, and thus their associated parameters are harder to correctly estimate a priori.

Behavior: Once the general architecture of the algorithm has been defined, the user needs to configure the behavior of these components through parameters. This has a direct impact on the performance and reliability of the algorithms. Indeed, its efficiency is strongly related to the way the Exploration versus Exploitation (EvE) dilemma is addressed, determining the ability of the EA to escape from local optima in order to sparsely visit interesting areas, while also focus on the most promising ones, to thus reaching global solutions. However, it is known that more exploration of the search space is necessary in the initial generations, while more exploitation should be done when the search is approaching to the optimum; thus, a static definition of the application rates for each operator will be always suboptimal.

We propose transparently including these two possible levels of control into the basic algorithmic process, as illustrated in Figure 7.1. Firstly, at the *design* level, the *Adaptive Operator Management* (AOM) aims at handling the suitable subset of operators that is made available to the algorithm, excluding the useless operators and including possibly useful ones. These operators are extracted from a set of potential operators (either automatically generated or predefined). Based on such subset of available operators, defined and maintained by the AOM, the *Adaptive Operator Selection* (AOS) is used to control the *behavior* of the algorithm, handling the problem of selecting the operators to be applied at every instant of the search. It clearly appears that the concepts of AOM and AOS are fully complementary, and very useful for making an EA more autonomous. Within the combination of both approaches, the design and the behavior of the algorithm are automatically adapted while solving the problem, from continuous observation of the performance of the operators during the search, based on its needs with regard to exploration and exploitation.

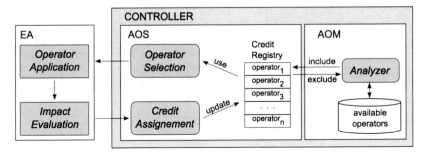

Fig. 7.1: General scheme of the controller, depicting its two main components, *Adaptive Operator Selection* and *Adaptive Operator Management*

This chapter will focus on these control mechanisms, by reviewing existing solutions previously proposed by the authors, and providing examples of their possible applications by means of cases studies. Firstly, an overview of the current state of the art in parameter setting in EAs and, more specifically, in adaptive control of operators, is presented in Section 7.2. Then, some efficient AOS combinations are presented in Section 7.3. A case study of the application of an AOM technique, combined with the AOS, is presented in Section 7.4. In both cases, experimental results are presented, applying the adaptive EA to the solution of SAT problems. Finally, conclusions and possible paths for further research are drawn in Section 7.5.

7.2 Parameter Setting in Evolutionary Algorithms

Slightly departing from the point of view adopted in previous surveys [14, 13], we shall firstly distinguish here between *external tuning* and *internal control* of the parameters.

Methods that perform *external tuning* consider the algorithm as a black box: some external process is run (e.g., another optimization algorithm, or some standard statistical analysis), to determine a good (range of) value(s) for the parameters at hand. Most of these methods are not limited to Evolutionary Algorithms and can be applied to any type of algorithm, although some of them have been designed in the evolutionary framework. The most straightforward approach is the complete factorial Design of Experiments, in which the same number of runs is executed for each setting, and the one that achieves the best performance on average is selected for the algorithm. Other approaches have been recently proposed, such as the SPO [3], in which the optimal values are refined after a few runs; the Racing methods [6, 46], in which settings are discarded once proven worst than others ***; and the recent Meta-EA approaches [8], in which some runs are actually stopped at some point of the evolution, and some of their parameters are modified before they are restarted.

On the other hand, *internal control* methods work directly on the values of the parameters while solving the problem, i.e., on-line. Such kind of mechanisms for modifying parameters during an algorithm execution were invented early in EC history, and most of them are still being investigated nowadays. Indeed, there are at least two strong arguments to support the idea of changing the parameters during an *Evolutionary Algorithms* run:

- As evolution proceeds, more information about the landscape is known by the algorithm, so it should be possible to take advantage of it. This applies to global properties (for example, knowing how rugged the landscape is) and to local ones (for example, knowing whether a solution has been improved lately or not).
- As the algorithm proceeds from a global (early) exploration of the landscape to a more focused, exploitation-like behavior, the parameters should be adjusted to take care of this new reality. This is quite obvious, and it has been empirically and theoretically demonstrated that different values of parameters might be op-

timal at different stages of the search process (see [13, p. 21] and references therein).

The different approaches that have been proposed to internally adapt the parameters can be classified into three categories, depending on the type of information that is used to adjust the parameters during the search (such classification, used by many authors in the 1990s, has been made in [14] and republished in the corresponding chapter of [13]).

Deterministic parameter control implements a set of deterministic rules, without any feedback from the search. This is, in general, hard to achieve, because of a simple reason: it relies heavily on knowing beforehand how long it will take to achieve a given target solution with the running algorithm and, obviously, this cannot be easily predictable. But even if it were, the way to balance exploration and exploitation can hardly be guessed. This situation is worsened by two facts: first, there is usually a big variance between different runs *on the very same problem*; and second, these methods often require additional parameters that are used to tune the deterministic parameter (starting with the total number of generations the algorithm will run), and even though these parameters can be considered second-order, their influence is nevertheless critical.

Since our knowledge about the way the search should behave is always limited, it is sometimes possible, and advantageous, to let evolution itself tune some of the parameters: such a *Self-Adaptive* parameter control adjusts parameters "for free", i.e., without any direct specification from the user. In other words, individuals in the population might contain "regulatory genes" that control some of the parameters, e.g., the mutation and recombination rates; and these regulatory genes are subject to the same evolutionary processes as the rest of the genome [23]. For quite some time in the 1990s, self-adaptation was considered as the royal road to success in Evolutionary Computation. First of all, the idea that the parameters are adapted *for free* is very appealing, and its parallel with the behavior of self-regulated genes is another appealing argument. On the practical side, as self-adaptive methods require little knowledge about the problem and, "what is probably more important", about the way the search should proceed, they sometimes are the only way to go when nothing is actually known about the problem at hand. However, by using such an approach, the algorithm needs to explore, in parallel, the search space of the variables of the problem and also the search space of the parameter values, which potentially increases the complexity of the search.

Then, it is clear that, whenever some decisions can be made to help the search follow an efficient path, this should be done. *Adaptive* or *Feedback-based* methods follow this rationale, being the most successful approaches today in on-line parameter control. These methods are based on the monitoring of particular properties of the evolutionary process, and use changes in these properties as an input signal to change the parameter values. The controllers presented in this work are included in this category.

Adaptive Operator Selection (AOS), presented in Section 7.3, is used to autonomously select which operator, among the available ones, should be applied by the EA at a given instant, based on how they have performed in the search so far. The set of operators available to the EA is usually static, defined *a priori* by an expert or an off-line method. However, since the performance of operators is continuously changing during the search, useless operators might be deleted, while possibly useful ones might be inserted in such a set. This is the underlying idea proposed by the *Adaptive Operator Management* (AOM), presented in Section 7.4.

To achieve the goals associated with AOS, two components are defined (as shown in Fig. 7.1): the *Credit Assignment* - how to assess the performance of each operator based on the impact of its application on the progress of the search; and the *Operator Selection* rule - how to select between the different operators based on the rewards that they have received so far. The rest of this section will survey existing AOS methods, looking at how they address each of these issues.

7.2.1 Learning by Credit Assignment

The AOS learns the performance of each operator based on a performance measure, assessed by what is usually referred to as the *Credit Assignment* mechanism. Starting with the initial work in AOS, from the late 1980's [12], several methods have been proposed to do this, with different ways of transforming the impact of the operator application into a quality measure.

Most of the methods use only the fitness improvement as a reward, i.e., the quality gain of the newborn offspring compared to a base individual, which might be (i) its parent [26, 42, 2], (ii) the current best individual [12], (iii) or the median individual [24] of the current population. If no improvement is achieved, usually a null credit is assigned to the operator.

Regarding which operator should be credited after the achievement of a given fitness improvement, the most common approach is to assign a credit to the operator that was directly responsible for the creation of the newborn offspring. Some authors, however, propose assigning credit to the operators used to generate the ancestors of the current individual (e.g., using some bucket brigade-like algorithm [12, 24]), based on the claim that the existence of efficient parents is indeed as important as the creation of improved offspring. Others, however, do not consider ancestors at all ([26, 42]), and some even suggest that this sometimes degrades the results [2].

The existing approaches also differ in the statistics that are considered in order to define the numerical reward. Most of the methods calculate their rewards based just on the most recent operator application, while others use the average of the quality achieved over a few applications. More recently, in [43], the utilization of extreme values over a few applications was proposed (statistical outliers), based on the idea that highly beneficial but rare events might be better for the search than regular but smaller improvements. The reported comparative results with other *Credit Assignment* mechanisms show the superiority of this approach over a set of continuous benchmark problems. Though used in a different context, the methods presented in this paper are also based on this idea.

7.2.2 Adaptation by Operator Selection Rule

The most common and straightforward way of doing *Operator Selection* is the so-called *Probability Matching* (PM) [20, 26, 2]. In brief, each operator is selected by a roulette-wheel-like process with a probability that is proportional to its known empirical quality (as defined by the *Credit Assignment* mechanism, e.g., the average of the received rewards). A user-defined α parameter might be used to introduce some relaxation in the update of this empirical estimation. Besides, a minimal probability (p_{min}) is usually applied, so that no operator is "lost" during the process: one operator that is currently bad might become useful at some further stage of the search. If an operator receives just null rewards (or the maximal reward) for some time, its expected reward will go to p_{min} (or $p_{max} = 1 - K * p_{min}$). However, this convergence is very slow, and, experimentally, mildly relevant operators keep being selected, which badly affects the performance of the algorithm [41].

Originally proposed for learning automata, *Adaptive Pursuit* (AP) [41] is an *Operator Selection* technique that partly addresses this drawback by implementing a winner-take-all strategy. Another user-defined parameter, β, is used to control the greediness of the strategy, i.e., how fast the probability of selecting the current best individual will converge to p_{max} while that of selecting the others will go to p_{min}. In its original proposal for the AOS problem [41], AP was applied in a set of artificial scenarios, in which the operator qualities were changing every Δt applications, achieving significant improvements over the performance of PM.

Different approaches, such as APGAIN [45], propose an *Operator Selection* divided into two periods. During a first learning stage (which is repeated several times during the run, so the changes can be followed), the operators are randomly selected, and the rewards gathered are used to extract initial knowledge about them. In a later stage, such knowledge is exploited by the technique in order to efficiently select the operators. The main drawback in this case is that roughly a quarter of the generations are dedicated to the learning, thus doing random selection, which may strongly affect the performance of the algorithm if disruptive operators are present in the set.

7.2.3 Meta-Parameters vs. Original Parameters

Clearly, the main objective of parameter setting in EAs is to automate some of the choices that should be made by the user. What happens in reality is that, although some of the parameters are efficiently defined by these autonomous controllers, the controllers themselves also have their own parameters that need to be tuned. This section considers this issue.

If one wants to define "manually" (or by the use of some off-line tuning technique) the parameters of an EA related to the variation operators, several choices need to be made. In addition to defining the list of operators that will be used by the algorithm, we need at least one other parameter to define *for each operator*: its application rate. Additionally, some operators also require some extra settings, e.g., the rate of flipping a bit in a bit-flip mutation operator, or the number of crossing points that will be used by an n-point crossover operator. Thus, considering a static definition of the operators and their respective rates, the number of original parameters to be defined is a multiple of the number of operators that is used by the EA.

All these parameters are automatically handled by the proposed adaptive controllers while solving the problem. The list of operators is managed by the *Adaptive Operator Management*, while the application rates are abstracted by the *Adaptive Operator Selection*. The extra parameters are abstracted as well, by considering different variants of the same operator as different operators, e.g., 1-point and 2-point crossover are treated as different operators instead of different (discrete) settings of the same operator.

As previously mentioned, to be able to do this, the controllers also introduce their own parameters, usually called hyper- or meta-parameters. But, firstly, there is a fixed number of meta-parameters instead of a multiple of the number of operators; secondly, these parameters are (supposed to be) less sensitive than the original ones; and finally, they are adapted online according to the needs of the search, while a static setting would always be suboptimal, as discussed in Section 7.2. For example, concerning the Operator Selection techniques reviewed in Section 7.2.2, Probability Matching needs the definition of the minimal probability p_{min} and the adaptation rate α; while the Adaptive Pursuit needs the definition of a third parameter, the learning rate β. For Credit Assignment, usually at least one parameter should be considered: the number of operator applications that are taken into account to calculate an empirical estimation that the controller keeps about each operator.

Even though the number of meta-parameters is fixed and small, given the initial lack of knowledge about the behavior of the controllers and what would be a good setting for them, off-line parameter tuning methods can be used to facilitate such definitions.

The cost of acquiring information about parameters is variable: whereas some parameters can be easily recognized as preponderant and easy to tune, others could have little effect over the search (i.e., different parameter values do not significantly affect the results), or could require many experiments to discover a correct value. For example, information theory could be used to characterize the relevance of the parameters, as in [34, 35]. The proposed method, called REVAC, uses Shannon and differential entropies to find the parameters with higher impact on the efficiency of the algorithm while estimating the utility of the possible parameters values.

Another possibility is the utilization of Racing methods: instead of performing m runs on each of the n instances in order to find the best configuration (i.e., a full factorial Design of Experiments), the candidate configurations are eliminated as soon as there is enough statistical evidence that they are worse than the current best one. Cycles of "execution/comparison/elimination" are repeated until there is only one configuration left, or some other stopping criterion is achieved. So, instead of wasting computing time to estimate with precision the performance of inferior candidates, Racing allocates the computational resources in a better way, by focusing on the most promising ones and consequently achieving lower variance estimates for them. Racing typically requires 10-20% of the computing time of a complete factorial Design of Experiments.

7.3 Adaptive Operator Selection

In this section, we review three AOS methods from our previous work. *Compass* [31] (Section 7.3.1) presents a *Credit Assignment* mechanism able to efficiently measure the impact of the operators application, while *Ex-DMAB* [15] (Section 7.3.2) proposes an *Operator Selection* that quickly adapts to the dynamics of the search. These two elements were combined into *ExCoDyMAB* [28], which is described and empirically compared to the original methods, respectively, in Sections 7.3.3 and 7.3.4.

7.3.1 Compass: Focusing on Credit Assignment

The *Adaptive Operator Selection* technique proposed in [31] uses a very simplistic *Operator Selection* mechanism, *Probability Matching*. However, its *Credit Assignment* mechanism is very efficient in measuring the impact of the operator's application, thus being the main contribution of this work. As is known, in case of multimodal functions, fitness improvement is not the only important criterion for the progress of the search; some level of diversity should also be maintained in the population; otherwise the search will quite probably converge prematurely and get trapped into a local optimum. Based on these assumptions, *Compass* provides a way to encourage improvements for both criteria, and works as follows.

Firstly, each time an operator is applied, three measures are collected: the variations in the population diversity and in the mean fitness of the population, respectively ΔD and ΔQ, and also its execution time T, as shown in Fig. 7.2(a). The average performance of each operator w.r.t. ΔD and ΔQ over the last τ applications is displayed in a "diversity vs. fitness" plot, represented by the points in Fig. 7.2(b). A user-defined angle θ defines the compromise between obtaining good results and maintaining diversity in the population, addressing the EvE dilemma. In practice, such an angle defines the plane from which perpendicular distances to the points are measured. Finally, these measures (δ_i) are divided by the operator's execution time to obtain the final aggregated evaluation of each one (Fig. 7.2(c)), which is the reward used to update the selection preferences.

This approach was used to control a steady state evolutionary algorithm applied to the well-known Boolean satisfiability problem (SAT) [9], automatically selecting from six ill-known operators. The SAT problem was chosen because it offers a large variety of instances with different properties and search landscapes, besides allowing the scaling of the instance difficulty. The experiments have demonstrated that this *AOS* method is efficient and provides good results when compared to other

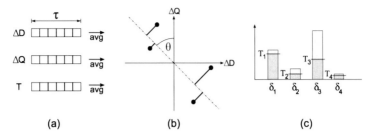

Fig. 7.2: *Compass* credit assignment: Sliding windows of three measures are maintained (a). Average measures of ΔD and ΔQ are plotted and distance of those points are measured from a plane with a slope of θ (b). Finally, those distances are divided by the execution time, resulting in the reward assigned to the operator (c)

existing mechanisms, such as Adaptive Pursuit [41] and APGAIN [45], based on the fitness improvement.

Such performance was achieved mainly due to the strength of the *Credit Assignment* mechanism proposed, which provides a robust measurement of the impact of the operator application by simultaneously considering several criteria. The *Operator Selection* rule used (PM) is rather simple, known being quite conservative and slow w.r.t. the adaptation, as already discussed in Section 7.2.2.

7.3.2 Ex-DMAB: Focusing on Operator Selection

Adaptive Operator Selection proposed in [15], uses extreme fitness improvement as *Credit Assignment*, based on the assumption that attention should be paid to extreme, rather than average, events [43]. This is implemented simply by assigning as a reward the maximum of the fitness improvements achieved by the application of a given operator over the last κ applications.

Concerning the *Operator Selection* mechanism, the idea is that such a task might be seen as another level of the Exploration vs. Exploitation (EvE) dilemma: there is the need to apply as much as possible the operator known to have brought the best results so far (exploitation), while nevertheless exploring the other options, as one of them might become the best at a further stage of the search. The EvE dilemma has been intensively studied in the context of *Game Theory*, and more specifically within the so-called Multi-Armed Bandit (MAB) framework. Among the existent MAB variants, the *Upper Confidence Bound* (UCB) [1] was chosen to be used as it provides asymptotic optimality guarantees, although it is described as "Optimism in front of the Unknown".

More formally, the UCB algorithm works as follows. Each variation operator is viewed as an *arm* of an MAB problem, and is associated with (i) its empirical reward $\hat{p}_j$, i.e., the average performance obtained by it so far; and (ii) a confidence interval, based on the number of times n_j such an operator has been tried. At each time step t, the algorithm selects the arm with the best upper bound w.r.t. the following quantity:

$$\hat{p}_{j,t} + C \sqrt{\frac{\log \sum_k n_{k,t}}{n_{j,t}}} \tag{7.1}$$

The first term of this equation favors the best empirical arm (exploitation) while the second term ensures that each arm is selected infinitely often (exploration). In the original setting [1], all rewards are Boolean, and hence their empirical means $\hat{p}_{i,t}$ are in $[0,1]$. As this is not the case in the context of AOS, a *Scaling factor* C is needed in order to properly correct the balance between the two terms.

The most important issue, however, is that the original MAB setting is static, while the AOS scenario is dynamic, i.e., the quality of the operators is likely to vary during the different stages of the search. Even though the exploration term in the UCB algorithm ensures that all operators will be tried infinitely many times, in case a change occurs (w.r.t. the operator's performance), it might take a long time before the new best operator catches up. To be able to efficiently follow the changes in dynamic environments, it has been proposed [21] to combine a statistical test with the UCB algorithm that efficiently detects changes in time series, the *Page-Hinkley* (PH) test [37]. Basically, as soon as a change in the reward distribution is detected (based on a threshold γ), e.g., the "best" operator is probably not the best anymore, the UCB algorithm is restarted from scratch, thus being able to quickly rediscover the new best operator.

The UCB algorithm involves one meta-parameter, the scaling factor C, while the PH test involves two parameters, the threshold γ for the change detection, and δ, which enforces the robustness of the test when dealing with slowly varying environments. Note that, according to initial experiments in [10], δ has been kept fixed at 0.15. The UCB algorithm, coupled with the Scaling factor and the PH change-detection test, has been termed as *Dynamic MAB (DMAB)*, with the complete *AOS* combination being denoted by *Ex-DMAB*.

Ex-DMAB has been used to adapt a $(1+\lambda)$-EA by efficiently choosing on-line from four mutation operators to solve the OneMax problem [10], and has been tried on yet another unimodal benchmark problem, the Long k-Path [16], this time effectively selecting from five mutation operators. In both cases, the optimal operator selection strategy was extracted by means of Monte Carlo simulations, and *Ex-DMAB* was shown to perform statistically equivalently to it while significantly improving over the naive (uniform selection) approach. It has also been used to adapt a $(100,100)$-EA with four crossover operators and one mutation operators in the Royal Road problem [17], also performing significantly better than the naive approach. For the three problems, we have also used other *AOS* combinations as baselines for comparison, namely Adaptive Pursuit, Probability Matching and static MAB (UCB without restarts), coupled with Extreme or Average rewards. *Ex-DMAB* was shown to be the best option.

However, as its *Credit Assignment* considers only the fitness improvement, *Ex-DMAB* would probably be less efficient on rougher landscapes, quickly converging to local optima. In order to solve multimodal problems, diversity should also be considered somehow. Besides being problem-dependent, another drawback of *Ex-DMAB* is that the variance of the fitness improvements changes as the search advances (in the beginning of the search larger improvements are more easily achieved, while in the fine-tuning final stage only very small improvements are possible). Thus, there do not exist any robust values for the meta-parameters C and γ that make it likely to perform well during the whole search. A possible solution to this is the utilization of comparison-based rewards (that, by definition, always keep the same variance, no matter what the problem and the search stage), e.g., a credit based on ranks or the use of some normalized measures.

7.3.3 *ExCoDyMAB = Compass + Ex-DMAB : An Efficient AOS Combination*

The previous sections showed that the strengths and weaknesses of both *Compass* and *Ex-DMAB* methods are complementary: *Compass* measures in a holistic way the effects of the operator applications over the population, but the operator selection rule is rather rudimentary, while *DMAB* has an effective way to adapt and select the operators, but its credit assignment mechanism is probably too simplistic. It seems hence natural to combine the *Credit Assignment* of the former with the *Operator Selection* mechanism of the latter.

However, even though merging both modules seems to be straightforward, some important issues need to be further explored:

- *Compass* uses sliding windows of size τ in the "measurement stage", with a unique reward value in its output, while *Ex-DMAB* stores in a sliding window the last κ outputs (rewards) of its *Credit Assignment* module. Should we keep both windows, or would it be redundant? And if only one is kept, which one should it be? From here on, these two windows will be referred to as $W1$ for the measurement window and $W2$ for the reward window.

- Another issue concerning the sliding windows is that of their usage: should the algorithm use their average (A), extreme (E), or simply instantaneous (I) value (equivalent to using no window at all)? The *Extreme* was shown to be stronger in unimodal problems, but how do such results hold in this completely different scenario?

- The last issue concerns the other meta-parameters. Besides tuning the size and type of $W1$ and $W2$, we also need to tune the values of the angle θ in *Compass*, and the scaling factor C and change detection threshold γ in *DMAB*. Since the idea is not to simply replace some parameters (the operator application probabilities) by other ones at a higher level of abstraction, we need to better understand their effects. One way to find answers to this

Table 7.1: Racing survivors

Name	W1 type, size	W2 type, size	C	γ
A	Extreme, 10	Instantaneous	7	1
B	Extreme, 10	Average, 10	7	1
C	Extreme, 10	Average, 50	7	1
D	Extreme, 10	Extreme, 10	7	3

questions is to experimentally study their influence on the performance of the *AOS* in the situation, and propose some robust default values whenever possible.

To analyze such issues, an empirical study was done, considering the following values for each of the meta-parameters: $C \in \{5, 7, 10\}$; $\gamma \in \{1, 3, 5\}$; and the window type(size) combinations $\in \{A(10), A(50), E(10), E(50), I(1)\}$ for both $W1$ and $W2$, summing up to a total of 225 possible configurations. The angle θ for *Compass* was fixed at 0.25, based on preliminary experiments (see [28] for further details). *ExCoDyMAB* was experimented within the same Evolutionary Algorithm used in [31], with the objective of selecting from six ill-known variation operators while solving different instances of the well-known combinatorial SAT problem [9].

Given the high number of possible configurations and the initial lack of knowledge about which values should be good for each of the mentioned meta-parameters, an off-line tuning technique was used for their setting, the F-Race [6], a Racing method that uses the Friedman's two-way Analysis of Variance by Ranks as a statistical test to eliminate the candidates. The stopping criteria for the Racing was set to 80 runs over all the instances (a "training set" of 7 instances was used), with eliminations taking place after each run, starting from the 11th.

At the end of the Racing process, four configurations were still "alive", presented in Table 7.1. This clearly indicates that the most important sliding window is $W1$, and it should be used in its Extreme configuration with a size of 10 (i.e., taking as *Compass* inputs the maximal of the last ten values), no matter which kind/size of $W2$ is being used. This fact emphasizes the need to identify rare but good improvements, greatly supporting the idea raised in [15]. Besides, the size of 10 for $W1$ could be interpreted as follows: with the Extreme policy, a larger τ would produce high durability of the extreme values, even when the behavior of the operator has changed. On the other hand, a smaller value $\tau = 1$ (i.e., the same as choosing Instantaneous policy) would forget those "rare but good" cases. One could suppose that an optimal size for $W1$ depends on the fitness landscape and the operators used. Further research is needed to better understand the setting of τ. The configuration "C" was found to be the best among them, and was thus used in the empirical comparison with other techniques, presented in the following.

Table 7.2: Comparative results on the 22 SAT instances: average (std dev) number of false clauses (over 50 runs)

Method Problem		Compass	Ex-DMAB	Uniform Choice
4blocks	2.8 (0.9)	6 (0.9)	6.2 (0.9)	13.4 (0.6)
aim	1 (0)	1 (0)	1.2 (0.3)	3.6 (1.8)
f1000	10.3 (2.3)	30.9 (6.2)	16.4 (2.6)	55.8 (8.6)
CBS	0.6 (0.6)	0.4 (0.5)	1 (0.9)	7 (2.7)
Flat200	7.2 (1.7)	10.6 (2.1)	10.7 (2.2)	37.7 (5.5)
logistics	6.5 (1.3)	7.6 (0.5)	8.8 (1.5)	17.9 (4.1)
medium	1.5 (1.5)	0 (0)	1.8 (1.6)	8.8 (3.4)
Par16	15.2 (3.1)	64 (10.2)	24.1 (5.7)	131.1 (14.5)
sw100-p0	9.2 (1.2)	12.8 (1.4)	12.5 (1.7)	25.9 (3.4)
sw100-p1	0 (0)	0.5 (0.6)	1.1 (0.8)	11.3 (3.5)
Uf250	0.9 (0.7)	1.8 (0.9)	1.7 (0.8)	9.1 (3.3)
Uuf250	2.5 (1)	4.5 (1.2)	3.1 (1.1)	12.7 (3.2)
Color	48 (2.5)	61.3 (2.2)	49.3 (3.4)	80.4 (6.6)
G125	8.8 (1.3)	20.6 (2)	13.5 (1.7)	28.8 (4.6)
Goldb-heqc	72.9 (8.5)	112.2 (15.2)	133.2 (15.9)	609.7 (96.2)
Grieu-vmpc	16.7 (1.7)	15.2 (1.7)	19.6 (1.8)	24.1 (3.3)
Hoons-vbmc	69.7 (14.5)	268.1 (44.6)	248.3 (24.1)	784.5 (91.9)
Manol-pipe	163 (18.9)	389.6 (37.2)	321 (38.1)	1482.4 (181.5)
Schup	306.6 (26.9)	807.9 (81.8)	623.7 (48.5)	1639.5 (169.9)
Simon	29.6 (3.3)	43.5 (2.7)	35.3 (6.3)	72.6 (11.3)
Velev-eng	18.3 (5.2)	29.5 (7.3)	118 (37.1)	394 (75.8)
Velev-sss	2 (0.6)	4.6 (1)	5.9 (3.9)	62.7 (25.2)
Comparison	-	18 - 2	21 - 0	22 - 0

7.3.4 Experimental Results

The performance of *ExCoDyMAB* (C) was compared with the baseline techniques, the original *Compass* and *Ex-DMAB*, and also with the *Uniform* (naive) selection, with 50 runs being performed on 22 SAT instances, obtained from [22] and from the SAT Race 2006. A preliminary off-line tuning of the meta-parameters by means of F-Race [6] was also done for the other techniques, in order to compare them at their best (more details in [28]).

The results are presented in Table 7.2. The columns show the mean number of false clauses after 5,000 function evaluations, averaged over 50 runs, and the standard deviation in parentheses, with the best results for each problem presented in boldface. The last line of the table summarizes the comparison, by showing the number of "wins - losses" of *ExCoDyMAB* compared to each of the baseline techniques, according to a Student t-test with 95% confidence.

Note that the purpose of this work was not to build an amazing SAT solver, but rather to experiment with a different *AOS* and validate *ExCoDyMAB* within an EA solving a general, difficult combinatorial problem. The results of Table 7.2 show that a basic EA using rather naive operators can indeed solve some instances. The main interesting result is that this set of benchmarks was difficult enough to highlight the benefit of using the proposed combination of *Compass* and *Ex-DMAB* rather than either separately – or a naive, blind choice. The deliberate choice of several non-specialized operators was also important for validating the control ability of *ExCoDyMAB* when facing variation operators of very different characteristics and performance. Competing for SAT races implies using highly specialized operators, such as the ones implemented in GASAT [25], and is left for further consideration.

This study highlights the importance of both control stages of *AOS*, namely credit assignment and operator selection rules. Both features, *Compass Credit Assignment* combined with *DMAB Operator Selection*, contribute to the overall performance of the proposed autonomous control method, explaining the efficiency gain over each previous method used alone. Additionally, using the Extreme values from the aggregated performance measure allows the algorithm to identify occasional but highly beneficial operators, while the inclusion of population diversity with the traditional fitness improvement measure contributes to escaping from local optima.

One main drawback of *ExCoDyMAB* is the tuning of its meta-parameters. Though the normalization of fitness improvements and diversity by *Compass* might result in a possible robust setting for the scaling parameter of the MAB balance (i.e., the value found here using Racing), further work is needed for a deeper understanding of how to tune the meta-parameter γ that triggers the change detection test. As previously mentioned, the off-line meta-parameter tuning using the F-Race paradigm can be done in a fraction (15%) of the time needed for a complete factorial DOE, but this is still quite costly.

Although its good performances rely expensive procedures, *ExCoDyMAB* was found to outperform the main options available to the naive EA user, namely (i) using a fixed or deterministic strategy (including the naive, uniform selection); and (ii) using a different *AOS* strategy, including the combinations previously proposed by the authors. Furthermore, *ExCoDyMAB* involves a fixed and limited number of parameters, whereas the number of operator rates increases with the number of operators, as discussed in Section 7.2.3.

7.4 Adaptive Operator Management

The design of the algorithm has great influence on the search performance. It can be seen as a parameter setting at a higher level of abstraction: one must decide, from several choices, the best configuration for the algorithm in order to maximize its performance. However, in this case "structural" parameters are considered instead of "behavioral" ones. In the case of operator control, instead of deciding on different values for the application rate of a given operator, it decides whether a given operator should be considered or not by the algorithm. It can be seen as a parameter optimization problem over a search space composed of algorithms.

This is usually done manually: the user decides, based mainly on his experience, which operators should be included in the search algorithm; or he tries a few configurations, assessing their performance in order to set down the definitive algorithm. As well as for the setting of the behavioral parameters, this design could also be assisted by parameter setting techniques, such as Racing or REVAC (see Section 7.2).

While the utilization of off-line tuning techniques is straightforward, providing the user a static "good" configuration, one might wonder how to do this in a dynamic way, while solving the problem. The motivations to do so would be the same as those that led us to propose the *AOS* methods: the efficiency of the operators changes according to the region of the search space that is currently being explored; thus a useless operator might be excluded from the algorithmic framework, while possibly useful ones might be included. The autonomous handling of the operators included in the algorithm while solving the problem is what we refer to as *Adaptive Operator Management* (AOM). Besides being yet another approach of doing parameter control, such an approach can also be described in the light of the recent field of hyper-heuristics [7].

7.4.1 Basic Concepts for Operator Management

There exist several criteria to determine whether an operator must be included or not in the search. The most straightforward one is the operator's recent success: if an operator has shown a good performance during the few last generations, it seems reasonable to keep it and exclude another with a lower performance. Another criterion could be based on a higher-level strategy of search, which encourages different levels of exploration and exploitation, as needed by the search. The latter implies the definition of a set of rules capable of guiding the search to efficiently balance intensification and diversification.

Regardless the criteria chosen to handle the operators, in general we can distinguish between the three following possible states for each operator, shown in Figure 7.3:

- **Unborn**, when the operators have never been used during the execution of the algorithm, or no information about their performance is available from previous runs.
- **Alive** refers to the operators that are currently being used by the algorithm. In this state, a trace of their performance is maintained.
- **Dead**, when the operators have been excluded from the search. The difference between this state and the *unborn* one is that here the controller has had the opportunity to assess the behavior of the operator; thus it could be eventually re-included later, with a clue about the effect it could produce.

It is arguable whether the observed performance of a given dead operator can be extrapolated to a later state of the search. Indeed, a dismissed "bad" operator could become useful in a different stage of the search, i.e., when the characteristics of the fitness landscape being currently explored by the population have changed and/or

Fig. 7.3: Operator states

the placement of the population over the search space has changed. This could mean that the information stored in the profiles of the dead operators is no longer useful; thus there are only two states: *dead* and *alive*. Nevertheless, we prefer to consider the tree-state model, which offers wider possibilities in terms of previous knowledge inclusion.

Such states are linked by the following three transitions:
- **Birth**, when an operator is included for the first time into the search.
- **Death**, when an operator has been considered useless for the current state of the search, and thus removed from it.
- **Revival**, when a previously dismissed operator is brought back to the search.

The core of *Adaptive Operator Management* lies in defining the conditions under which the transitions must take place. These conditions correspond to the different criteria described above, which define whether and when an operator should be included and eliminated from the search.

7.4.2 Blacksmith: A Generic Framework for AOM

The framework presented here, *Blacksmith*, aims to control the design by modifying the set of operators available to the algorithm during its execution. The structure of the entire controller is shown in Figure 7.4. In addition to comprising the *Adaptive Operator Selection* component, *Adaptive Operator Management* is composed of a set of operator definitions from which the operators are extracted. This set could be simply a list of possible operators that can be included in the search, or an entire module that mixes subcomponents to create operators in an intelligent manner.

Blacksmith manages operator states by defining a couple of data structures that are attached to the operator definition. The rules that define the transition from one state to another can be summarized as follows:

1. Initially, when an operator is *Unborn* state, Blacksmith only knows its name.
2. Once operators are born (i.e., they pass from *unborn* to *alive* state) two data structures are associated with the operator name: *data*, which stores recent measures of performance, and *profile*, which summarizes the information in *data*, by calculating meaningful statistics. A fixed number of operators is kept in the registry, and they are evaluated at regular intervals.
3. Operators that have been applied a sufficient number of times are considered for elimination. This condition is required to ensure that the operator has low

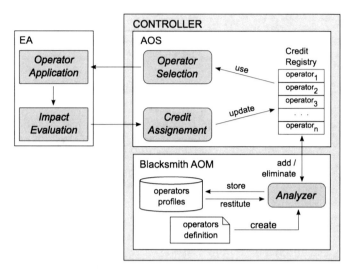

Fig. 7.4: Scheme of the controller used in this work, depicting its two main components, *Adaptive Operator Selection* and *Adaptive Operator Management*

performance. The weakest of those "known enough" operators is deleted (pass-
ing to the *dead* state) and a new one is inserted in the registry. In order to give
all operators a chance to show their skills, all *unborn* operators are tried before
dead ones are revived.

4. When there is no unborn operator left, replacement of unsuccessful operators
   is performed by including dead ones. Such reinsertion of the operator is not
   performed in a blind way, since the *profile* of the dead operators is maintained
   (*data*, in contrast, is dismissed).

Note that this implementation is generic enough to manage any operator: all that
the AOM knows about the operators is their name and a summary of their performance.

In the implementation presented here, ten operators are kept concurrently in the
algorithm, and they are evaluated every 50 generations. An operator's performance
is considered to be well known after five applications, and is thus eligible for deletion. An operator is deleted if its performance lies in the lower third of the alive
operators. The performance is assessed by a *Credit Assignment* mechanism (indeed,
the same information is shared between the AOS and AOM), presented in the following section.

This approach was applied to control an evolutionary algorithm based on GASAT
[25], which solves the SAT problem [19, 5]. The controller deals with crossover
operators that are created by combining four different criteria, summing up to 307
possible crossover operators. The criteria are related to how the clauses are selected
and what action is performed on them, as follows (further details can be found in
[29]):

1. **Selection of clauses that are false in both parents**: select none, in chronological order, randomly, randomly from the smallest and randomly from the biggest.
2. **Action to perform on the selected clauses**: either take no action or consider one of four ways to flip variables with different criteria.
3. **Selection of clauses that are true in both parents**: same as in (1).
4. **Action to perform on the selected clauses**: same as in (2).

7.4.3 Credit Assignment

In order to improve the Credit Assignment, beyond Compass (described in Section 7.3.1), we compare two other schemes of evaluation, both based on the concept of Pareto dominance [38]. In n-dimensional space, we say that a point $a = (a_1, a_2, \ldots, a_n)$ dominates another point $b = (b_1, b_2, \ldots, b_n)$ if a_i *is better than* $b_i, \forall i = 1 \ldots n$. Here the word *"better"* is used in the context of optimization: if we consider a maximization problem in dimension i, then a dominates b if $a_i > b_i$; on the other hand, if the objective is to minimize a given function, then a dominates b if $a_i < b_i$. When neither of the two points dominates the other, they are said to be incomparable. In our case, we have a two-dimensional space $(\Delta Diversity, \Delta Quality)$ with two criteria that we want to maximize.

The first scheme is called *Pareto Dominance (PD)*, and it counts the number of operators dominated by other operators (see Figure 7.5(b)). The purpose here is to obtain a high value. The second evaluation method, called *Pareto Rank (PR)*, computes the number of operators that dominate a given operator (Figure 7.5(c)). Here the objective is to obtain low values. Operators with a PR value of 0 belong to the Pareto frontier. There exists an important difference between these two evaluations: whereas PR will prefer only operators which are not dominated, PD also rewards operators which are in strong competition with the others.

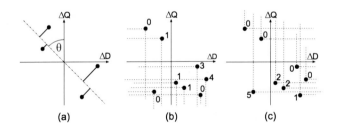

(a) (b) (c)

Fig. 7.5: Credit Assignment schemes. Compass (a), Pareto Dominance (b), Pareto Rank (c)

After the application of an operator, the values of $\Delta Diversity$ and $\Delta Quality$ are sent to the controller. The credit assignment module computes then the evaluation

(C, PD or PR, depending on the scheme selected), and normalizes the resulting values across all operators (normalization is achieved by dividing values by the highest possible value). The normalized values are stored in the Credit Registry as the assigned rewards. A list of the last m rewards of each operator (corresponding to its last m applications) is recorded in the registry in order to provide to the operator selection module an updated history of the performances of each operator.

7.4.4 Operator Selection

The operator selection module selects the next operator to be applied based on its past performances, but without neglecting an exploration of the possible available operators. As mentioned in Section 7.3.2, *Ex-DMAB* is inspired by the multi-armed bandit approach, used in game theory. The strategy always chooses the operator that maximizes expression (7.1).

However, this expression relies on the assumption that all operators are present in the evolutionary algorithm from the beginning of the run. If an operator is inserted during execution, its value of $n_{i,t}$ would be so low that the bandit-based AOS would have to apply it many times in order to adjust its exploration term with regards to the rest of the operators.

Since we are interested in operators that enter and exit the evolutionary algorithm during the search process, we have reformulated expression (7.1) in order to deal with a dynamic set of operators. This is mainly achieved by replacing the measure corresponding to the number of times that an operator has been applied with another criterion that corresponds to the number of generations elapsed since the last application of the operator (i.e., its idle time). This allows a new operator to immediately increase its evaluation by applying it once. The new evaluation of performance is then defined as follows:

$$MAB2_{o,t} = r_{o,t} + 2 \times exp(p \times i_{o,t} - p \times x \times NO_t) \qquad (7.2)$$

where $i_{o,t}$ is the idle time of operator o at time t, NO_t is the number of operators considered by the *Operator Selection* at time t, and x is the number of times the controller must wait before compulsorily applying the operator o. The behavior of the exploration component is better understood by looking at Figure 7.6. The value stays close to zero except when $i_{o,t}$ is close to $x \times NO_t$. Since the values of $r_{o,t}$ are normalized in $[0, 1]$, when an operator has not been applied for a long time, its application becomes mandatory. p is a positive parameter that adjusts the slope of the exponential.

Besides *MAB2*, referred to as **M2** from here on, the following three different *Operator Selection* methods were also analyzed:

- **Random (R)**, simply chooses randomly which among the operators currently available in the EA.

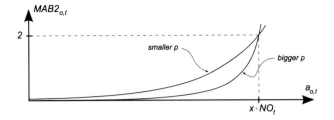

Fig. 7.6: behavior of exploratory component of expression (7.2)

- **Probability Matching (PM)**, which chooses the operator with a probability proportional to the reward values stored by the credit assignment module.
- **MAB2 + Stuck detection (M2D)**, which adds to M2 a method in order to detect the population is trapped in a local optimum. The detection is performed thanks to the linear regression of the values of the mean quality of the population during the last generations. If the value of the slope is close to zero and the difference between the maximum and minimum values of mean quality is small enough (i.e., almost a flat line), a diversification stage is performed, using only operators that have an exploration profile. This diversification stage is maintained until (i) the diversity reaches a range over the original value, or (ii) there are no exploration operators, or (iii) a predefined number of generations has passed without the desired diversity being reached.

7.4.5 Experimental Results

The combination of the three mentioned *Credit Assignment* mechanisms, namely Compass (C), Pareto Dominance (PD) and Pareto Rank (PR), with the four considered *Operator Selection* techniques (R, PM, M2, M2D), resulted in a set of 12 *Adaptive Operator Selection* combinations to be coupled with the proposed *Blacksmith Adaptive Operator Management*. These controllers were tested in the resolution of the satisfiability problem SAT, with the AOM continuously managing the operational set of ten operators by including and excluding them from a superset of 307 operators (described in Section 7.4.2); and the AOS autonomously selecting from this subset of ten available operators defined by the AOM which of them should be applied at each time instant, with both using the same assessment information provided by the *Credit Assignment* module.

The results were compared with state-of- the art crossovers FF, CC and CCTM [25, 18]. Besides, the Uniform crossover and a controller that simply applies one of the 307 possible combinations randomly (called *R307*) were used as baseline methods. Comparisons have been made on instances from different families and types of benchmarks, including nine crafted, ten random-generated and seven industrial instances. One algorithm execution refers to 100,000 applications of crossover operators.

These combinations are identified by the notation $X - Y$, where $X \in \{C, PD, PR\}$ denotes the Credit assignment mechanism used, and $Y \in \{M2, R, M2D, PM\}$ is the mechanism for operator selection. The parameters of the controller are those of Blacksmith and those of M2 and M2D. The registry has a fixed size of 20 operators. Every 50 generations[1], the Analyzer is invoked in order to find a weak operator to replace it with a fresh one. If an operator has been sufficiently applied (half of the size of the registry, i.e., ten times) and if its reward is in the lower third of the operators, it is selected to be deleted. The parameters of M2 are $p = 0.2$ and $x = 1.5$. M2D uses the data of the last 100 generations to compute the linear regression. The diversification stage is triggered when the value of the slope is within ± 0.0001 and the difference between maximal and minimal values is less than 0.001.

Figure 7.7 shows the convergence of the EA algorithm coupled with the different controllers, as well as those of state-of-the-art and baseline crossovers, solving an industrial instance. It can be seen that the worst performance is obtained by the Uniform Crossover, and most of the 12 controllers do not overcome the state-of-the-art crossovers CC and CCTM. However, two of them (PD-R and PD-PM) are able to converge quite quickly to better solutions.

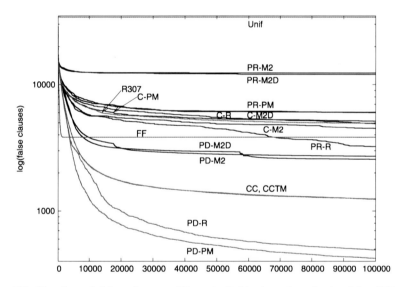

Fig. 7.7: Number of false clauses of best individual so far, obtained by different controllers and state-of-the-art crossovers when solving an industrial instance

---

[1] According to the usual taxonomy, this algorithm is a steady-state evolutionary algorithm; thus a generation corresponds to the application of one operator.

There is a clear advantage with *Credit Assignment* based on the Pareto Dominance (PD), since it appears in the most efficient controllers, followed by C and PR. One possible explanation is that PR considers equally all the operators placed on the Pareto frontier (points in the Figure 7.5(c) with value 0). This induces a balance between the exploration and the exploitation tendencies, thus preventing the evolutionary algorithm from leaning to one side or to the other. A similar behavior could be observed when using Compass according to its performance measure method. On the other hand, when using PD, the better evaluation of the operators that follow the general tendency (points in the Figure 7.5(b) with higher values) allows the evolutionary algorithm to break the status quo and finally improve the quality of the population. This "flexible balance" is the main asset of this *Credit Assignment* method.

It is interesting to notice that the most exploratory *Operator Selection* methods (PM and R) have produced some of the best results. It could seem surprising that a random operator selection could be able to overcome sophisticated methods that carefully try to balance EvE at the operator selection level, such as the bandit-based approaches. A possible hypothesis for this could be that a mix of different crossover operators works better than a single one due the diversity of behavior the former can manifest; however, the poor results obtained by R307 show that simply mixing different operators does not produce good results.

Another, more plausible, hypothesis is that an exploratory *Operator Selection* method will provide updated information to AOM, since the rate of application of "bad" operators is higher than when an exploitative *Operator Selection* method is used. In this manner, R and PM rapidly produce the five applications needed to evaluate the unsuccessful operators and move them from *alive* to *dead* state.

Another factor to consider is that *Adaptive Operator Management* uses the same criterion as *Adaptive Operator Selection* to choose which operator will survive. Since AOM continually dismisses the worst operators, an important part of exploitation is done by AOM; thus AOS can focus exclusively on exploration. Differently stated, AOS performs exploration among the operators that AOM has decided to keep by using an exploitative criteria.

Table 7.3 shows the mean (standard deviation) of the performance of the algorithms for each problem over 25 executions. We show here only the two most successful controllers (PD-PM and PD-R). The best results (and those that are indistinguishable from them, using a Student t-test with 95% confidence) are in boldface. Here we focus on the most successful controllers, PD-R and PD-PM.

PD-R obtained top results on 19 instances and PD-PM on 17, while CC and CCTM were the best ones only five times. Even though the controller configurations obtained the worst results on two industrial instances, we observed the best improvements on this family of instances, especially on *I5* and *I6*, in which the adaptively controlled evolutionary algorithms obtained up to 260 times fewer false clauses than the best state-of-the-art crossover. The results presented in this comparison show the generality of PD-PM and PD-R on different families and types of instances. The overhead of this improvement is really moderate since the time dedicated to control represents less that 10% of the total execution time.

Table 7.3: Comparison of PD-PM and PD-R with state-of-the-art crossovers over crafted, random and industrial instances. Number of false clauses and standard deviation

	PD-PM		PD-R		FF		CC		CCTM	
C1	**35.4**	(5.4)	**34.8**	(2.8)	503.2	(41.0)	44.7	(5.2)	42.1	(4.7)
C2	**35.8**	(2.6)	38.0	(4.2)	509.4	(31.6)	46.0	(4.4)	47.6	(4.9)
C3	35.4	(3.7)	**35.6**	(3.6)	490.0	(37.7)	48.4	(4.1)	47.1	(3.3)
C4	45.1	(3.8)	**43.4**	(4.6)	491.6	(36.5)	48.7	(3.0)	48.2	(3.4)
C5	10.5	(1.8)	**9.8**	(2.8)	47.9	(4.2)	11.6	(1.8)	10.2	(1.5)
C6	8.6	(1.9)	8.3	(1.7)	36.9	(3.3)	**8.4**	(1.6)	**8.7**	(1.4)
C7	8.8	(1.8)	8.0	(1.9)	38.7	(4.2)	**8.4**	(1.2)	**8.7**	(1.7)
C8	10.0	(2.4)	**9.7**	(2.5)	48.2	(4.1)	11.3	(1.4)	11.6	(1.6)
C9	150.9	(31.2)	**123.3**	(28.8)	973.2	(77.4)	214.7	(15.9)	217.0	(14.8)
R1	7.5	(1.5)	**7.2**	(1.1)	34.2	(5.4)	9.5	(1.9)	9.7	(1.8)
R2	6.4	(1.3)	**5.7**	(1.4)	30.6	(3.8)	7.3	(1.4)	7.7	(1.6)
R3	8.4	(1.4)	**8.2**	(1.5)	32.1	(3.8)	10.6	(1.6)	10.9	(1.9)
R4	4.2	(1.5)	**3.5**	(1.4)	26.3	(3.8)	7.4	(1.2)	7.4	(1.8)
R5	8.2	(2.1)	7.8	(1.8)	40.0	(6.0)	**8.4**	(1.5)	9.1	(1.4)
R6	**6.7**	(1.6)	7.9	(1.6)	44.2	(6.4)	8.7	(1.5)	8.8	(1.4)
R7	6.1	(1.7)	**5.8**	(2.1)	39.4	(5.5)	7.6	(1.6)	7.8	(1.4)
R8	9.0	(1.2)	**8.8**	(1.6)	49.2	(5.3)	10.3	(1.9)	9.9	(1.7)
R9	9.1	(1.6)	9.0	(1.7)	41.9	(5.7)	10.0	(1.7)	**9.0**	(1.5)
R10	**110.1**	(5.7)	115.1	(8.3)	654.0	(39.5)	153.0	(9.2)	150.0	(7.9)
I1	**123.6**	(11.4)	167.6	(32.3)	439.3	(27.3)	354.4	(11.4)	349.6	(11.7)
I2	**99.7**	(8.2)	134.7	(22.5)	469.1	(26.3)	372.0	(35.5)	367.8	(32.2)
I3	2.7	(2.9)	8.6	(7.7)	216.5	(18.9)	**1.0**	(0.0)	**1.0**	(0.0)
I4	2.8	(2.4)	6.3	(4.8)	116.2	(11.2)	**1.0**	(0.2)	**1.1**	(0.4)
I5	59.4	(98.5)	**38.0**	(1.4)	12567.6	(547.1)	10044.2	(384.4)	9928.1	(382.0)
I6	127.8	(317.1)	**35.1**	(1.5)	9736.2	(404.9)	7567.7	(238.0)	7521.2	(272.9)
I7	**44.2**	(1.3)	48.4	(2.2)	1877.6	(195.1)	61.8	(1.7)	61.6	(1.8)

7.5 Conclusions

Finding an appropriate parameter setting has a great effect on the performance of search algorithms. This parameterization can be fixed during the whole execution, or it can be continuously modified while the problem is being solved, adapting its features to the needs of each stage of the search, which is commonly known as parameter control.

Parameter control requires good knowledge of the mechanics of the search, and in many cases of the details of the problem. Actually, it is mostly achieved in a customized way, making it difficult to transfer the knowledge obtained through the control of one algorithm to another one. As a consequence, one of the main issues in parameter control lies in the creation of generic schemes that could be used transparently by the user of a search algorithm, without requiring prior knowledge about the problem scope and the details of the search algorithm.

In an intuitive way, we can distinguish betweentwo main types of parameters: those that control the *behavior* of the algorithm, such as the application rates of a given variation operator, and those that affect the core structure of the algorithm itself, called *structural*. In this chapter, we have discussed two approaches for dealing with both behavioral and structural parameters, namely *Adaptive Operator Selection* and *Adaptive Operator Management*.

ExCoDyMAB AOS, presented in Section 7.3.3, combines *Compass*, an efficient *Credit Assignment* approach proposed in [31] that associates two performance measures (quality and diversity) with a given operator, with *Ex-DMAB*, an engineered *Operator Selection* mechanism based on the multi-armed bandit approach, originally proposed in [15]. Such an AOS combination has been shown to be very efficient in selecting between ill-known operators within an EA applied to SAT problems. Although some meta-parameters still need to be tuned, off-line techniques such as the F-Race [6] can be used, which reduces the tuning time to around 15% of the total time that would be required by a complete factorial design of experiments.

To deal with *structural* parameters, we have defined a general framework in which operators can be in one of the following three states: unborn, alive or dead. The inclusion or exclusion of operators defines the transitions between these states. We have presented an *Adaptive Operator Management* called Blacksmith, which builds operators in a combinatorial way and attaches a profile to each one of them, in order to evaluate whether an operator must be kept, dismissed or reincorporated into the search. This AOM technique has been shown to have the ability to autonomously and continuously manage an efficient set of ten operators within an EA applied to SAT problems, extracted from a superset of 307 unknown operators. Even if one well-performing static configuration could be found, such a definition if done by hand would be very time-consuming.

Both AOS and AOM controllers are based on the same assessment of operators, referred to as *Credit Assignment*. In order to attain generality with regard to the implementation of the search algorithm, it is necessary to measure the performance of the controlled items in a generic way. This measure must make sense to a wide set of search algorithms and be compatible with the search itself. Two high-level criteria are used to evaluate the performance of operators: mean fitness and population diversity. These two measures are respectively aligned with exploitation and exploration, that constitute the two main goals of any search algorithm. Note that even though population diversity is closely related to population-based algorithms, an equivalent measure of exploration, such as a time-diversity diversity measure (e.g., one considering the amount of exploration that the point has performed in recent operations) could be used instead by single-point based algorithms.

Given this generality, the combination of both controllers, *Adaptive Operator Selection* and *Adaptive Operator Management*, could be used as a framework to manage and select variation operators not only for evolutionary algorithms, but for search algorithms in general.

A possible improvement to the approach presented in this chapter would be the inclusion of a high-level strategy to guide the behavior of both controllers. Notice that both AOM and AOS currently consider the same criteria to guide their choices, i.e., Pareto optimality (or a similar idea) of both mean fitness and population diversity. In the way it is currently being done, the same weight is given to both trends, the exploration/diversification and the exploitation/intensification criteria; as is known, the needs of the search w.r.t. such trends vary as the search goes on; thus keeping the same balance might lead to overall suboptimal behavior. Intuitively, one idea would be to consider some strategy that starts by encouraging the exploration behavior, and gradually shifts to the exploitation; or one that searches for a certain compromise level of EvE [30, 32].

Another path of further research concerns the identification of the best operators used during one or more runs. Even though our approach can choose appropriate operators given a search state, no information about this knowledge is delivered at the end. This automatically extracted knowledge could be used to guide the selection of the better operators for the next run, or to establish a map between operators and the search states for which they are best suited.

Some further analysis should be done for both *Adaptive Operator Selection* and *Adaptive Operator Management* concerning the relevance and sensitivity of their meta-parameters, which would help us to better understand their behavior.

One could also remark that, in this architecture, bad operators should be tried before assessing their poor performances. Another architecture could be envisaged to separate the evaluation of the operators and their effective use. For instance, in a distributed solving scheme, some controllers could be devoted to the evaluation of the performance according to given state of the search while other controllers could concentrate on the effective solving of the problem, the controllers exchanging information about the current state of the search.

References

[1] Auer, P., Cesa-Bianchi, N., Fischer, P.: Finite-time analysis of the multiarmed bandit problem. Machine Learning **47**(2/3), 235–256 (2002)
[2] Barbosa, H. J. C., Sá, A. M.: On adaptive operator probabilities in real coded genetic algorithms. In: Proc. XX Intl. Conf. of the Chilean Computer Science Society (2000)
[3] Bartz-Beielstein, T., Lasarczyk, C., Preuss, M.: Sequential parameter optimization. In: Proc. CEC'05, pp. 773–780. IEEE Press (2005)
[4] Battiti, R., Brunato, M., Mascia, F.: Reactive Search and Intelligent Optimization, *Operations Research/Computer Science Interfaces*, vol. 45. Springer (2008)
[5] Biere, A., Heule, M., van Maaren, H., Walsh, T. (eds.): Handbook of Satisfiability, *Frontiers in Artificial Intelligence and Applications*, vol. 185. IOS Press (2009)
[6] Birattari, M., Stützle, T., Paquete, L., Varrentrapp, K.: A racing algorithm for configuring metaheuristics. In: W. B. Langdon et al. (ed.) Proc. GECCO '02, pp. 11–18. Morgan Kaufmann (2002)

[7] Burke, E. K., Hyde, M., Kendall, G., Ochoa, G., Ozcan, E., Woodward, J.: Handbook of Meta-heuristics, chap. A Classification of Hyper-heuristics Approaches. Springer (2009)

[8] Clune, J., Goings, S., Punch, B., Goodman, E.: Investigations in Meta-GA: Panaceas or Pipe Dreams? In: H. G. Beyer et al. (ed.) Proc. GECCO'05, pp. 235–241. ACM Press (2005)

[9] Cook, S. A.: The complexity of theorem-proving procedures. In: Proc. ACM Symposium on Theory of Computing, pp. 151–158. ACM Press (1971)

[10] Da Costa, L., Fialho, A., Schoenauer, M., Sebag, M.: Adaptive operator selection with dynamic multi-armed bandits. In: M. Keijzer et al. (ed.) Proc. GECCO'08, pp. 913–920. ACM Press (2008)

[11] Da Costa, L., Schoenauer, M.: GUIDE, a Graphical User Interface for Evolutionary Algorithms Design. In: J. H. Moore et al. (ed.) GECCO Workshop on Open-Source Software for Applied Genetic and Evolutionary Computation (SoftGEC). ACM Press (2007)

[12] Davis, L.: Adapting operator probabilities in genetic algorithms. In: J. D. Schaffer (ed.) Proc. ICGA'89, pp. 61–69. Morgan Kaufmann (1989)

[13] Eiben, A., Michalewicz, Z., Schoenauer, M., Smith, J.: Parameter Setting in Evolutionary Algorithms, chap. Parameter Control in Evolutionary Algorithms, pp. 19–46. Vol. 54 of Lobo et al. [27] (2007)

[14] Eiben, A. E., Hinterding, R., Michalewicz, Z.: Parameter control in evolutionary algorithms. IEEE Trans. Evolutionary Computation 3(2), 124–141 (1999)

[15] Fialho, A., Da Costa, L., Schoenauer, M., Sebag, M.: Extreme value based adaptive operator selection. In: G. Rudolph et al. (ed.) Proc. PPSN'08, *LNCS*, vol. 5199, pp. 175–184. Springer (2008)

[16] Fialho, A., Da Costa, L., Schoenauer, M., Sebag, M.: Dynamic multi-armed bandits and extreme value-based rewards for adaptive operator selection in evolutionary algorithms. In: T. Stuetzle et al. (ed.) Proc. LION'09 (2009)

[17] Fialho, A., Schoenauer, M., Sebag, M.: Analysis of adaptive operator selection techniques on the royal road and long k-path problems. In: F. Rothlauf et al. (ed.) Proc. GECCO'09, pp. 779–786. ACM Press (2009)

[18] Fleurent, C., Ferland, J. A.: Object-oriented implementation of heuristic search methods for graph coloring, maximum clique, and satisfiability. In: Cliques, Coloring, and Satisfiability: Second DIMACS Implementation Challenge, *DIMACS Series in Discrete Mathematics and Theoretical Computer Science*, vol. 26, pp. 619–652 (1996)

[19] Garey, M. R., Johnson, D. S.: Computers and Intractability: A Guide to the Theory of NP-Completeness. W.H. Freeman & Company, San Francisco (1979)

[20] Goldberg, D.: Probability Matching, the Magnitude of Reinforcement, and Classifier System Bidding. Machine Learning 5(4), 407–426 (1990)

[21] Hartland, C., Gelly, S., Baskiotis, N., Teytaud, O., Sebag, M.: Multi-armed bandit, dynamic environments and meta-bandits. In: Online Trading of Exploration and Exploitation Workshop, NIPS (2006)

[22] Hoos, H., Stützle, T.: SATLIB: An Online Resource for Research on SAT, pp. 283–292. IOS Press, www.satlib.org (2000)

[23] Jong, K. D.: Parameter Setting in Evolutionary Algorithms, chap. Parameter Setting in EAs: a 30 Year Perspective, pp. 1–18. Vol. 54 of Lobo et al. [27] (2007)

[24] Julstrom, B. A.: What have you done for me lately? Adapting operator probabilities in a steady-state genetic algorithm on genetic algorithms. In: L. J. Eshelman (ed.) Proc. ICGA'95, pp. 81–87. Morgan Kaufmann (1995)

[25] Lardeux, F., Saubion, F., Hao, J. K.: GASAT: A genetic local search algorithm for the satisfiability problem. Evolutionary Computation **14**(2), 223–253 (2006)

[26] Lobo, F., Goldberg, D.: Decision making in a hybrid genetic algorithm. In: T. Bäck et al. (ed.) Proc. ICEC'97, pp. 121–125. IEEE Press (1997)

[27] Lobo, F., Lima, C., Michalewicz, Z. (eds.): Parameter Setting in Evolutionary Algorithms, *Studies in Computational Intelligence*, vol. 54. Springer (2007)

[28] Maturana, J., Fialho, A., Saubion, F., Schoenauer, M., Sebag, M.: Extreme compass and dynamic multi-armed bandits for adaptive operator selection. In: Proc. CEC'09 (2009)

[29] Maturana, J., Lardeux, F., Saubion, F.: Autonomous operator management for evolutionary algorithms. Journal of Heuristics **16**, 881–909 (2010)

[30] Maturana, J., Saubion, F.: On the design of adaptive control strategies for evolutionary algorithms. In: Proc. EA'07, *LNCS*, vol. 4926. Springer (2007)

[31] Maturana, J., Saubion, F.: A compass to guide genetic algorithms. In: G. Rudolph et al. (ed.) Proc. PPSN'08, *LNCS*, vol. 5199, pp. 256–265. Springer (2008)

[32] Maturana, J., Saubion, F.: From parameter control to search control: Parameter control abstraction in evolutionary algorithms. Constraint Programming Letters 4, Special Issue on Autonomous Search, 39–65 (2008)

[33] Moore, G. A.: Crossing the Chasm: Marketing and Selling High-Tech Products to Mainstream Customers. Collins Business Essentials (1991)

[34] Nannen, V., Eiben, A. E.: A method for parameter calibration and relevance estimation in evolutionary algorithms. In: Proc. GECCO'06, pp. 183–190. ACM Press (2006)

[35] Nannen, V., Eiben, A. E.: Relevance estimation and value calibration of evolutionary algorithm parameters. In: Proc. IJCAI'07, pp. 975–980 (2007)

[36] Nannen, V., Smit, S. K., Eiben, A. E.: Costs and benefits of tuning parameters of evolutionary algorithms. In: G. Rudolph et al. (ed.) Proc. PPSN'08, pp. 528–538. Springer (2008)

[37] Page, E.: Continuous inspection schemes. Biometrika **41**, 100–115 (1954)

[38] Pareto, V.: Cours d'économie politique. In: Vilfredo Pareto, Oeuvres complètes, Genève: Librairie Droz (1896)

[39] Rice, J. R.: The algorithm selection problem. Advances in Computers **15**, 65–118 (1976)

[40] Sywerda, G.: Uniform crossover in genetic algorithms. In: Proc. ICGA'89, pp. 2–9. Morgan Kaufmann Publishers Inc., San Francisco, CA, USA (1989)

[41] Thierens, D.: An adaptive pursuit strategy for allocating operator probabilities. In: H. G. Beyer (ed.) Proc. GECCO'05, pp. 1539–1546. ACM Press (2005)

[42] Tuson, A., Ross, P.: Adapting operator settings in genetic algorithms. Evolutionary Computation **6**(2), 161–184 (1998)

[43] Whitacre, J. M., Pham, T. Q., Sarker, R. A.: Use of statistical outlier detection method in adaptive evolutionary algorithms. In: M. Cattolico (ed.) Proc. GECCO'06, pp. 1345–1352. ACM Press (2006)

[44] Wolpert, D. H., Macready, W. G.: No free lunch theorems for optimization. IEEE Transactions on Evolutionary Computation **1**(1), 67–82 (1997)

[45] Wong, Y. Y., Lee, K. H., Leung, K. S., Ho, C. W.: A novel approach in parameter adaptation and diversity maintenance for genetic algorithms. Soft Computing **7**(8), 506–515 (2003)

[46] Yuan, B., Gallagher, M.: Statistical racing techniques for improved empirical evaluation of evolutionary algorithms. In: X. Yao et al. (ed.) Proc. PPSN'04, pp. 172–181. Springer (2004)

Chapter 8
Parameter Adaptation in Ant Colony Optimization

Thomas Stützle, Manuel López-Ibáñez, Paola Pellegrini, Michael Maur,
Marco Montes de Oca, Mauro Birattari, and Marco Dorigo

8.1 Introduction

Ant colony optimization (ACO) is a metaheuristic inspired by the foraging behavior of ants [13, 9, 14, 11, 15, 8]. In ACO algorithms, artificial ants are probabilistic solution construction procedures that are biased by artificial pheromones and heuristic information. Heuristic information can be derived from a problem instance to guide ants in the solution construction process. Pheromones are represented as numerical information that is modified iteratively to reflect the algorithm's search experience. Modifications bias the search towards good quality solutions [45].

The behavior of ACO algorithms depends strongly on the values given to parameters [11, 18]. In most ACO applications, parameter values are kept constant throughout each run of the algorithm. However, varying the parameters at computation time, either in prescheduled ways or depending on the search progress, can enhance the performance of an algorithm. Parameter control and parameter adaptation are recurring themes in the field of evolutionary algorithms (EAs) [31]. The adaptation of parameters while solving a problem instance by exploiting machine learning techniques is also the unifying theme in the research area of reactive search [3]. In the ACO literature as well, several strategies have been proposed and tested for modifying parameters while solving a problem instance.

Thomas Stützle, Manuel López-Ibáñez, Marco Montes de Oca, Mauro Birattari, Marco Dorigo
IRIDIA, CoDE, Université Libre de Bruxelles, Brussels, Belgium
e-mail: {stuetzle,manuel.lopez-ibanez,mmontes,mbiro,mdorigo}@ulb.ac.be

Michael Maur
Fachbereich Rechts- und Wirtschaftswissenschaften, TU Darmstadt, Darmstadt, Germany
e-mail: maur@stud.tu-darmstadt.de

Paola Pellegrini
Dipartimento di Matematica Applicata, Università Ca' Foscari Venezia, Venezia, Italia
e-mail: paolap@unive.it

Y. Hamadi et al. (eds.), *Autonomous Search*,
DOI 10.1007/978-3-642-21434-9_8,
© Springer-Verlag Berlin Heidelberg 2011

In this chapter, we first review available research results on parameter adaptation in ACO algorithms. We follow the three classes of parameter control strategies discussed by Eiben et al. [17]. Then, we analyze the development of the solution quality reached by ACO algorithms over computation time for specific parameter values. The goal is to identify situations where the best fixed parameter settings depend strongly on the available computation time because it is exactly in such situations that prescheduled variation of parameter values can improve strongly the anytime performance [44] of an algorithm. We observe strong dependencies for the MAX-MIN Ant System (MMAS) [42], and we show that prescheduled parameter variation actually leads to much improved behavior of MMAS over computation time without compromising the final solution quality obtained.

This chapter is structured as follows. In Section 8.2 we give an introductory description of the main ACO variants, which are also mentioned later in the chapter. After a short review of parameter adaptation in Section 8.3, we discuss relevant literature concerning ACO in Section 8.4. The experimental part in this chapter is divided into two sections: Section 8.5 describes experiments with fixed parameter settings, whereas Section 8.6 describes experiments with prescheduled parameter variation. From the review of the literature and our experimental study, we provide some conclusions and suggestions for future research in Section 8.7.

8.2 Ant Colony Optimization

The earliest ACO algorithms used the traveling salesman problem (TSP) as an example application. The TSP is typically represented by a graph $G = (V, E)$, V being the set of $n = |V|$ vertices, representing the cities, and E being the set of edges that fully connect the vertices. A distance d_{ij} is associated with each edge (i, j). The objective is to find a Hamiltonian cycle of minimum total cost. The TSP is a computationally hard problem, but the application of ACO algorithms to it is simple, which explains why ACO applications to the TSP have played a central role in the development of this algorithmic technique.

8.2.1 Ant Colony Optimization for the TSP

When applying ACO to the TSP, a pheromone value τ_{ij} is associated with each edge $(i, j) \in E$. The pheromone value represents the attractiveness of a specific edge for the ants, according to the experience gained at runtime: the higher the amount of pheromone on an edge, the higher the probability that ants choose it when constructing solutions. Pheromone values are iteratively updated by two mechanisms: pheromone evaporation and pheromone deposit. In addition to the pheromone trails, the ants' solution construction process is also biased by a heuristic value $\eta_{ij} = 1/d_{ij}$,

procedure ACOMetaheuristic
ScheduleActivities
ConstructSolutions
DaemonActions   //*optional*
UpdatePheromones
end-ScheduleActivities
end-procedure

Fig. 8.1: ACO metaheuristic for NP-hard problems in pseudo-code

which represents the attractiveness of each edge (i, j) from a greedy point of view.

The algorithmic outline of an ACO algorithm (see Figure 8.1) contains three main procedures: ConstructSolutions, DaemonActions, and UpdatePheromones. The main characteristics of these procedures are as follows.

- ConstructSolutions. This procedure includes the routines needed for ants to construct solutions incrementally. After the selection of the starting city, one node at a time is added to an ant's path. An ant's decision about where to go next is biased by the pheromone trails τ_{ij} and the heuristic information η_{ij}. In general, the higher the two values, the higher the probability of choosing the associated edge. Typically, two parameters, $\alpha > 0$ and $\beta \geq 0$, are used to weigh the relative influence of the pheromone and the heuristic values, respectively. The rule that defines the ant's choice is specific to each ACO variant.

- DaemonActions. This procedure comprises all problem-specific operations that may be considered for boosting the performance of ACO algorithms. The main example of such operations is the introduction of a local search phase. In addition, daemon actions implement centralized tasks that cannot be performed by an individual ant. This type of procedures is for optional use, but typically several daemon actions are very useful for significantly improving the performance of ACO algorithms [11].

- UpdatePheromones. This procedure updates the pheromone trail values in two phases. First, pheromone evaporation is applied to decrease pheromone values. The degree of decrement depends on a parameter $\rho \in [0, 1]$, called *evaporation rate*. The aim of pheromone evaporation is to avoid an unlimited increase of pheromone values and to allow the ant colony to forget poor choices made previously. A pheromone deposit is then applied to increase the pheromone values that belong to good solutions the ants have generated. The amount of pheromone deposited and the solutions considered are peculiar to each ACO variant.

ACO algorithms involve a number of parameters that need to be set appropriately. Of these, we already have mentioned α and β, which are used to weigh the relative influence of the pheromone and heuristic values in the ants' solution construction. The role of these parameters in biasing the ants' search is intuitively similar. Higher values of α emphasize differences in the pheromone values, and higher values of β have the same effect on the heuristic values. The initial value of the pheromones, τ_0, has a significant influence on the convergence speed of the algorithm; however, its recommended setting depends on the particular ACO algorithm. The evaporation rate parameter, ρ, $0 \leq \rho \leq 1$, regulates the degree of the decrease in pheromone trails. If ρ is low, the influence of the pheromone values will persist longer, while high values of ρ allow fast forgetting of previously very attractive choices and, hence, allow faster focus on new information that incorporate into the pheromone matrix. Another parameter is the number of ants in the colony, m. For a given computational budget, such as maximum computation time, the number of ants is a critical parameter for determining the trade-off between the number of iterations that can be made and how broad the search is at each of the iterations. In fact, the larger the number of ants per iteration, the fewer the iterations of the ACO algorithm.

8.2.2 ACO Variants

The ACO framework can be implemented in many different ways. In the literature, several ACO algorithms have been proposed, which differ in some choices characterizing the ConstructSolutions and UpdatePheromones procedures. Three of the main variants are described next. For a description of other variants we refer the interested reader to Dorigo and Stützle [11].

8.2.3 Ant System

Ant System (AS) is the first ACO algorithm proposed in the literature [12, 13]. In AS, an ant k in node i chooses the next node j with probability given by the random proportional rule, defined as

$$p_{ij} = \frac{[\tau_{ij}]^\alpha \cdot [\eta_{ij}]^\beta}{\sum_{h \in N^k} [\tau_{ih}]^\alpha \cdot [\eta_{ih}]^\beta}, \tag{8.1}$$

where N^k is its feasible neighborhood. The feasible neighborhood excludes nodes already visited in the partial tour of ant k, and it may be further restricted to a candidate set of the nearest neighbors of a city i. Once an ant has visited all nodes, it returns to its starting node.

During the execution of the UpdatePheromones procedure in AS, all m ants deposit pheromones at each iteration. The pheromone trail values are updated as

$$\tau_{ij} \leftarrow (1 - \rho) \cdot \tau_{ij} + \sum_{k=1}^{m} \Delta \tau_{ij}^k, \tag{8.2}$$

where $\Delta \tau_{ij}^k$ is defined as

$$\Delta \tau_{ij}^k = \begin{cases} F(k) & \text{if edge } (i,j) \text{ is part of the solution constructed by ant } k, \\ 0 & \text{otherwise,} \end{cases} \tag{8.3}$$

where $F(k)$ is the amount of pheromone deposited on the edges of the solution constructed by ant k. $F(k)$ is equal to the reciprocal of the cost of the solution constructed by ant k, possibly multiplied by a constant Q. Hence, the better the solution, the higher the amount of pheromone deposited by an ant.

8.2.4 MAX–MIN Ant System

MAX–MIN Ant System (MMAS) is an improvement over the AS algorithm [42]. The main difference is the handling of the pheromone trails update. Firstly, only one solution is used for the pheromone deposit. This is typically either the iteration-best solution or the best-so-far solution, that is, the best solution since the start of the algorithm. Secondly, all pheromone values are bounded by the interval $[\tau_{\min}, \tau_{\max}]$. The pheromone update rule used is

$$\tau_{ij} \leftarrow \max\{\tau_{\min}, \min\{\tau_{\max}, (1-\rho) \cdot \tau_{ij} + \Delta \tau_{ij}^{\text{best}}\}\} \tag{8.4}$$

where $\Delta \tau_{ij}^{\text{best}}$ is defined as

$$\Delta \tau_{ij}^{\text{best}} = \begin{cases} F(s^{\text{best}}) & \text{if edge } (i,j) \text{ is part of the best solution } s^{\text{best}}, \\ 0 & \text{otherwise.} \end{cases} \tag{8.5}$$

$F(s^{\text{best}})$ is the reciprocal of the cost of the solution considered for the deposit. For more details on the pheromone initialization and the use of occasional pheromone trail reinitializations, we refer the reader to Stützle and Hoos [42].

8.2.5 Ant Colony System

Ant Colony System (ACS) [10] differs in several ways from AS and MMAS. ACS uses the *pseudorandom proportional rule* in the solution construction: with a probability q_0 the next city to visit is chosen as

$$j = \arg\max_{h \in N^k} \{\tau_{ih} \cdot \eta_{ih}^{\beta}\}, \tag{8.6}$$

that is, the most attractive edge is selected greedily with fixed probability. With probability $1 - q_0$, the AS random proportional rule defined by Equation 8.1 is used. In ACS, the parameter α is fixed to 1, and, therefore, it is often omitted.

The pheromone deposit of ACS modifies only the pheromone values of edges from one solution. As in MMAS, this solution is either the iteration-best or the best-so-far solution. The ACS pheromone update formula is

$$
\tau_{ij} \leftarrow
\begin{cases}
(1 - \rho) \cdot \tau_{ij} + \rho \cdot \Delta \tau_{ij} & \text{if } (i, j) \text{ is part of the best solution } s^{\text{best}}, \\
\tau_{ij} & \text{otherwise},
\end{cases}
\tag{8.7}
$$

with $\Delta \tau_{ij} = F(s^{\text{best}})$.

A local pheromone update rule is applied during the solution construction of the ants. Each time an ant traverses an edge (i, j), τ_{ij} is modified as

$$
\tau_{ij} \leftarrow (1 - \xi) \cdot \tau_{ij} + \xi \cdot \tau_0,
\tag{8.8}
$$

where $\xi \in (0, 1)$ is a parameter called pheromone decay coefficient, and τ_0 is the initial value of the pheromones. In ACS, τ_0 is a very small constant with value $1/(n \cdot L_{nn})$, where L_{nn} is the length of a nearest neighbor tour. The local pheromone update aims at avoiding stagnation: it decreases the pheromone value on the previously used edges and makes them less attractive for other ants.

8.3 Overview of Parameter Adaptation Approaches

The dependency of the performance of metaheuristics on the settings of their parameters is well known. In fact, finding appropriate settings for an algorithm's parameters is considered to be a nontrivial task and a significant amount of work has been devoted to it. The approaches for tackling this task can roughly be divided into offline versus online procedures.

Offline tuning has the goal of finding appropriate settings for an algorithm's parameters before the algorithm is actually deployed. Traditionally, offline tuning has mostly been done in a trial-and-error fashion. This process is time-consuming, human-intensive and error-prone, and it often leads to the uneven tuning of different algorithms. Moreover, tuning by trial and error depends much on the intuition and experience of the algorithm developer, and it is typically undocumented and therefore not reproducible. More recently, increasing effort has been devoted to methods that allow for search-based, hands-off tuning of algorithm parameters. The study of techniques for automatic algorithm configuration is currently a rather active research area. Still, these methods typically come up with one specific parameter setting which then remains the same while the algorithm solves a particular instance.

An alternative to offline tuning is online tuning. Typically, this consists of the modification of an algorithm's parameter settings while solving a problem instance. A potential advantage of online modification of parameters is that algorithms may adapt better to the particular instance's characteristics. When instances are relatively heterogeneous, parameter settings that result in good performance on average across all instances may lead to much worse results on some instances. Online parameter adaptation may also be useful for achieving the best performance for a stage of search. It is often possible to identify an algorithm's explorative and exploitative search phases, and good parameter settings in these phases may again be very different. Finally, if algorithms are applied in situations that are very different from the context in which they have been developed or tuned offline, allowing parameters to change online may increase an algorithm's robustness.

There are different ways of modifying parameters during the run of an algorithm. Many strategies have been widely studied in the context of EAs, and Eiben et al. [17] give a possible classification of these strategies. Perhaps the simplest way is to define the parameter variation rule before actually running the algorithm. In such an approach, the problem is observed from an offline perspective: Static parameters are substituted with (deterministic or randomized) functions depending on the computational time or the number of algorithm iterations. Eiben et al. [17] call such strategies *deterministic* parameter control; however, we prefer the term *prescheduled* parameter variation, because the adjective "deterministic" does not correctly characterize this way of controlling parameters, given that the schedule could also allow randomized choices. Even if prescheduled parameter variation is an online tuning method, it does not make the offline tuning problem disappear since also the parameters that define the schedule need to be appropriately set.

An alternative is to use *adaptive parameter settings*, where the parameter modification scheme is defined as a function of some statistics on the algorithm behavior. Various measures can be used for this online adaptation. They can be grouped depending on whether they are based on absolute or relative evidence. In the first case, the adaptation strategy monitors the occurrence of some events during the run, for example, some fixed threshold of the distance between the solutions visited. Surpassing the threshold then triggers a set of rules for parameter variation. In the second case, the adaptation strategy considers the relative difference between the performance achieved with different parameter settings, and adapts the parameter values to resemble the most successful ones. For parameter adaptation to work, some additional decisions need to be made beyond the ones strictly related to the implementation of the algorithm. In particular, the equations that describe the parameter update need to be defined a priori and it is hoped that the mechanisms used are very robust with respect to their definitions.

A further possibility that has been the object of studies consists of having the parameters modified at run time by the algorithm itself. Specifically, dimensions that represent parameters of exploration strategies are added to the search space of the problem. The optimization process is then executed in this new space. Eiben et al. [17] name this approach *self-adaptation*. Taking the notion of self-adaptation a step further, we call *search-based adaptation* strategies that use a search algorithm different from the underlying algorithm for parameter adaptation. This class of strategies includes techniques such as local search and EAs for adapting the parameters of ACO algorithms.

Table 8.1: Schematic description of the literature on adaptive ACO. Some of the articles propose general adaptation strategies that could be used for several parameters. Here we only indicate the parameters that have been adapted experimentally

Authors	Adaptation strategy	ACO variant	Parameters
Merkle and Middendorf [34]	pre-scheduled	variant of AS	β
Merkle et al. [35]	pre-scheduled	variant of AS	β, ρ
Meyer [36]	pre-scheduled	AS	α
Randall and Montgomery [40]	adaptive	ACS	candidate set
Chusanapiputt et al. [6]	adaptive	AS	α, β
Li and Li [28]	adaptive	new variant	α, β
Hao et al. [24]	adaptive	ACS	ρ
Kovářík and Skrbek [27]	adaptive	variant of MMAS	β
Li et al. [29]	adaptive	variant of ACS	q_0, pheromone update
Cai et al. [5]	adaptive	ACS	ρ
Randall [39]	self-adaptation	ACS	β, ρ, q_0, ξ
Martens et al. [32]	self-adaptation	MMAS	α, β
Förster et al. [19]	self-adaptation	new variant	pheromone update
Khichane et al. [26]	self-adaptation	MMAS	α, β
Pilat and White [38]	search-based adaptation	ACS	β, ξ, q_0
Gaertner and Clark [20]	search-based adaptation	AS–ACS combination	β, ρ, q_0
Hao et al. [23]	search-based adaptation	variant of ACS	β, ρ, q_0
Garro et al. [22]	search-based adaptation	variant of ACS	algorithm specific
Ling and Luo [30]	search-based adaptation	variant of ACS	α, ρ, Q
Amir et al. [1]	search-based adaptation	ACS	β, q_0
Anghinolfi et al. [2]	search-based adaptation	variant of ACS	β, q_0
Melo et al. [33]	search-based adaptation	multi-colony ACS	α, β, ρ, q_0

8.4 Parameter Adaptation in ACO

The study of the impact of various parameters on the behavior of ACO algorithms has been an important subject since the first articles [12, 13]. We schematically summarize in Table 8.1 the main approaches that have been used in the ACO literature to adapt parameter values, following roughly the classes defined by Eiben et al. [17]. This summary shows that the most frequently chosen parameters for adaptation are α, β, q_0 (in the case of ACS), and those that control the pheromone update. We describe these approaches in the following sections.

8.4.1 Prescheduled Variation of Parameter Settings

There is surprisingly little work on prescheduled parameter variation for ACO algorithms. Merkle and Middendorf [34] describe a contribution that considers an ACO algorithm for the resource-constrained project scheduling problem. Its authors propose decreasing the value of the parameter β linearly over the run of an algorithm from an initial value of 2 to 0. In a subsequent study, Merkle et al. [35] consider the same problem and its authors propose modifying the parameter β and the evaporation rate ρ. For β they propose a schedule as described before. For ρ, they propose starting with a small value to increase the initial exploration of the search space, and later setting the evaporation rate to a high value for an intensive search around the best solutions found by the algorithm.

Meyer [36] proposes a variant of AS called α-annealing. The idea at the basis of its author's work is to change the value of α according to some annealing schedule. Increasing α slowly throughout the search can keep diversity in the beginning and gradually increase the selective pressure to cover better regions of the search space in the later phases.

8.4.2 Adaptive Approaches

Many of the approaches proposed in the literature can be classified as adaptive. In these approaches, some parameters are modified according to some rules that take into account the search behavior of the ACO algorithm. The average λ-branching factor [21] is one of the first proposed measures of ACO behavior. Other measures include entropy-based measures for the pheromone, dispersion of solutions generated by the algorithm, or simply the quality of the solutions generated [7, 37]. Favaretto et al. [18] propose a technique for measuring the effect of parameter variation on the exploration performed; this technique may also serve as an indicator for defining parameter adaptation strategies.

In an early discussion of the usefulness of parameter adaptation in ACO algorithms, Merkle and Middendorf [34] propose a decomposition of the search into different phases to allow for the development of parameter adaptation strategies. Several strategies have been proposed later and we divide these in two groups: adaptations based on the dispersion of the pheromone trails, and adaptations based on the quality of solutions. Within the first group, Li and Li [28] introduce an ACO algorithm that varies the parameters α and β over time. Their parameter adaptation strategy considers a measure of the entropy on the action choice probabilities of the ants during solution construction and they aim at obtaining a schedule for α and β. During the early stage of the search, the value of α is small enough to allow extensive exploration of the search space; the value of α increases over time to improve the local search ability of the algorithm. They suggest the opposite schedule for β. Li et al. [29] propose a variant of ACS that uses a "cloud-based fuzzy strategy" for choosing the solution to be used in the global pheromone update. The main idea is that, as the pheromones get more concentrated around a single solution, tours other than the best-so-far one have a good chance of depositing pheromone. Additionally, they adapt the parameter q_0 with the goal of decreasing it as soon as the pheromone trails concentrate on very few edges. Chusanapiputt et al. [6] propose a variation of AS for dealing with the unit commitment problem. Three of the algorithm's parameters are adapted based on pheromone dispersion.

In a second group of papers, the driver of adaptation is the quality of the solutions generated. Hao et al. [24] and Cai et al. [5] propose a variant of ACS for the TSP. In their implementation, a different value of the parameter ρ is associated with each ant depending on the quality of its solution. This mechanism aims at having high-quality solutions contribute more pheromone than low-quality ones. Amir et al. [1] add a fuzzy logic controller module to the ACS algorithm for the TSP for adapting the value of β and q_0. The adaptive strategy uses two performance measures: the difference between the optimal solution (or the best known solution) and the best one found and the variance of the solutions visited by a population of ants. Kovářík and Skrbek [27] describe an approach that divides the ant colony into groups of ants using different parameter settings. They adapt the number of ants in each of the groups depending on the improvement of the solution quality obtained by each group; however, they do not give all details of the adaptation strategy. An analysis of the experimental results of adapting the values of β indicates that better initial performance is obtained with high values of β while towards the end of the run low values of β are preferable. Randall and Montgomery [40] apply an adaptation mechanism to an ACS algorithm that uses a *candidate set strategy* as a speedup procedure. At each step, the component to be added to the partial solution under construction is chosen from the ones belonging to such a set. The composition of the set is modified throughout the run. Elements that give low probability values are eliminated temporarily from the search process (they become tabu). After a number of iterations, they are added again to the candidate set. The threshold for establishing which elements are tabu is varied throughout the search process, depending on the quality of solutions being produced.

8.4.3 Search-Based Parameter Adaptation

Various adaptive ACO strategies fall into the category of self-adaptive strategies [17], where an algorithm tunes itself by integrating its parameters into its search task. We first present strategies that are "purely self-adaptive" in the original meaning used by Eiben et al. [17]. Later, we discuss approaches that use other search algorithms than ACO for adapting parameters. Given that these strategies are search-based, most of the approaches discussed in the following use as feedback the quality of the solutions generated.

8.4.3.1 Pure Self-Adaptive Approaches

The first self-adaptive approach to ACO is by Randall [39]. He suggests discretizing the parameter range and associating with each resulting value of a parameter a pheromone trail that gives the desirability of choosing it. In his approach, each ant adapts its own parameter settings and chooses them at each iteration before solution construction. This mechanism is tested by adapting the parameters β, q_0, ρ, and ξ for ACS applied to the TSP and the quadratic assignment problem. The comparison of the results to the default parameter settings is somehow inconclusive.

Martens et al. [32] propose a self-adaptive implementation of MMAS applied to the generation of decision rule-based classifiers. In their AntMiner+ algorithm, ants choose suitable values for the parameters α and β. This is done by introducing for each parameter a new vertex group in the construction graph. The values of α and β are limited to integers between 1 and 3. Unlike in the previous paper, here parameters are treated as interdependent.

Förster et al. [19] apply the same idea to an ACO approach for a multi-objective problem. The parameters adapted are specific to the algorithm proposed, but they are mostly related to pheromone deposit. As in Randall [39], the dependence among parameters is neglected, that is, no new nodes are added to the construction graph. A separate pheromone matrix is recorded, each column representing a parameter to be adapted. Before starting the solution construction, each ant selects probabilistically its own parameter settings based on the pheromone matrix.

Khichane et al. [26] study a self-adaptive mechanism to tune the parameters α and β of their implementation of MMAS and apply it to constraint satisfaction problems. However, differently from the previous works, they do not define parameter settings at the level of an individual ant. For each iteration one common parameter setting for the whole ant colony is defined. The two parameters are considered independent of each other. The authors propose two variants of the parameter adaptation mechanism. In the first one, called global parameter learning ant-solver (GPL-Ant-solver), the colony uses the same parameter setting during the solution construction of each ant. In the second one, called distributed parameter learning ant-solver (DPL-Ant-solver), at each step of the solution construction the colony chooses new values for α and β; hence, in this case the pheromones that encode specific parameter settings refer to the desirability of choosing a specific parameter value for a specific construction step. In an experimental evaluation of the two variants, both are shown to reach similar performance levels. A comparison with an offline tuned version of their ant-solver shows that for some instances the adaptive version performs better while for others the opposite is true.

8.4.3.2 Other Search Algorithms for Adapting Parameters

Pilat and White [38] test two ways of using an EA to adjust parameters of an ACS algorithm; one of them has to do online tuning. Their approach to online tuning uses an EA to determine, at each ACS iteration, the parameter settings of four ants before constructing solutions. The EA in turn uses the constructed solutions to further evolve a set of good parameter settings. The authors choose three parameters for adaptation, namely, β, q_0, and ξ. Their results are somewhat inconclusive. This approach is similar to the mechanism used in an earlier paper by White et al. [43], where the authors evolve the parameters α and β in an ACO algorithm for a telecommunications routing problem. As an alternative to the online tuning of ACO parameters by an EA, Pilat and White [38] explore the use of an EA as an offline tuning mechanism, analogously to Botee and Bonabeau [4].

Gaertner and Clark [20] propose a similar adaptive approach. In their work, every ant is initialized with a random parameter combination, where the parameter values are chosen from a predefined range. Over time, the entire population of ants evolves, breeding ants with parameter combinations which find improved solutions. In their experiments, the authors consider an algorithm based on a combination of AS and ACS for the TSP. They test their approach on three parameters: β, ρ and q_0.

Hao et al. [23] propose a variant of ACS in which each ant is characterized by its own parameter setting. The usual random-proportional rule is applied for selecting subsequent moves. After each iteration, the parameter configurations are modified using a particle swarm optimization (PSO) approach. Three parameters are adapted throughout the algorithm's run, namely β, ρ and q_0. If the PSO mechanism assigns a value outside a predefined range to a parameter, then the parameter is randomly reinitialized. Following a similar idea, Ling and Luo [30] propose using an artificial fish swarm algorithm for exploring the parameter space. Its authors also consider a variant of ACS and vary the three parameters α, ρ and Q, a parameter that influences the amount of pheromone an ant deposits. The main difference between this work and the one of Hao et al. [23] is that Ling and Luo use the same parameter setting for all ants.

Garro et al. [22] present an algorithm that evolves parameters using an EA. An individual in the EA represents an ant characterized by specific parameter values. The authors study a variant of ACS for automatically determining the path a robot should follow from its initial position to its goal position. They adapt three parameters of a newly proposed state transition rule.

Anghinolfi et al. [2] adapt two parameters using a local search in the parameter space. They define the neighborhood of the current parameter setting as all possible combinations of parameter settings that can be obtained by increasing or decreasing each parameter value by a fixed amount. Therefore, in the case of two parameters, at each iteration five parameter configurations are tested: the incumbent one and its four resulting neighbors. The test is done by assigning each parameter combination to one of five equal-sized groups of ants; each group then uses its parameters to generate solutions. After each iteration, the incumbent parameter setting is changed to the one that produced the iteration-best solution. In their experiments, the authors adapt two parameters of a variant of ACS, namely β and q_0. They observe better performance of the adaptive strategy than a tuned, fixed parameter setting.

Finally, Melo et al. [33] propose a multi-colony ACS algorithm, where several colonies of ants try to solve the same problem simultaneously. Each colony uses different parameter settings for α, β, ρ and q_0. Apart from exchanging solutions among the colonies, their proposal includes a mutation operator that replaces the parameter settings of the worst colony with the value of the same parameter in the best colony modified by a small, uniformly random value.

8.4.4 *Conclusions from the Review*

The review above shows that there is ongoing interest in parameter adaptation in the ACO literature. However, we also observe that several of the contributions apply adaptive techniques without prior in-depth understanding of the effect of individual parameters. Without such an understanding, decisions about which parameters to adapt and how to adapt them are mostly arbitrary. In particular, we did not find in our review any systematic study of the effect of different parameter settings on the anytime behavior of ACO algorithms. It is our intuition that such an analysis can inform decisions not only about which parameters may be worth varying during runtime, but also about how to perform such variations. Moreover, the anytime behavior of fixed parameter settings provides a baseline for evaluating the performance of parameter adaptations. In the following sections, we first provide a systematic study of the anytime behavior of ACO algorithms, and we use the knowledge acquired from it to design successful schemes for prescheduled parameter variation.

8.5 Experimental Investigation of Fixed Parameter Settings

In this section, we examine the effect that various parameters have on the performance of ACS and MMAS. In particular, we are interested in the development of the best-so-far solution over time when varying one parameter at a time. Our goal is to identify which parameter settings produce the best results at any moment during the run of the algorithm. Clearly, a parameter setting that produces very good solutions during the initial stages of a run but leads to much worse results later is an interesting candidate for having its settings varied online. In other words, we are interested in the anytime behavior [44] of specific parameter settings to clearly identify opportunities for the adaptation of parameter values over the computation period.

Our experimental analysis is based on the publicly available ACOTSP software [41], which we compiled with gcc, version 3.4. Experiments are carried out on a cluster of Intel Xeon™ E5410 quad-core processors running at 2.33 GHz with 6 MB L2 Cache and 8 GB RAM under Rocks Cluster GNU/Linux. Due to the sequential implementation of the code, only one core is used for running the executable.

We test two ACO algorithms, ACS and MMAS. Table 8.2 gives the default values for the parameters under study. In each experiment where one parameter is varied, the others are kept fixed at their default values. For the remaining parameters, that is, τ_0, ξ (for ACS), the choice of iteration-best or best-so-far update, and so on, we use the default values given by Dorigo and Stützle [11], which are also the default values of the ACOTSP package. We test the algorithms with and without the use of the first-improvement 2-opt local search provided by the ACOTSP package. For the experiments, TSP instances are randomly generated using the instance generator provided for the 8th DIMACS challenge on the TSP; in particular, points are generated uniformly at random in a square of side length 1,000,000. When using ACO

Table 8.2: Default settings of the parameters under study for MMAS and ACS without local search and with 2-opt local search

Algorithm	β	ρ	m	q_0
ACS	2	0.1	10	0.9
ACS + 2-opt	2	0.1	10	0.98
MMAS	2	0.02	n	0.00
MMAS + 2-opt	2	0.2	25	0.00

algorithms without local search, the tests are done on instances of size 100, 200, 400 and 800; because of the much higher performance of the ACO algorithms when local search is used, we use with local search larger instances of size $1,500$, $3,000$ and $6,000$ to minimize possible floor effects. The presentation of the experimental results is based on the development of the relative deviation of the best solution found by an ACO algorithm from the optimal solution (or the best known solution for the instance of size $6,000$). Each of the curves of the solution quality over time, or SQT curves [25], is the average of 25 executions of each parameter setting. Since we only present plots, we give for each setting results on only one instance. However, the trends are the same on all instances and, hence, the plots are representative of the general results.

8.5.1 Fixed Parameter Settings for Ant Colony System

In the case of ACS, we study the effect of β, which regulates the influence of the heuristic information; m, the number of ants; ρ, the evaporation factor; and q_0, the probability of making a deterministic choice in Equation 8.6. Here, we present SQT curves only for the case where ants' solutions are improved by a local search for the instance of size $3,000$. The final conclusions concerning the usefulness of the variation of parameters at run-time were the same on the other instances and when using ACS without local search. Figures 8.2 to 8.5 report the results on parameters β, m, ρ, and q_0, respectively.

The main overall conclusion we obtain from these results is that very often there is a single parameter setting that performs best during most of the available run-time. Hence, there does not appear to be a clear benefit to varying the parameter settings at run-time. This conclusion remains the same if ACS is run without local search; the main difference is that the performance is more variable and more dependent on specific parameter values. In more detail, the observations and conclusions that arise for the single parameters from the presented results are the following.

β, *Figure 8.2:* Medium range values of β equal to 2 or 5 produce the best results during most of the runtime. Smaller values of β are initially worse but, after enough computation time, eventually match the results of the default value. Much larger values (e.g., $\beta = 10$) are quickly overshadowed by smaller ones.

m, Figure 8.3: The default value of ten ants results in very good anytime performance. Interestingly, very small values (notably $m = 1$) make the algorithm perform slightly worse during the whole runtime, whereas much larger values ($m = 100$) lead to much worse results. The latter effect is probably due to too much diversification because of the application of the local pheromone update rule in ACS.

ρ, Figure 8.4: Surprisingly, the differences among different settings of ρ are almost imperceptible. Without local search (not shown here), large ρ values produce faster convergence. However, after a short time small values close to the default ($\rho = 0.1$) produce progressively better results.

q_0, Figure 8.5: As suggested in the literature, good values of q_0 tend to be close to 1. In extreme cases, a value of 1 quickly leads to search stagnation, while values smaller than 0.75 produce very slow convergence towards good solutions. Similar results are obtained when local search is disabled.

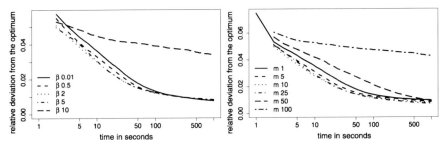

Fig. 8.2: ACS with various values of β Fig. 8.3: ACS with various numbers of ants (m)

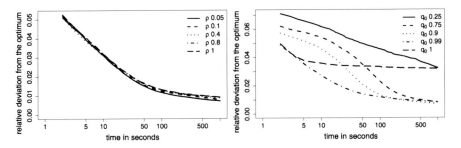

Fig. 8.4: ACS with various values of ρ Fig. 8.5: ACS with various values of q_0

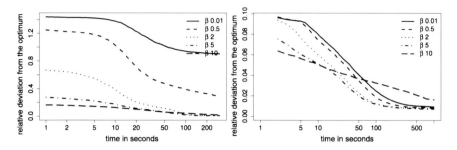

Fig. 8.6: MMAS with various fixed values of β; left plot without local search and right plot with local search

8.5.2 Fixed Parameter Settings for MAX–MIN Ant System

We now study the impact of the same parameters, β, m, ρ, and q_0, on the anytime behavior of MMAS. For MMAS, the behavior is more interesting from a parameter adaptation point of view. We therefore present results for the cases with and without local search. Results without local search are for one instance with 400 nodes, whereas with local search, they are for an instance with 3,000 nodes.

β, *Figure 8.6:* Without local search (upper part of Figure 8.6), MMAS requires relatively large values of β, which produce a significant advantage, especially during the initial part of the run, over the default setting of $\beta = 2$. While results with the default setting eventually match the results obtained with higher settings, values of β less than 2 lead to quite poor performance. With local search, the differences are much smaller and a setting of $\beta = 10$ is quickly overshadowed by lower ones. This suggests that starting with a high value of β may enhance the performance of MMAS at the beginning of the run, but a value close to the default may produce better results for larger computation times.

m, *Figure 8.7:* With and without local search, the number of ants shows a clear trade-off between early and late performance. In particular, a low number of ants (for example, $m = 5$) produces the best results during the early stages of the algorithm run. However, a higher number of ants (for example, $m = 100$) obtains much better results towards the end of the run. Without local search, the fast initial progress with few ants soon levels off and apparently leads to search stagnation. In this case, the default setting of $m = 400$ appears to be already too high, and it slows down the algorithm compared to using 100 ants without improving the final result. With local search, the SQT curves cross for the different parameter settings and those with few ants ($m = 1$ and $m = 5$) result in worse final solution quality. In fact, a larger number of ants ($m \geq 25$) pays off if the algorithm is allowed to run for enough time. This result suggests that increasing the number of ants from an initially low value may lead to better anytime behavior of MMAS.

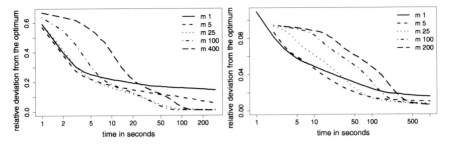

Fig. 8.7: MMAS with various fixed numbers of ants (*m*); left plot without local search and right plot with local search

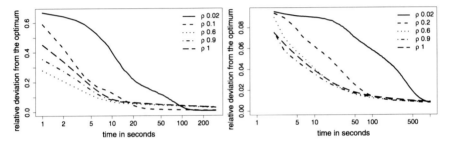

Fig. 8.8: MMAS with various fixed values of ρ; left plot without local search and right plot with local search

ρ, *Figure 8.8:* There is some degree of trade-off between large and small values of ρ. Large values (for example, $\rho = 0.6$) converge faster than the default values ($\rho = 0.02$ without local search, $\rho = 0.2$ with local search). Nevertheless, low values of ρ are able to achieve the same performance, and, given sufficient time, produce the best final results. This effect is most noticeable in the case without local search. Hence, starting with a high evaporation factor and then reducing it over time to its default value appears to be a promising strategy.

q_0, *Figure 8.9:* Finally, we test the use of the pseudorandom proportional rule of ACS (Equation 8.6) in MMAS. Here, we study the effect of different values of q_0 as we previously did for ACS. In this case, a clear trade-off is observed: high values of q_0 perform best for a short runtime, whereas low values of q_0 ($q_0 = 0$ effectively reverts to the standard MMAS) generally result in better final performance.

Summarizing the above experiments, in MMAS a strong trade-off exists for various parameters between the performance of fixed settings for short and long computation times, making the behavior of MMAS very different from that of ACS. In particular, β, *m* and q_0 seem good candidates for the use of variable settings to achieve good anytime performance. Therefore, in the next section, we examine a few simple ways of varying the parameter settings of MMAS during the run.

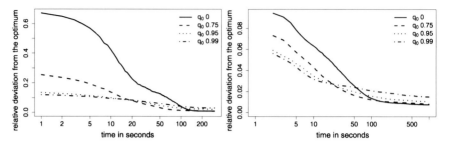

Fig. 8.9: MMAS using the pseudorandom proportional rule with various fixed values of q_0; left plot without local search and right plot with local search

8.6 Prescheduled Parameter Variation for MMAS

Given that MMAS was a clear candidate for the varying of parameters during the computation period, we examine various schedules for changing the parameter settings. In this section, we give exemplary results for prescheduled parameter variation. In particular, we show results concerning the adaptation of the parameters β, m, and q_0. These illustrate the types of improvement in the anytime behavior of MMAS that may be obtained. We do not consider varying the evaporation factor, ρ, since we did not find schedules that significantly improve performance over a fixed, high setting (such as $\rho = 0.6$).

First, we study the variation of β. We tested schedules that decrease the value of β linearly with the iteration counter as well as schedules where a switch from a high value to a low value occurs at a fixed iteration number. The latter type of schedule resulted in better anytime performance and, hence, we focus on these here. The procedure of these schedules is to start with the high value of $\beta = 20$, which was shown to yield good performance at the start of the run, and to later set it directly to a lower value close to the default value. Figure 8.10 shows the results with local search for three scenarios that differ in the number of iterations after which β is changed, from 20 to 3; in particular, we consider 50 (a_β 1), 100 (a_β 2) and 200 (a_β 3) iterations. The schedule a_β 1 obtained the best SQT curve, and delaying the change of β produces worse results. In the case without local search (not shown here), delaying the switch from the high to the low value of β showed some improvement. Nonetheless, for simplicity, we choose strategy a_β 1 for further comparison. Figure 8.11 compares the use of strategy a_β 1 to the use of the default value and a large value ($\beta = 20$) of β without (left) and with (right) local search. In both cases, the prescheduled parameter variation is able to combine the best results of both fixed settings, achieving a better anytime performance.

In the case of the number of ants, m, the strategies studied here start with a single ant and increase the number of ants as the algorithm progresses. Figure 8.12 shows that there is progressive degradation of the quality of the results as the rate at which ants are added increases. The best results are obtained with the lowest rate

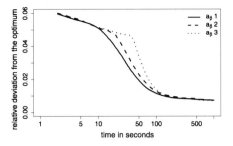

Fig. 8.10: MMAS, scheduled variation of parameter β; the three strategies a_β 1 to a_β 3 start each with β equal to 20 and set β to 3 after 50 (a_β 1), 100 (a_β 2), and 200 (a_β 3) iterations, respectively

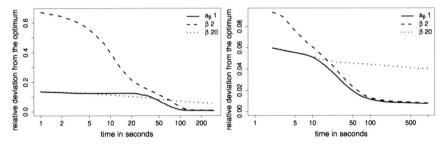

Fig. 8.11: MMAS, comparison of fixed and varying parameter settings for β; left side: case without local search; right side: case with local search. The adaptation strategy used is a_β 1 (see caption of Figure 8.10 for details)

(a_m 1, which adds one ant after every ten iterations) for both cases, with and without local search (only the local search case is shown for conciseness). The comparison between the fixed and prescheduled settings (Figure 8.13) shows a clear benefit with the use of prescheduled variation of m, which matches the good performance obtained for short runtimes by only one ant and for long runtimes by a large number of ants.

For varying q_0 for MMAS, we tested strategies that start with a high value of $q_0 = 0.99$ and decrease it until they reach a setting of q_0 equal to 0. Figure 8.14 shows four strategies that decrease q_0 at different rates, namely, by 0.001 every 15 iterations (a_{q_0} 1), by 0.001 every two iterations (a_{q_0} 2), by 0.005 every iteration (a_{q_0} 3), and by 0.02 every iteration (a_{q_0} 4). Without local search, the strategies that decrease q_0 more slowly result in faster convergence to good solutions (not shown here). However, with local search there is a trade-off between the slowest and the fastest decrease of q_0, with the former being better at the start of the algorithm, and the latter performing best for higher computation times. This suggests that more sophisticated strategies may be able to further enhance the performance of the algorithm. Nevertheless, the comparison of the schedule a_{q_0} 2 with fixed pa-

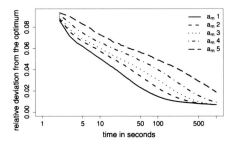

Fig. 8.12: MMAS, scheduled variation of the number of ants (m). All strategies a_m 1 to a_m 5 start with one ant and iteratively increase the number of ants. In particular, a_m 1 adds one ant every ten iterations, a_m 2 adds one ant every second iteration, a_m 3 adds one ant each iteration, a_m 4 adds two ants each iteration, and a_m 5 adds five ants each iteration

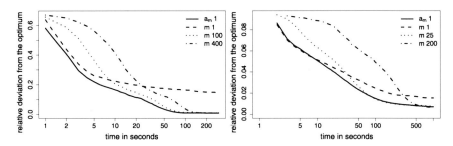

Fig. 8.13: MMAS, comparison of fixed and varying parameter settings for parameter m; left side: case without local search; right side: case with local search. The adaptation strategy used is a_m1 (see caption of Figure 8.12 for details)

rameter settings shows that prescheduled parameter variation is able to match the best results of both fixed parameter settings during the execution time of the algorithm.

A general observation from our study of prescheduled parameter variation is that considerable improvements of the anytime behavior of MMAS are possible without their substantially affecting the final performance achieved by the algorithm. In some additional experiments, we verified that the same conclusion is also true for the parameter α, which weights the influence of the pheromone trails. In fact, for similar simple schedules as proposed previously, we could observe strong improvements in the anytime behavior compared to performance with fixed settings for α. Further studies need to verify whether the same observations on the usefulness of simple prescheduled parameter variations hold for other problems.

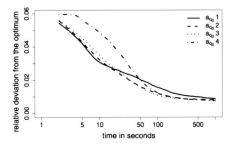

Fig. 8.14: MMAS, scheduled variation of the parameter q_0. All strategies (a_{q_0} 1 to a_{q_0} 4) start at $q_0 = 0.99$ and decrease q_0 to 0. In particular, a_{q_0} 1 decreases q_0 by 0.001 every 15 iterations, a_{q_0} 2 by 0.001 every 2 iterations, a_{q_0} 3 by 0.005 every iteration, and a_{q_0} 4 by 0.02 every iteration

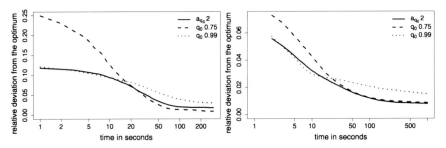

Fig. 8.15: MMAS, comparison of fixed and varying parameter settings for parameter q_0; left side: case without local search; right side: case with local search. The adaptation strategy used is a_{q_0} 2 (see caption of Figure 8.14 for details)

8.7 Conclusions and Future Work

In this chapter, we have given an overview of the literature on parameter adaptation in ACO algorithms. A variety of approaches have been proposed but the overall impression from the research results is that further efforts are required to determine the most suitable strategies for parameter adaptation and their role and importance in ACO algorithms that perform at the state-of-the-art level. Only few of the presented publications have shown clear computational advantages, for example, in the form of better average solution quality reachable in highly effective ACO algorithms.

In the second part of the chapter, we have given an experimental analysis of the impact that specific parameters have on the anytime behavior of ACO algorithms. For the application of ACS and MMAS to the TSP we could determine very different behavior of these algorithms. While the anytime behavior of ACS was rather insensitive to parameter variation, the analysis of the anytime behavior of MMAS has identified clear opportunities for prescheduled parameter variation. We tested a number of fairly straightforward schedules of the values for the parameters β, m, and q_0 in MMAS. As a result, we could observe that the anytime behavior of MMAS can be greatly improved without significant loss in the final solution quality.

Our computational study can clearly be extended in different directions. An interesting extension is the study of interactions between different parameter settings. Such a study may hint at combined variations of at least two parameters that can further improve the anytime behavior of MMAS or even its final performance. Finally, it is certainly worthwhile studying in more detail the contribution of adaptive strategies that take into account the internal state of the algorithm in order to adapt to different classes of instances. For their study, it is probably preferable to consider problems where the algorithm parameters depend more strongly on specific instance classes than they do in the TSP.

For our main conclusion we can state that parameter adaptation is a relatively large field that is receiving strong attention by the research community. Many techniques have already been proposed in the context of other heuristic methods, but their adoption in ACO algorithms still opens up a number of research opportunities with potentially significant impact.

Acknowledgements This work was supported by the META-X project, an *Action de Recherche Concertée* funded by the Scientific Research Directorate of the French Community of Belgium. Mauro Birattari, Thomas Stützle and Marco Dorigo acknowledge support from the Belgian F.R.S.-FNRS, of which they are Research Associates and Research Director, respectively. The authors also acknowledge support from the FRFC project "*Méthodes de recherche hybrides pour la résolution de problèmes complexes*".

References

[1] Amir C., Badr A., Farag I.: A fuzzy logic controller for ant algorithms. Computing and Information Systems 11(2):26–34 (2007)

[2] Anghinolfi D., Boccalatte A., Paolucci M., Vecchiola C.: Performance evaluation of an adaptive ant colony optimization applied to single machine scheduling. In: Li X., et al. (eds.) Simulated Evolution and Learning, 7th International Conference, SEAL 2008, Lecture Notes in Computer Science, vol. 5361, Springer, Heidelberg, Germany, pp. 411–420 (2008)

[3] Battiti R., Brunato M., Mascia F.: Reactive Search and Intelligent Optimization, Operations Research/Computer Science Interfaces, vol. 45. Springer, New York, NY (2008)

[4] Botee H. M., Bonabeau E.: Evolving ant colony optimization. Advances in Complex Systems 1:149–159 (1998)

[5] Cai Z., Huang H., Qin Y., Ma X.: Ant colony optimization based on adaptive volatility rate of pheromone trail. International Journal of Communications, Network and System Sciences 2(8):792–796 (2009)

[6] Chusanapiputt S., Nualhong D., Jantarang S., Phoomvuthisarn S.: Selective self-adaptive approach to ant system for solving unit commitment problem.

In: Cattolico M., et al. (eds.) GECCO 2006, ACM press, New York, NY, pp. 1729–1736 (2006)

[7] Colas S., Monmarché N., Gaucher P., Slimane M.: Artificial ants for the optimization of virtual keyboard arrangement for disabled people. In: Monmarché N., et al. (eds.) Artificial Evolution - 8th International Conference, Evolution Artificielle, EA 2007, Lecture Notes in Computer Science, vol. 4926, Springer, Heidelberg, Germany, pp. 87–99 (2008)

[8] Dorigo M.: Ant colony optimization. Scholarpedia 2(3):1461 (2007)

[9] Dorigo M., Di Caro G.: The Ant Colony Optimization meta-heuristic. In: Corne D., Dorigo M., Glover F. (eds.) New Ideas in Optimization, McGraw Hill, London, UK, pp. 11–32 (1999)

[10] Dorigo M., Gambardella L. M.: Ant Colony System: A cooperative learning approach to the traveling salesman problem. IEEE Transactions on Evolutionary Computation 1(1):53–66 (1997)

[11] Dorigo M., Stützle T.: Ant Colony Optimization. MIT Press, Cambridge, MA (2004)

[12] Dorigo M., Maniezzo V., Colorni A.: The Ant System: An autocatalytic optimizing process. Tech. Rep. 91-016 Revised, Dipartimento di Elettronica, Politecnico di Milano, Italy (1991)

[13] Dorigo M., Maniezzo V., Colorni A.: Ant System: Optimization by a colony of cooperating agents. IEEE Transactions on Systems, Man, and Cybernetics - Part B 26(1):29–41 (1996)

[14] Dorigo M., Di Caro G., Gambardella L. M.: Ant algorithms for discrete optimization. Artificial Life 5(2):137–172 (1999)

[15] Dorigo M., Birattari M., Stützle T.: Ant colony optimization: Artificial ants as a computational intelligence technique. IEEE Computational Intelligence Magazine 1(4):28–39 (2006)

[16] Dorigo M., et al. (eds.): Ant Algorithms: Third International Workshop, ANTS 2002, Lecture Notes in Computer Science, vol. 2463. Springer, Heidelberg, Germany (2002)

[17] Eiben A. E., Michalewicz Z., Schoenauer M., Smith J. E.: Parameter control in evolutionary algorithms. In: [31], pp. 19–46 (2007)

[18] Favaretto D., Moretti E., Pellegrini P.: On the explorative behavior of MAX–MIN Ant System. In: Stützle T., Birattari M., Hoos H. H. (eds.) Engineering Stochastic Local Search Algorithms. Designing, Implementing and Analyzing Effective Heuristics. SLS 2009, Lecture Notes in Computer Science, vol. 5752, Springer, Heidelberg, Germany, pp. 115–119 (2009)

[19] Förster M., Bickel B., Hardung B., Kókai G.: Self-adaptive ant colony optimisation applied to function allocation in vehicle networks. In: Thierens D., et al. (eds.) GECCO'07: Proceedings of the 9th Annual Conference on Genetic and Evolutionary Computation, ACM, New York, NY, pp. 1991–1998 (2007)

[20] Gaertner D., Clark K.: On optimal parameters for ant colony optimization algorithms. In: Arabnia H. R., Joshua R. (eds.) Proceedings of the 2005 International Conference on Artificial Intelligence, ICAI 2005, CSREA Press, pp. 83–89 (2005)

[21] Gambardella L. M., Dorigo M.: Ant-Q: A reinforcement learning approach to the traveling salesman problem. In: Prieditis A., Russell S. (eds.) Proceedings of the Twelfth International Conference on Machine Learning (ML-95), Morgan Kaufmann Publishers, Palo Alto, CA, pp. 252–260 (1995)

[22] Garro B. A., Sossa H., Vazquez R. A.: Evolving ant colony system for optimizing path planning in mobile robots. In: Electronics, Robotics and Automotive Mechanics Conference, IEEE Computer Society, Los Alamitos, CA, pp. 444–449 (2007)

[23] Hao Z., Cai R., Huang H.: An adaptive parameter control strategy for ACO. In: Proceedings of the International Conference on Machine Learning and Cybernetics, IEEE Press, pp. 203–206 (2006)

[24] Hao Z., Huang H., Qin Y., Cai R.: An ACO algorithm with adaptive volatility rate of pheromone trail. In: Shi Y., van Albada G. D., Dongarra J., Sloot P. M. A. (eds.) Computational Science – ICCS 2007, 7th International Conference, Proceedings, Part IV, Lecture Notes in Computer Science, vol. 4490, Springer, Heidelberg, Germany, pp. 1167–1170 (2007)

[25] Hoos H. H., Stützle T.: Stochastic Local Search–Foundations and Applications. Morgan Kaufmann Publishers, San Francisco, CA (2005)

[26] Khichane M., Albert P., Solnon C.: An ACO-based reactive framework for ant colony optimization: First experiments on constraint satisfaction problems. In: Stützle T. (ed.) Learning and Intelligent Optimization, Third International Conference, LION 3, Lecture Notes in Computer Science, vol. 5851, Springer, Heidelberg, Germany, pp. 119–133 (2009)

[27] Kovářík O., Skrbek M.: Ant colony optimization with castes. In: Kurkova-Pohlova V., Koutnik J. (eds.) ICANN'08: Proceedings of the 18th International Conference on Artificial Neural Networks, Part I, Lecture Notes in Computer Science, vol. 5163, Springer, Heidelberg, Germany, pp. 435–442 (2008)

[28] Li Y., Li W.: Adaptive ant colony optimization algorithm based on information entropy: Foundation and application. Fundamenta Informaticae 77(3):229–242 (2007)

[29] Li Z., Wang Y., Yu J., Zhang Y., Li X.: A novel cloud-based fuzzy self-adaptive ant colony system. In: ICNC'08: Proceedings of the 2008 Fourth International Conference on Natural Computation, IEEE Computer Society, Washington, DC, vol. 7, pp. 460–465 (2008)

[30] Ling W., Luo H.: An adaptive parameter control strategy for ant colony optimization. In: CIS'07: Proceedings of the 2007 International Conference on Computational Intelligence and Security, IEEE Computer Society, Washington, DC, pp. 142–146 (2007)

[31] Lobo F., Lima C. F., Michalewicz Z. (eds.): Parameter Setting in Evolutionary Algorithms. Springer, Berlin, Germany (2007)

[32] Martens D., Backer M. D., Haesen R., Vanthienen J., Snoeck M., Baesens B.: Classification with ant colony optimization. IEEE Transactions on Evolutionary Computation 11(5):651–665 (2007)

[33] Melo L., Pereira F., Costa E.: MC-ANT: A multi-colony ant algorithm. In: Artificial Evolution - 9th International Conference, Evolution Artificielle, EA

2009, Lecture Notes in Computer Science, vol. 5975, Springer, Heidelberg, Germany, pp. 25–36 (2009)

[34] Merkle D., Middendorf M.: Prospects for dynamic algorithm control: Lessons from the phase structure of ant scheduling algorithms. In: Heckendorn R. B. (ed.) Proceedings of the 2000 Genetic and Evolutionary Computation Conference - Workshop Program. Workshop "The Next Ten Years of Scheduling Research", Morgan Kaufmann Publishers, San Francisco, CA, pp. 121–126 (2001)

[35] Merkle D., Middendorf M., Schmeck H.: Ant colony optimization for resource-constrained project scheduling. IEEE Transactions on Evolutionary Computation 6(4):333–346 (2002)

[36] Meyer B.: Convergence control in ACO. In: Genetic and Evolutionary Computation Conference (GECCO), Seattle, WA, late-breaking paper available on CD (2004)

[37] Pellegrini P., Favaretto D., Moretti E.: Exploration in stochastic algorithms: An application on MAX–MIN Ant System. In: Nature Inspired Cooperative Strategies for Optimization (NICSO 2008), Studies in Computational Intelligence, vol. 236, Springer, Berlin, Germany, pp. 1–13 (2009)

[38] Pilat M. L., White T.: Using genetic algorithms to optimize ACS-TSP. In: [16], pp. 282–287 (2002)

[39] Randall M.: Near parameter free ant colony optimisation. In: Dorigo M., et al. (eds.) Ant Colony Optimization and Swarm Intelligence: 4th International Workshop, ANTS 2004, Lecture Notes in Computer Science, vol. 3172, Springer, Heidelberg, Germany, pp. 374–381 (2004)

[40] Randall M., Montgomery J.: Candidate set strategies for ant colony optimisation. In: [16], pp. 243–249 (2002)

[41] Stützle T.: ACOTSP: A software package of various ant colony optimization algorithms applied to the symmetric traveling salesman problem. URL http://www.aco-metaheuristic.org/aco-code/ (2002)

[42] Stützle T., Hoos H. H.: MAX–MIN Ant System. Future Generation Computer Systems 16(8):889–914 (2000)

[43] White T., Pagurek B., Oppacher F. Connection management using adaptive mobile agents. In: Arabnia H. R. (ed.) Proceedings of the International Conference on Parallel and Distributed Processing Techniques and Applications (PDPTA'98), CSREA Press, pp. 802–809 (1998)

[44] Zilberstein S.: Using anytime algorithms in intelligent systems. AI Magazine 17(3):73–83 (1996)

[45] Zlochin M., Birattari M., Meuleau N., Dorigo M.: Model-based search for combinatorial optimization: A critical survey. Annals of Operations Research 131(1–4):373–395 (2004)

Part III
New Directions and Applications

Chapter 9
Continuous Search in Constraint Programming

Alejandro Arbelaez, Youssef Hamadi, and Michèle Sebag

9.1 Introduction

In Constraint Programming, properly crafting a constraint model which captures all the constraints of a particular problem is often not enough to ensure acceptable runtime performance. Additional tricks, e.g., adding redundant and channeling constraints, or using some global constraint (depending on your constraint solver) which can efficiently do part of the job, are required to achieve efficiency. Such tricks are far from being obvious, unfortunately; they do not change the solution space, and users with a classical mathematical background might find it hard to see why adding redundancy helps.

For this reason, users are often left with the tedious task of tuning the search parameters of their constraint solver, and this again is both time consuming and not necessarily straightforward. Parameter tuning indeed appears to be conceptually simple (i/ try different parameter settings on representative problem instances, ii/ pick up the setting yielding the best average performance). Still, most users easily consider instances which are not representative of their problem, and get misled.

The goal of the presented work is to allow any user to eventually get his or her constraint solver to achieve top performance on his or her problems. The proposed

Alejandro Arbelaez
Microsoft-INRIA joint lab, Orsay, France
e-mail: alejandro.arbelaez@inria.fr

Youssef Hamadi
Microsoft Research, Cambridge, CB3 0FB, UK
LIX, École Polytechnique, F-91128 Palaiseau, France
e-mail: youssefh@microsoft.com

Michèle Sebag
Project-Team TAO, INRIA Saclay, Ile-de-France
LRI (UMR CNRS 8623), Orsay, France
Microsoft-INRIA joint lab, Orsay, France
e-mail: michele.sebag@inria.fr

Y. Hamadi et al. (eds.), *Autonomous Search*,
DOI 10.1007/978-3-642-21434-9_9,
© Springer-Verlag Berlin Heidelberg 2011

approach is based on the concept of Continuous Search (CS), on gradually building a heuristics model tailored to the user's problems, and on mapping a problem instance onto some appropriate parameter setting. An important contribution to the state of the art (see [42] for a recent survey and Section 9.6 for more) is the relaxing of the requirement that a large set of representative problem instances be available beforehand to support off-line training. The heuristics model is initially empty (set to the initial default parameter setting of the constraint solver) and it is enriched along the course of lifelong learning approach, exploiting the problem instances submitted by the user to the constraint solver.

Formally, CS interleaves two functioning modes. In production or exploitation mode, the instance submitted by the user is processed by the constraint solver; the current heuristics model is used to parameterize the constraint solver based on the instance at hand. In learning or exploration mode, CS reuses the last submitted instance, running other heuristics than the one used in production mode in order to find which heuristics would have been most efficient for this instance. CS thus gains some expertise with regard to this particular instance, which is used to refine the general heuristics model through Machine Learning (Section 9.2.3). During the exploration mode, new information is thus generated and exploited in order to refine the heuristics model in a transparent manner, without requiring the user's input and by only using the idle computer's CPU cycles.

Our claim is that the CS methodology is realistic (most computational systems are always on, especially production ones) and compliant with real-world settings, where the solver is critically embedded within large and complex applications. The CS computational cost must be balanced with the huge computational cost of off-line training [19, 25, 24, 35, 46, 47]. Finally, lifelong learning appears a good way to construct an efficient and agnostic heuristics model, and able to adapt to new modeling styles or new classes of problem.

This chapter is organized as follows. Background material is presented in Section 9.2. Section 9.3 introduces the Continuous Search paradigm. Section 9.4 details the proposed algorithm. Section 9.5 reports on its experimental validation. Section 9.6 discusses related work, and the chapter concludes with some perspectives on further studies.

9.2 Background and Notations

This section briefly introduces definitions used in the rest of the chapter.

9.2.1 Constraint Satisfaction Problems

Definition 1 *A Constraint Satisfaction Problem (CSP) is a triple $\langle X, D, C \rangle$ where,* $X = \{X_1, X_2, \ldots, X_n\}$ *represents a set of n variables.*

$D = \{D_1, D_2, \ldots, D_n\}$ *represents the set of associated domains, i.e., possible values for the variables.*

$C = \{C_1, C_2, \ldots, C_m\}$ *represents a finite set of constraints.*

Each constraint C_i involves a set of variables and is used to restrict the combinations of values between these variables. The degree $deg(X_i)$ of a variable is the number of constraints involving X_i and $dom(X_i)$ denotes the current domain of X_i.

Solving a CSP involves finding a solution, i.e., an assignment of values to variables such that all constraints are satisfied. If a solution exists the problem is satisfiable, and it is unsatisfiable otherwise. A depth-first search backtracking algorithm can be used to tackle CSPs. At each step of the search process, an unassigned variable X and a valid value v for X ($v \in dom(X)$) are selected and a constraint $X = v$ is added to the search process. In case of infeasibility, the search backtracks and can undo previous decisions with new constraints, e.g., $X \neq v$. The search thus explores a so-called search tree, where each leaf node corresponds to a solution. In the worst-case scenario the search process has to explore exponential space. Therefore, it is necessary to combine the exploration of variables and values with a look-ahead strategy able to narrow the domains of the variables and reduce the remaining search space through constraint propagation. Restarting the search engine [23, 29, 31] helps to reduce the effects of early mistakes in the search process. A restart is done when some cutoff limit in the number of failures (backtracks) is reached (i.e., at some point in the search tree).

9.2.2 Search Heuristics

In this section, we briefly review the basic ideas and principles of the last generation of CSP heuristics. As we pointed out above, a CSP heuristic includes a variable/value selection procedure. Most common value selection strategies can be summarized as follows: *min-value* selects the minimum value, *max-value* selects the maximum value, *mid-value* selects the median value and *rand-value* selects a random value from the remaining domain. On the other hand, variable selection heuristics are usually more important and include more sophisticated algorithms.

In [11], Boussemart et al. proposed *wdeg* and *dom-wdeg* to focus the search on difficult constraints. The former selects the variable that is involved in most failed constraints. A weight is associated with each constraint and incremented each time the constraint fails. Using this information *wdeg* selects the variable whose weight is maximal. The latter, *dom-wdeg*, is a mixture of the current domain and the weighted degree of a variable, choosing the variable that minimizes the ratio $\frac{dom}{wdeg}$.

In [36], Refalo proposed the *impact* dynamic variable-value selection heuristic. The approach of this strategy is to measure the *impact* on the search space reduction, given by the Cartesian product of the variables (i.e., $|v_1| \times \ldots \times |v_n|$); using this information the impact of a variable is averaged over all previous decisions in the search tree and the variable with the highest impact is selected.

It is also worth mentioning another category of dynamic variable heuristics, which corresponds to *min-dom* [26] and *dom-deg*. The former, "first-fail principle: try first where you are more likely to fail", chooses the variable with minimum size domain; the latter selects the variable with the smallest domain that is involved in most constraints (i.e., minimizes the ratio $\frac{dom}{deg}$). While only deterministic heuristics will be considered in this chapter, the proposed approach can be extended to randomized algorithms by following the approach proposed in [27].

9.2.3 Supervised Machine Learning

Supervised Machine Learning exploits data labelled by the expert to automatically build hypotheses that emulate the expert's decisions [44]. Only the binary classification case will be considered in the following. Formally, a learning algorithm processes a training set $\mathscr{E} = \{(x_i, y_i), x_i \in \Omega, y_i \in \{1, -1\}, i = 1 \ldots n\}$ made of n examples (x_i, y_i), where x_i is the example description (e.g., a vector of values $\Omega = \mathbf{R}^d$) and y_i is the associated label; example (x, y) is referred to as positive (or negative) iff y is 1 (or minus 1). The learning algorithm outputs a hypothesis $f : \Omega \mapsto Y$ associating with each example description x a label $y = f(x)$ in $\{1, -1\}$. Among ML applications are pattern recognition, from computer vision to fraud detection [30], protein function prediction [2], game playing [21], and autonomic computing [38].

Among the prominent ML algorithms are *Support Vector Machines* (SVMs) [16]. Linear SVM considers real-valued positive and negative instances ($\Omega = \mathbf{R}^d$) and constructs the separating hyperplane that maximizes the margin (Figure 9.1), i.e., the minimal distance between the examples and the separating hyperplane. The margin maximization principle provides good guarantees about the stability of the solution and its convergence towards the optimal solution when the number of examples increases.

The linear SVM hypothesis $f(x)$ can be described from the sum of the scalar products of the current instance x and some of the training instances x_i, called support vectors:

$$f(x) = <w, x> + b = \sum \alpha_i <x_i, x> + b$$

The SVM approach can be extended to non-linear spaces by mapping the instance space Ω into a more expressive feature space $\Phi(\Omega)$. This mapping is made implicit through the so-called *kernel trick*, by defining $K(x, x') = < \Phi(x), \Phi(x') >$; it preserves all good SVM properties provided the kernel is positive definite. Among the most widely used kernels are the Gaussian kernel ($K(x, x') = exp\{-\frac{\|x-x'\|^2}{\sigma^2}\}$) and the polynomial kernel ($K(x, x') = (<x, x'> + c)^d$). More complex separating hypotheses can be built on such kernels

$$f(x) = \sum \alpha_i K(x_i, x) + b$$

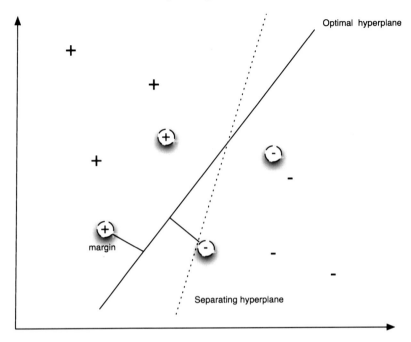

Fig. 9.1: Linear Support Vector Machine. The optimal hyperplane is the one maximizing the minimal distance to the example

using the same learning algorithm core as in the linear case. In all cases, a new instance x is classified as positive (or negative) if $f(x)$ is positive (or negative).

9.3 Continuous Search in Constraint Programming

The Continuous Search paradigm, illustrated on Figure 9.2, considers a functioning system governed from a heuristics model (which could be expressed as a set of rules, a knowledge base or a neural net). The core of continuous search is to exploit the problem instances submitted to the system in a three step process:

1. unseen problem instances are solved using the current heuristics model;
2. the same instances are solved with other heuristics, yielding new information. This information associates with the description x of the example (accounting for the problem instance and the heuristics) a Boolean label y (the heuristics either improves or does not improve on the current heuristics model);
3. the training set $\mathcal{E}$, augmented with these new examples (x, y), is used to revise or relearn the heuristics model.

The exploitation or production mode (step 1) aims at solving new problem instances as quickly as possible. The exploration or learning mode (steps 2 and 3) aims at learning a more accurate heuristics model.

Definition 2 *A continuous search system is endowed with a heuristics model, which is used as is to solve the current problem instance in production mode, and which is improved using instances previously seen in learning mode.*

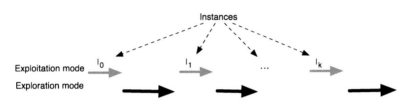

Fig. 9.2: Continuous Search scenario

Initially, the heuristics model of a continuous search system is empty, that is, it is set to the default settings of the search system. In the proposed CS-based constraint programming, the default setting is a given heuristic, *DEF* in the following (Section 9.4). Arguably, *DEF* is a reasonably good strategy on average; the challenge is to improve on *DEF* for the types of instances encountered in production mode.

9.4 Dynamic Continuous Search

The Continuous Search paradigm is applied to a restart-based constraint solver, defining the *dyn-CS* algorithm. After giving a general overview of *dyn-CS*, this section details the different modules thereof.

Figure 9.3 depicts the general scheme of *dyn-CS*. The constraint-based solver involves several restarts of the search. A restart is launched after the number of backtracks in the search tree reaches a user-specified threshold. The search stops after a given time limit. Before starting the tree-based search and after each restart, the description x of the problem instance is computed (Section 9.4.1). We will call these descriptions checkpoints.

The global picture of the Continuous Search paradigm is described in Figure 9.4. In production (or exploitation) mode, the heuristic model f is used to compute the heuristic $f(x)$ to be applied to the entire checkpoint window, i.e., until the next restart. Not to be confused with the *choice point*, which selects a variable/value pair at each node in the search tree, *dyn-CS* selects the most promising heuristic at a given

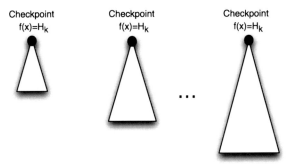

Fig. 9.3: *dyn-CS*: selecting the best heuristic at each restart point

checkpoint and uses it for the whole checkpoint window. In learning (or exploration) mode, other combinations of heuristics are applied (Section 9.4.4) and the eventual result (depending on whether the other heuristics improved on heuristic $f(x)$) leads to the building of training examples (Section 9.4.3). The augmented training set is used to relearn $f(x)$.

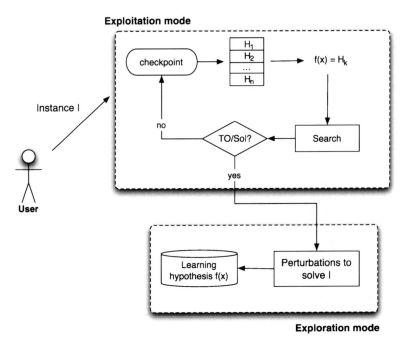

Fig. 9.4: Continuous Search in Constraint Programming

9.4.1 Representing Instances: Feature Definition

At each checkpoint (or restart), the description of the problem instance is computed, including its static and dynamic features.

While a few of these descriptors had already been used in SAT portfolio solvers [27, 47], many descriptors had to be added as CSPs are more diverse than SAT instances: SAT instances only involve Boolean variables and clauses, in contrast with CSPs, which use variables with large domains, and a variety of constraints and pruning rules [6, 8, 34].

9.4.1.1 Static Features

Static features encode the general description of a given problem instance; they are computed once for each instance as they are not modified during the resolution process. The static features also allow one to discriminate between types of problems and different instances.

- **Problem definition** (four features): Number of variables, constraints, variables assigned/not assigned at the beginning of the search.
- **Variables size information** (six features): Size prod, sum, min, max, mean and variance of all variable domain sizes.
- **Variables degree information** (eight features): min, max, mean and variance of all variable degrees (or variable domains/degrees).
- **Constraints information** (six features): The degree (or arity) of a given constraint c is represented by the total number of variables involved in c. Likewise the size of c is represented by the product of its corresponding variable domain sizes. Taking into account this information, the following features are computed: min, max, and mean of constraint size and degree.
- **Filtering cost category** (eight features): Each constraint c is associated with a category[1]. In this way, we compute the number of constraints for each category. Intuitively, each category represents the implementation cost of the filtering algorithm. $Cat = \{Exponential, Cubic, Quadratic, Linear expensive, Linear cheap, Ternary, Binary, Unary\}$, where *Linear expensive* (or *cheap*) indicates the complexity of a linear equation constraint and the last three categories indicate the number of variables involved in the constraint. More information about the filtering cost category can be found in [20].

9.4.1.2 Dynamic Features

Two kinds of dynamic features are used to monitor the performance of the search effort at a given checkpoint: global statistics describe the progress of the overall search process; local statistics check the evolution of a given strategy.

---

[1] Out of eight categories, detailed in
http://www.gecode.org/doc-latest/reference/classGecode_1_1PropCost.html.

- **Heuristic criteria** (15 features): each heuristic criterion (e.g., *wdeg*, *dom-wdeg*, *impacts*) is computed for each variable; their `prod`, `min`, `max`, `mean` and `variance` over all variables are used as features.
- **Constraint weights** (12 features): the `min`, `max`, `mean` and `variance` of all constraint weights (i.e., constraints *wdeg*). Additionally, the `mean` for each filtering cost category is used as a feature.
- **Constraints information** (three features): the `min`, `max` and `mean` of a constraint's *run-prop*, where *run-prop* indicates the number of times the propagation engine has called the filtering algorithm of a given constraint.
- **Checkpoint information** (33 features): for every checkpoint$_i$ relevant information from the previous checkpoint$_{i-1}$ (when available) is included into the feature vector. From checkpoint$_{i-1}$ we include the total number of nodes and maximum search depth. From the latest non-failed node, we consider the total number of assigned variables, satisfied constraints, sum of variables *wdeg* (or size and degree) and product of variable degrees (or *domain*, *wdeg* and *impacts*) of non-assigned variables. Finally, using the previous 11 features the `mean` and `variance` are computed taking into account all visited checkpoints.

The attributes listed above include a collection of 95 features.

9.4.2 Feature Preprocessing

Feature preprocessing is a very important step in Machine Learning [45], which can significantly improve the prediction accuracy of the learned hypothesis. Typically, the descriptive features detailed above are on different scales; the number of variables and/or constraints can be high while the impact of (variable, value) is between 0 and 1. A data normalization step is performed using the *min-max normalization* [41] formula:

$$v'_i = \left(\frac{v_i - min_i}{max_i - min_i} \right) \times (max_{new} - min_{new}) + min_{new}$$

Where $min_{new} = -1$, $max_{new} = 1$ and min_i (or max_i) corresponds to the normalization value for the *ith* feature. In this way, feature values are scaled down to $[-1, 1]$. Although selecting the most informative features might improve the performance, in this chapter we do not consider any feature selection algorithm, and only features that are constant over all examples are removed as they offer no discriminant information.

9.4.3 Learning and Using the Heuristics Model

The selection of the best heuristic for a given problem instance is formulated as a binary classification problem, as follows. Let $\mathcal{H}$ denote the set of k candidate heuristics, two particular elements in $\mathcal{H}$ being *DEF* (the default heuristic yielding reasonably good results on average) and *dyn-CS*, the (dynamic) ML-based heuristic model initially set to *DEF*.

Definition 3 *Each training example* $p_i = (x_i, y_i)$ *is generated by applying some heuristic h* ($h \in \mathcal{H}, h \neq dyn\text{-}CS$) *at some checkpoint in the search tree of a given problem instance. Description* x_i ($\in \mathbf{R}^{97}$) *is made of the static feature values describing the problem instance, the dynamic feature values computed at this checkpoint and describing the current search state, and two additional features: checkpoint-id, which gives the number of checkpoints up to now, and cutoff-information, which gives the cutoff limit of the next restart. The associated label* y_i *is positive iff the associated runtime (using heuristic h instead of dyn-CS at the current checkpoint) improves on the heuristic model-based runtime (using dyn-CS at every checkpoint); otherwise, label* y_i *is negative.*

If the problem instance cannot be solved whatever the heuristic used, i.e., times out during the exploration or exploitation mode, it is discarded (since the associated training examples do not provide any relevant information).

In production mode, the hypothesis f learned from the above training examples (their generation is detailed in next subsection) is used as follows:

Definition 4 *At each checkpoint, for each* $h \in \mathcal{H}$, *the description* x_h *and the associated value* $f(x_h)$ *are computed.*
If there exists a single h such that $f(x_h)$ *is positive, it is selected and used in the subsequent search effort.*
If there exists several heuristics with positive $f(x_h)$, *the one with maximal value is selected[2].*
If $f(x_h)$ *is negative for all h, the default heuristic* DEF *is selected.*

9.4.4 Generating Examples in Exploration Mode

The Continuous Search paradigm uses the idle computer's CPU cycles to explore different heuristic combinations on the last seen problem instance, and to see whether one could have done better than the current heuristics model on this instance. The rationale for this exploration is that improving on the last seen instance (albeit meaningless from a production viewpoint since the user already got a solution) will deliver useful indications as to how to best deal with other similar instances. In this way, the heuristics model will be tailored to the distribution of problem instances actually dealt with by the user.

---

[2] The rationale for this decision is that the margin, i.e., the distance of the example from the separating hyperplane, is interpreted as the confidence of the prediction [44].

The CS exploration proceeds by slightly perturbing the heuristics model. Let *dyn-CS* $^{-i,h}$ denote the policy defined as: use heuristics model *dyn-CS* at all checkpoints except the *i*th one, and use heuristic *h* at *i* checkpoint.

Algorithm 1: Exploration time (instance: $\mathscr{I}$)

 1: $\mathscr{E} = \{\}$ //initialize the training set
 2: **for all** *i* in checkpoints($\mathscr{I}$) // loop over checkpoints ($\mathscr{I}$) **do**
 3:     **for all** *h* in $\mathscr{H}$ // loop over all heuristics **do**
 4:         Compute *x* describing the current checkpoint and *h*
 5:         **if** $h \neq dyn\text{-}CS$ **then**
 6:             Launch *dyn-CS* $^{-i,h}$
 7:             Define $y = 1$ iff *dyn-CS* $^{-i,h}$ improves on *dyn-CS* and -1 otherwise
 8:             $\mathscr{E} \leftarrow \mathscr{E} \cup \{x,y\}$
 9:         **end if**
10:     **end for**
11: **end for**
12: return $\mathscr{E}$

Algorithm 1 describes the proposed Exploration mode for Continuous Search. A limited number (ten in this work) of checkpoints in the *dyn-CS* based resolution of instance $\mathscr{I}$ are considered (line 2); for each checkpoint and each heuristic *h* (distinct from the *dyn-CS*), a lesion study is conducted, applying *h* instead of *dyn-CS* at the *i*th checkpoint (heuristics model *dyn-CS* $^{-i,h}$); the example (described from the *i*th checkpoint and *h*) is labelled positive iff *dyn-CS* $^{-i,h}$ improves on *dyn-CS*, and added to the training set $\mathscr{E}$; once the exploration mode for a given instance is finished the hypothesis model is updated by retraining the SVM, including the feature preprocessing, as stated in Section 9.4.2.

9.4.5 Imbalanced Examples

It is well known that one of the heuristics often performs much better than the others for a particular distribution of problems [15]. Accordingly, negative training examples considerably outnumber the positive ones (it is difficult to improve on the winning heuristics). This phenomenon, known as *Imbalanced distribution*, might severely hinder the SVM algorithm [1]. Simple ways of enforcing a balanced distribution in such cases, intensively examined in the literature and considered in earlier work [4], are oversampling examples in the minority class (generating additional positive examples by perturbing the available ones according to a Gaussian distribution) and/or undersampling examples in the majority class.

Another option is to use prior knowledge to rebalance the training distribution. Formally, instead of labeling an example positive (or negative) iff its associated runtime is strictly less (or greater) than that of the heuristic model, we consider the difference between the runtimes. If the difference is less than some tolerance value *dt*, then the example is relabeled as positive.

The number of positive examples and hence the coverage of the learned heuristics model increases with *dt*; in the experiments (Section 9.5), *dt* is set to 20% of the time limit iff *time-exploitation* (time required to solve a given instance in production mode) is greater than 20% of the time limit; otherwise *dt* is set to *time-exploitation*.

9.5 Experimental Validation

This section reports on the experimental validation of the proposed Continuous Search approach. All tests were conducted on Linux Mandriva-2009 boxes with 8 GB of RAM and 2.33 Ghz Intel processors.

9.5.1 Experimental Setting

The presented experiments consider 496 CSP instances taken from different repositories.

- **nsp**: 100 *nurse-scheduling* instances from the MiniZinc[3] repository.
- **bibd**: 83 *Balanced Incomplete Block Design* instances from the XCSP [39] repository, translated into Gecode using Tailor [22].
- **js**: 130 *Job Shop* instances from the XCSP repository.
- **geom**: 100 *Geometric* instances from the XCSP repository.
- **lfn**: 83 *Langford number* instances, translated into Gecode using global and channelling constraints.

The learning algorithm used in the experimental validation of the proposed approach is a Support Vector Machine with a Gaussian kernel using the libSVM implementation with default parameters [14]. All considered CSP heuristics (Section 9.2) are homemade implementations integrated in the Gecode 2.1.1 [20] constraint solver. *dyn-CS* was used as a heuristics model on top of the heuristics set $\mathscr{H} = \{dom\text{-}wdeg, wdeg, dom\text{-}deg, min\text{-}dom, impacts\}$, with *min-value* as the value selection heuristic. The cutoff value used to restart the search was initially set to 1,000 and the cutoff increase policy to ×1.5; the same cutoff policy is used in all the experimental scenarios.

Continuous Search was assessed against the best two dynamic variable ordering heuristics on the considered problems, namely *dom-wdeg* and *wdeg*. It must be noted that Continuous Search, being a lifelong learning system, will depend on the curriculum, that is, the order of the submitted instances. If the user "pedagogically" starts by submitting informative instances first, the performance in the first

[3] http://www.g12.cs.mu.oz.au/minizinc/download.html.

stages will be better than if untypical and awkward instances are considered first. For the sake of fairness, the performance reported for Continuous Search on each problem instance is the median performance over ten random orderings of the CSP instances.

9.5.2 Practical Performances

The first experimental scenario involves a timeout of five minutes. Figure 9.5 highlights the Continuous Search results on Langford number problems, compared to *dom-wdeg* and *wdeg*. The x-axis gives the number of problems solved and the y-axis presents the cumulated runtime. The (median) *dyn-CS* performance (gray line) is satisfactory as it solves 12 more instances than *dom-wdeg* (black line) and *wdeg* (light gray line). The dispersion of the *dyn-CS* results depending on the instance ordering is depicted through the set of dashed lines. Indeed, traditional portfolio approaches such as [27, 40, 47] do not present such performance variations as they assume a complete set of training examples to be available beforehand.

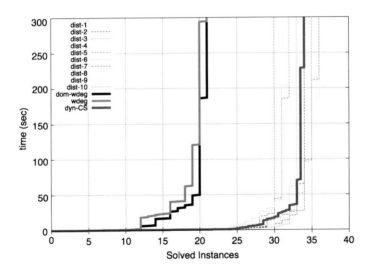

Fig. 9.5: Langford number (lfn)

Figures 9.6 to 9.9 depict the performance of *dyn-CS*, *dom-wdeg* and *wdeg* on all other problem families, respectively (bibd, js, nsp, and geom). On the bibd (Figure 9.7) and js (Figure 9.8) problems, the best heuristic is *dom-wdeg*, solving three more instances than *dyn-CS*. Note that *dom-wdeg* and *wdeg* coincide on bibd since all decision variables are Boolean.

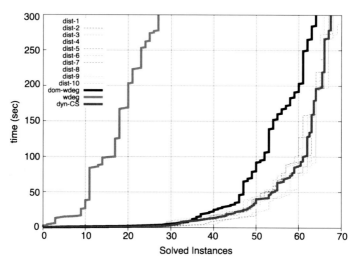

Fig. 9.6: Geometric (geom)

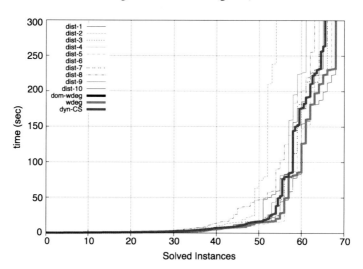

Fig. 9.7: Balanced incomplete block designs (bibd)

On nsp (Figure 9.9), *dyn-CS* solves nine more problems than *dom-wdeg*, but is outperformed by *wdeg* by 11 problems. On geom (Figure 9.6), *dyn-CS* improves on the other heuristics, solving respectively three more instances and 40 more instances than *dom-wdeg* and *wdeg*.

These results suggest that *dyn-CS* is most often able to pick up the best heuristic on a given problem family, and sometimes able to significantly improve on the best of the available heuristics.

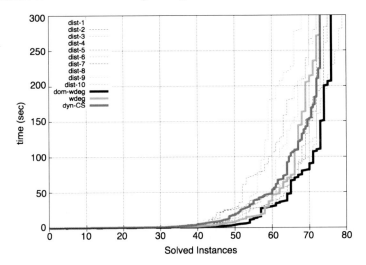

Fig. 9.8: Job Shop (js)

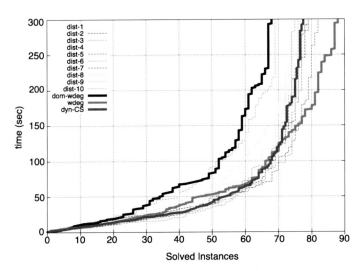

Fig. 9.9: Nurse Scheduling (nsp)

All experimental results concerning the first scenario are summarized in Table 9.1, reporting for each considered heuristic the number of instances solved (#sol), the total computational cost for all instances (time, in hours), and the average time per instance (avg-time, in minutes), over all problem families. These results confirm that *dyn-CS* outperforms *dom-wdeg* and *wdeg*, solving respectively 18 and 41 instances more out of 315. Furthermore, it shows that *dyn-CS* is slightly faster than the other heuristics, with an average time of 2.11 minutes, against 2.39 for *dom-wdeg* and 2.61 for *wdeg*.

Problem	dom-wdeg			wdeg			dyn-CS		
	#sol	time(h)	avg-time(m)	#sol	time(h)	avg-time(m)	#sol	time(h)	avg-time(m)
nsp	68	3.9	2.34	**88**	**2.6**	**1.56**	77	2.9	1.74
bibd	**68**	**1.8**	**1.37**	68	1.8	1.37	65	2.0	1.44
js	**76**	**4.9**	**2.26**	73	5.1	2.35	73	5.2	2.4
lfn	21	5.2	3.75	21	5.3	3.83	**33**	**4.1**	**2.96**
geom	64	3.9	2.34	27	6.8	4.08	**67**	**3.3**	**1.98**
Total	297	19.7	2.39	274	21.6	2.61	**315**	**17.5**	**2.11**

Table 9.1: Total solved instances (five minutes)

The second experimental results using a timeout of three minutes are presented in table 9.2; as can be observed, decreasing the time limit drastically reduces the total number of solved instances for *dom-wdeg* and *wdeg*. Therefore, selecting the right heuristic becomes critical. Here *dyn-CS* is able to solve 24 and 45 more instances than *dom-wdeg* and *wdeg*.

Problem	dom-wdeg			wdeg			dyn-CS		
	#sol	time(h)	avg-time(m)	#sol	time(h)	avg-time(m)	#sol	time(h)	avg-time(m)
nsp	61	2.8	1.68	**81**	**2.1**	**1.26**	75	2.2	1.32
bibd	**62**	**1.3**	**0.94**	62	1.3	0.94	60	1.4	1.01
js	**74**	**3.1**	**1.43**	69	3.3	1.52	67	3.4	1.57
lfn	20	3.2	2.31	20	3.2	2.31	**32**	**2.5**	**1.81**
geom	56	2.6	1.56	20	4.3	2.58	**63**	**2.2**	**1.32**
Total	273	13.0	1.57	252	14.2	1.72	**297**	**11.7**	**1.42**

Table 9.2: Total solved instances (three minutes)

Another interesting lesson learnt from the experiments concerns the difficulty of the underlying learning problem and the generalization error of the learnt hypothesis. The generalization error in the Continuous Search framework is estimated by 10-fold Cross-Validation [5] on the whole training set (including all training examples generated in the exploration mode). Table 9.3 reports on the predictive accuracy of the SVM algorithm (with the same default setting) on all problem families, with an average accuracy of 67.8%. As could have been expected, the predictive accuracy is correlated to the performance of Continuous Search: the problems with best accuracy and best performance improvement are geom and lfn.

To give an idea of order, 62% predictive accuracy was reported in the context of SATzilla [47], aimed at selecting the best heuristic in a portfolio.

A direct comparison of the predictive accuracy might however be biased. On the one hand SATzilla errors are attributed, according to its authors, to the selection of some near-optimal heuristics; on the other hand, Continuous Search involves several selection steps (in each checkpoint) and can thus compensate for earlier errors.

Timeout	bibd	nsp	geom	js	lfn	Total
3 Min	64.5%	64.2%	79.2%	65.6%	68.2%	68.3%
5 Min	63.2%	58.8%	76.9%	63.6%	73.8%	67.3%
Average	63.9%	61.5%	78.0%	64.6%	71.0%	**67.8%**

Table 9.3: Predictive accuracy of the heuristics model (10-fold Cross-Validation)

9.5.3 Exploration Time

Now we turn our attention to the CPU time required to complete the exploration mode. Table 9.4 shows the total exploration time considering a timeout of five and three minutes for each problem family; the `median` value is computed taking into account all instance orderings, and `instance` estimates the total exploration time for a single problem instance.

Problem	TO = 5 Min		TO = 3 Min	
	Median	Instance	Median	Instance
bibd	73.6	0.9	48.3	0.6
nsp	151.1	1.5	106.8	1.1
geom	71.6	0.7	37.6	0.4
lfn	161.8	1.9	100.3	1.0
js	215.6	1.7	135.6	1.0

Table 9.4: Exploration time (in Hours)

As can be seen the time required to complete the exploration mode after solving a problem-instance is on average no longer than two hours. On the other hand, since the majority of the instances for the bibd and geom problems can be quickly solved by *dyn-CS*, it is not surprising that their required time is significantly less than that of nsp, js and lfn.

9.5.4 The Power of Adaptation

Our third experimental test combines instances from different domains in order to show how CS is able to adapt to changing problem distributions. Indeed, unlike classical portfolio-based approaches which can only be applied if the training and exploitation sets come from the same domain, CS can adapt to changes and provide top performance even if the problems change.

In this context, Table 9.5 reports the results on the geom (left) and bibd (right) problems by considering the following two scenarios. In the first scenario, we are going to emulate a portfolio-based search which would use the wrong domain to train. In *nsp-geom*[‡], CS incrementally learns while solving the 100 nsp instances,

Problem	#Sol	time (h)	Problem	#Sol	time (h)
nsp-geom‡	55	4.1	lfn-bibd‡	23	5.3
nsp-geom†	67	3.4	lfn-bibd†	63	2.3

Table 9.5: Total solved instances (five minutes)

and then solves one by one the 100 geom instances. However, when switching to this second domain, incremental learning is switched off, and checkpoint adaptation uses the model learnt on nsp. In the second scenario, *nsp-geom†*, we solve nsp, and then geom instances one by one, but this time we keep the incremental learning on when switching from the first domain to the second one, as if CS were not aware of the transition.

As we can see in the first line of the table, training on the wrong domain gives poor performance (55 instances solved in 4.1 hours). In contrast, the second line shows that CS can recover from training on the wrong domain thanks to its incremental adaptation (solving 67 instances in 3.4 hours). The right hand of the table reports similar results for the bibd problem.

As can be observed in nsp-geom† and lfn-bibd†, CS successfully identifies the new distribution of problems solving, respectively the same number and two fewer instances than geom and bibd when CS is only applied to this domain starting from scratch. However, the detection of the new distribution introduces an overhead in the solving time (see results for a single domain in Table 9.1).

9.6 Related Works

This section briefly reviews and discusses some related works, devoted to heuristic selection within CP and SAT solvers. This section is divided into three main categories: *per-instance learning*, *per-class learning* and *other work*.

9.6.1 Per-Instance Learning

This work is devoted to using supervised machine learning in a portfolio solver. The portfolio builds a model that maps a feature vector to the desired output (i.e., the best algorithm for a given instance). The two main considerations to take into account are the definition of a vector of features and the choice of a representative set of training examples.

SATzilla [47, 48] is a well-known SAT portfolio solver built upon a set of features. Roughly speaking, SATzilla includes two kinds of basic features: general features such as number of variables, number of propagators and local search features which actually probe the search space in order to estimate the difficulty of each problem instance for a given algorithm. The goal of SATzilla is to learn a run-time prediction function by using a linear regression model. Notice that this portfolio solver follows a traditional batch learning scenario in which all training examples are provided at once in order to learn a machine learning model.

CPHydra [32], the overall best solver in the 2008 CSP competition,[4] is a portfolio approach based on case-based reasoning [37]; it maintains a database with all solved instances (so-called *cases*). When a new instance *I* arrives a set of similar cases *C* is computed, and based on *C* CPHydra builds a switching policy which selects (and executes) a set of solvers that maximize the possibility of solving *I* within a given amount of time.

The approach most similar to the presented one is that of [40], which applies Machine Learning techniques to combine on-line heuristics into search tree procedures. Unfortunately, this work requires an important number of training instances to build a model with good generalization property. The Continuous Search approach was instead designed to take into account the fact that such a large and representative training set is missing in general, and should be constructed during the lifetime of the system.

In [12, 13] low-knowledge is used to select the best algorithm in the context of optimization problems; this work assumes a black-box optimization scenario where the user has no information about the problem or even about the domain of the problem, and the only known information is about the output (i.e., solution cost for each algorithm in the portfolio). Unfortunately, this mechanism is only applicable to optimization problems and cannot be used to solve CSPs.

9.6.2 Per-Class Learning

So far, we have presented machine learning approaches dealing with the heuristic selection problem. We switch our attention to the class-based methodologies dealing with the algorithm configuration problem whose objective is to find the best parameter configuration for a given algorithm in order to efficiently solve a given class of problems.

In [7], Birattari et al. presented F-Race, a racing algorithm in which candidates (i.e., parameter configurations) compete with each other, poor candidates are quickly discarded, and the search is focused on high-quality candidates. The race is repeated until one candidate remains or a given timeout is reached.

paramILS [28] is an Iterative Local search procedure to select the best parameter configurations for a given algorithm; it starts with an initial parameter configuration and then applies a set of local search operators to focus the search on promising candidates, avoiding the overtunning problem, which is the side-effect of considering a fixed number of training instances. It is worth mentioning that paramILS

---

[4] http://www.cril.univ-artois.fr/CPAI08/.

has been successfully used to tune algorithms from different domains, ranging from SAT to mixed integer programming problems.

Finally, Ansótegui et al. recently proposed GGA [3], a genetic algorithm to overcome the parameter algorithm configuration problem; this approach uses a gender separation strategy to reduce the number of executions of a given configuration candidate. An empirical comparison showed that GGA is able to find parameter configurations as good as the ones found by paramILS.

9.6.3 Other Work

To conclude the related work section, we highlight some strategies that have been developed to combine CSP heuristics during the resolution process.

The quickest first principle (QFP) [10] is a methodology for combining CSP heuristics. QFP relies on the fact that easy instances can frequently be solved by simple algorithms, while exceptionally difficult instances will require more complex heuristics. In this context, it is necessary to predefine an execution order for heuristics, and the switching mechanism is set according to the thrashing indicator; once the thrashing value of the current strategy reaches some cutoff value, it becomes necessary to continue the search procedure with the following heuristic in the sequence. Although QFP is a very elegant approach, this static methodology does not actually learn any information about solved instances.

The purpose of the *Adaptive Constraint Engine* (ACE) [18] is to unify the decision of several heuristics in order to guide the search process. Each heuristic votes for a possible variable/value decision to solve a CSP. Afterwards, a global controller selects the most appropriate variable/value pair according to previously (off-line) learnt weights associated with each heuristic. The authors however did not present any experimental scenario taking into account any restart strategy, although this nowadays is an essential part of constraint solvers. In the same way, Petrovic and Epstein in [33] showed that using a subset of powerful heuristics is more useful than using all available ones.

Finally, Combining Multiple Heuristics Online [43] and Portfolios with deadlines [46] are designed to build scheduler policies to adapt the execution of *blackbox* solvers during the resolution process. However, in these papers the switching mechanism is learnt or defined beforehand, while our approach relies on the use of machine learning to change the execution of heuristics on the fly.

9.7 Discussion and Perspectives

The main contribution of the presented approach, the Continuous Search framework, aims at designing a heuristics model tailored to the user's problem distribution, allowing him or her to get top performance from the constraint solver. The representative instances needed to train a good heuristics model are not assumed to be available beforehand; they are gradually built and exploited for improving the current heuristics model by stealing the idle CPU cycles of the computing system. Metaphorically speaking, the constraint solver uses its spare time to play against itself and gradually improve its strategy over time; further, this expertise is relevant to the real-world problems considered by the user, all the more so as it directly relates to the problem instances submitted to the system.

The experimental results suggest that Continuous Search is able to pick up the best of a set of heuristics on a diverse set of problems, by exploiting the incoming instances; in two out of five problems, Continuous Search swiftly builds up a mixed strategy, significantly overcoming all baseline heuristics. With the other classes of problems, its performance is comparable to that of the best two single heuristics. Our experiments also showed the capacity of adaptation of CS. Moving from one problem domain to another is possible thanks to its incremental learning capacity. This capacity is a major improvement over classical portfolio-based approaches, which only work when off-line training and exploitation use instances from the same domain.

Further work will investigate the use of Active Learning [9, 17] in order to select the most informative training examples and focus the exploration mode on the most promising heuristics. Another point regards the feature design; better features would be needed to get higher predictive accuracy, which governs the efficiency of the approach.

References

[1] Akbani, R., Kwek, S., Japkowicz, N.: Applying support vector machines to imbalanced datasets. In: J. F. Boulicaut, F. Esposito, F. Giannotti, D. Pedreschi (eds.) 15th European Conference on Machine Learning, *Lecture Notes in Computer Science*, vol. 3201, pp. 39–50. Springer, Pisa, Italy (2004)

[2] Al-Shahib, A., Breitling, R., Gilbert, D. R.: Predicting protein function by machine learning on amino acid sequences – a critical evaluation. BMC Genomics **78**(2) (2007)

[3] Ansótegui, C., Sellmann, M., Tierney, K.: A gender-based genetic algorithm for the automatic configuration of algorithms. In: I. P. Gent (ed.) 15th International Conference on Principles and Practice of Constraint Programming, *Lecture Notes in Computer Science*, vol. 5732, pp. 142–157. Springer, Lisbon, Portugal (2009)

[4] Arbelaez, A., Hamadi, Y., Sebag, M.: Online heuristic selection in constraint programming. In: International Symposium on Combinatorial Search. Lake Arrowhead, USA (2009)

[5] Bailey, T. L., Elkan, C.: Estimating the accuracy of learned concepts. In: IJ-CAI, pp. 895–901 (1993)

[6] Beldiceanu, N., Carlsson, M., Demassey, S., Petit, T.: Global constraint catalogue: Past, present and future. Constraints **12**(1), 21–62 (2007)

[7] Birattari, M., Stützle, T., Paquete, L., Varrentrapp, K.: A racing algorithm for configuring metaheuristics. In: W. B. Langdon, E. Cantú-Paz, K. E. Mathias, R. Roy, D. Davis, R. Poli, K. Balakrishnan, V. Honavar, G. Rudolph, J. Wegener, L. Bull, M. A. Potter, A. C. Schultz, J. F. Miller, E. K. Burke, N. Jonoska (eds.) Proceedings of the Genetic and Evolutionary Computation Conference, pp. 11–18. Morgan Kaufmann, New York, USA (2002)

[8] Bordeaux, L., Hamadi, Y., Zhang, L.: Propositional satisfiability and constraint programming: A comparative survey. ACM Comput. Surv. **38**(4) (2006)

[9] Bordes, A., Ertekin, S., Weston, J., Bottou, L.: Fast kernel classifiers with online and active learning. Journal of Machine Learning Research **6**, 1579–1619 (2005)

[10] Borrett, J. E., Tsang, E. P. K., Walsh, N. R.: Adaptive constraint satisfaction: The quickest first principle. In: W. Wahlster (ed.) 12th European Conference on Artificial Intelligence, pp. 160–164. John Wiley and Sons, Chichester, Budapest, Hungary (1996)

[11] Boussemart, F., Hemery, F., Lecoutre, C., Sais, L.: Boosting systematic search by weighting constraints. In: R. L. de Mántaras, L. Saitta (eds.) Proceedings of the 16th European Conference on Artificial Intelligence, pp. 146–150. IOS Press, Valencia, Spain (2004)

[12] Carchrae, T., Beck, J. C.: Low-knowledge algorithm control. In: D. L. McGuinness, G. Ferguson (eds.) Proceedings of the Nineteenth National Conference on Artificial Intelligence, Sixteenth Conference on Innovative Applications of Artificial Intelligence, pp. 49–54. AAAI Press / The MIT Press, San Jose, California, USA (2004)

[13] Carchrae, T., Beck, J. C.: Applying machine learning to low-knowledge control of optimization algorithms. Computational Intelligence **21**(4), 372–387 (2005)

[14] Chang, C. C., Lin, C. J.: LIBSVM: A library for support vector machines (2001). Software from `http://www.csie.ntu.edu.tw/~cjlin/libsvm`

[15] Correira, M., Barahona, P.: On the efficiency of impact based heuristics. In: P. J. Stuckey (ed.) 14th International Conference on Principles and Practice of Constraint Programming, *Lecture Notes in Computer Science*, vol. 5202, pp. 608–612. Springer, Sydney, Australia (2008)

[16] Cristianini, N., Shawe-Taylor, J.: An introduction to support vector machines and other kernel-based learning methods. Cambridge University Press (2000)

[17] Dasgupta, S., Hsu, D., Monteleoni, C.: A general agnostic active learning algorithm. In: J. C. Platt, D. Koller, Y. Singer, S. T. Roweis (eds.) Proceedings of the Twenty-First Annual Conference on Neural Information Processing Systems. MIT Press, Vancouver, British Columbia, Canada (2007)

[18] Epstein, S. L., Freuder, E. C., Wallace, R., Morozov, A., Samuels, B.: The Adaptive Constraint Engine. In: P. Van Hentenryck (ed.) 8th International Con-

ference on Principles and Practice of Constraint Programming, *Lecture Notes in Computer Science*, vol. 2470, pp. 525–542. Springer, NY, USA (2002)

[19] Gebruers, C., Hnich, B., Bridge, D. G., Freuder, E.C.: Using CBR to select solution strategies in constraint programming. In: H. Muñoz-Avila, F. Ricci (eds.) 6th International Conference on Case-Based Reasoning, Research and Development, *Lecture Notes in Computer Science*, vol. 3620, pp. 222–236. Springer, Chicago, IL, USA (2005)

[20] Gecode Team: Gecode: Generic constraint development environment (2006). Available from http://www.gecode.org

[21] Gelly, S., Silver, D.: Combining online and offline knowledge in UCT. In: Z. Ghahramani (ed.) Proceedings of the Twenty-Fourth International Conference on Machine Learning, *ACM International Conference Proceeding Series*, vol. 227, pp. 273–280. ACM, Corvalis, Oregon, USA (2007)

[22] Gent, I. P., Miguel, I., Rendl, A.: Tailoring solver-independent constraint models: A case study with essence' and minion. In: I. Miguel, W. Ruml (eds.) 7th International Symposium on Abstraction, Reformulation, and Approximation, *Lecture Notes in Computer Science*, vol. 4612, pp. 184–199. Springer, Whistler, Canada (2007)

[23] Gomes, C., Selman, B., Kautz, H.: Boosting combinatorial search through randomization. In: AAAI/IAAI, pp. 431–437 (1998)

[24] Gomes, C. P., Selman, B.: Algorithm portfolios. Artif. Intell. **126**(1-2), 43–62 (2001)

[25] Haim, S., Walsh, T.: Restart strategy selection using machine learning techniques. In: O. Kullmann (ed.) 12th International Conference on Theory and Applications of Satisfiability Testing, *Lecture Notes in Computer Science*, vol. 5584, pp. 312–325. Springer, Swansea, UK (2009)

[26] Haralick, R. M., Elliott, G. L.: Increasing tree search efficiency for constraint satisfaction problems. In: IJCAI, pp. 356–364. San Francisco, CA, USA (1979)

[27] Hutter, F., Hamadi, Y., Hoos, H. H., Leyton-Brown, K.: Performance prediction and automated tuning of randomized and parametric algorithms. In: F. Benhamou (ed.) 12th International Conference on Principles and Practice of Constraint Programming, *Lecture Notes in Computer Science*, vol. 4204, pp. 213–228. Springer, Nantes, France (2006)

[28] Hutter, F., Hoos, H. H., Stützle, T.: Automatic algorithm configuration based on local search. In: Proceedings of the Twenty-Second AAAI Conference on Artificial Intelligence, pp. 1152–1157. AAAI Press, Vancouver, British Columbia, Canada (2007)

[29] Kautz, H. A., Horvitz, E., Ruan, Y., Gomes, C. P., Selman, B.: Dynamic restart policies. In: AAAI/IAAI, pp. 674–681 (2002)

[30] Larochelle, H., Bengio, Y.: Classification using discriminative restricted Boltzmann machines. In: W. W. Cohen, A. McCallum, S. T. Roweis (eds.) Proceedings of the Twenty-Fifth International Conference on Machine Learning, *ACM International Conference Proceeding Series*, vol. 307, pp. 536–543. ACM, Helsinki, Finland (2008)

[31] Luby, M., Sinclair, A., Zuckerman, D.: Optimal speedup of Las Vegas algorithms. In: ISTCS, pp. 128–133 (1993)

[32] O'Mahony, E., Hebrard, E., Holland, A., Nugent, C., O'Sullivan, B.: Using case-based reasoning in an algorithm portfolio for constraint solving. In: Proceedings of the 19th Irish Conference on Artificial Intelligence and Cognitive Science (2008)

[33] Petrovic, S., Epstein, S. L.: Random subsets support learning a mixture of heuristics. International Journal on Artificial Intelligence Tools **17**(3), 501–520 (2008)

[34] Prasad, M. R., Biere, A., Gupta, A.: A survey of recent advances in SAT-based formal verification. STTT **7**(2), 156–173 (2005)

[35] Pulina, L., Tacchella, A.: A multi-engine solver for quantified Boolean formulas. In: C. Bessiere (ed.) 13th International Conference on Principles and Practice of Constraint Programming, *Lecture Notes in Computer Science*, vol. 4741, pp. 574–589. Springer, Providence, RI, USA (2007)

[36] Refalo, P.: Impact-based search strategies for constraint programming. In: M. Wallace (ed.) 10th International Conference on Principles and Practice of Constraint Programming, *Lecture Notes in Computer Science*, vol. 2004, pp. 557–571. Springer, Toronto, Canada (2004)

[37] Richter, M. M., Aamodt, A.: Case-based reasoning foundations. Knowledge Eng. Review **20**(3), 203–207 (2005)

[38] Rish, I., Brodie, M., Ma, S., et al.: Adaptive diagnosis in distributed systems. IEEE Trans. on Neural Networks **16**, 1088–1109 (2005)

[39] Roussel, O., Lecoutre, C.: XML representation of constraint networks: Format XCSP 2.1. CoRR **abs/0902.2362** (2009)

[40] Samulowitz, H., Memisevic, R.: Learning to solve QBF. In: Proceedings of the Twenty-Second AAAI Conference on Artificial Intelligence. AAAI Press, Vancouver, British Columbia (2007)

[41] Shalabi, L. A., Shaaban, Z., Kasasbeh, B.: Data mining: A preprocessing engine. Journal of Computer Science **2**, 735–739 (2006)

[42] Smith-Miles, K.: Cross-disciplinary perspectives on meta-learning for algorithm selection. ACM Comput. Surv. **41**(1) (2008)

[43] Streeter, M., Golovin, D., Smith, S. F.: Combining multiple heuristics online. In: Proceedings of the Twenty-Second AAAI Conference on Artificial Intelligence, pp. 1197–1203. AAAI Press, Vancouver, British Columbia, Canada (2007)

[44] Vapnik, V.: The Nature of Statistical Learning. Springer Verlag, New York, NY, USA (1995)

[45] Witten, I. H., Frank, E.: Data Mining, Practical Machine Learning Tools and Techniques. Elsevier (2005)

[46] Wu, H., Beek, P. V.: Portfolios with deadlines for backtracking search. International Journal on Artificial Intelligence Tools **17**, 835–856 (2008)

[47] Xu, L., Hutter, F., Hoos, H. H., Leyton-Brown, K.: The design and analysis of an algorithm portfolio for SAT. In: C. Bessiere (ed.) 13th International Conference on Principles and Practice of Constraint Programming, *Lecture Notes*

in Computer Science, vol. 4741, pp. 712–727. Springer, Providence, RI, USA (2007)

[48] Xu, L., Hutter, F., Hoos, H. H., Leyton-Brown, K.: Satzilla: Portfolio-based algorithm selection for sat. Journal of Artificial Intelligence Research **32**, 565–606 (2008)

Chapter 10
Control-Based Clause Sharing in Parallel SAT Solving

Youssef Hamadi, Said Jabbour, and Lakhdar Sais

10.1 Introduction

The recent successes of SAT solvers in traditional hardware and software applications have extended to important new domains. Today, they represent essential low-level reasoning components used in general theorem proving, computational biology, AI, and so. This popularity gain is related to their breakthrough in real-world instances involving millions of clauses and hundreds of thousands of variables. These solvers, called modern SAT solvers [22, 11], are based on a nice combination of (i) clause learning [21, 22], (ii) activity-based heuristics [22], and (iii) restart policies [14] enhanced with efficient data structures (e.g., watched literals [22]). This architecture is now standard, and today only minor improvements have been observed (cf. SAT competitions). Therefore, it seems difficult to bet on other order of magnitude gains without a radical algorithmic breakthrough.

Fortunately, the recent generalization of multicore hardware gives parallel processing capabilities to standard PCs. This represents a real opportunity for SAT researchers, who can now consider parallel SAT solving as an obvious way to substantial efficiency gains.

Recent works on parallel SAT are all based on the modern SAT architecture [22], and therefore systematically exploit clause learning as an easy way to extend the cooperation between processing units. When a unit learns a new clause, it can share

Youssef Hamadi
Microsoft Research, Cambridge, CB3 0FB, UK
LIX, École Polytechnique, F-91128 Palaiseau, France
e-mail: youssefh@microsoft.com

Said Jabbour
Université Lille-Nord de France, CRIL – CNRS UMR 8188, Artois, F-62307 Lens
e-mail: jabbour@cril.fr

Lakhdar Sais
Université Lille-Nord de France, CRIL – CNRS UMR 8188, Artois, F-62307 Lens
e-mail: sais@cril.fr

Y. Hamadi et al. (eds.), *Autonomous Search*,
DOI 10.1007/978-3-642-21434-9_10,
© Springer-Verlag Berlin Heidelberg 2011

it with all other units in order to prune their search spaces. Unfortunately, since the number of potential conflicts is exponential, the systematic sharing of learnt clauses is not practically feasible. The solution is to exchange up to some predefined size limit. This has the advantage of reducing the overhead of the cooperation while focusing the exchange on short clauses, recognized as more powerful in terms of pruning.

In this work, our goal is to improve the clause sharing scheme of modern parallel SAT solvers. Indeed, the approach based on some predefined size limit has several flaws, the first and most apparent being that an overestimated value might induce a very large cooperation overhead, while an underestimated one might completely inhibit the cooperation. The second flaw comes from the observation that the size of learnt clauses tends to increase over time (see Section 10.4.1), leading to an eventual halt in the cooperation. The third flaw is related to the internal dynamics of modern solvers, which tend to focus on particular subproblems thanks to the activity/restart mechanisms. In parallel SAT, this can lead two search processes toward completely different subproblems such that clause sharing becomes pointless.

We propose a dynamic clause sharing policy which uses pairwise size limits to control the exchange between any pair of processing units. Initially, high limits are used to enforce the cooperation, and allow pairwise exchanges. On a regular basis, each unit considers the number of foreign clauses received from other units. If this number is below/above a predefined threshold, the pairwise limits are increased/decreased. This mechanism allows the system to maintain a throughput. It addresses the first and second flaws. To address the last flaw, related to the poor relevance of the shared clauses, we extend our policy to integrate the quality of the exchanges. Each unit evaluates the quality of the received clauses, and the control is able to selectively increase/decrease the pairwise limits based on the underlying quality of the recently communicated clauses, the rationale being that the information recently received from a particular source is qualitatively linked to the information which could be received from it in the very near future. The evolution of the pairwise limits w.r.t. the throughput or quality criterion follows an AIMD (Additive-Increase-Multiplicative-Decrease) feedback control-based algorithm [6].

The chapter is organized as follows. After the presentation of previous works in Section 10.2, Section 10.3 gives some technical background. Our dynamic control-based clause sharing policies are motivated and presented in Section 10.4. Section 10.5 presents an extensive experimental evaluation of the new policies, as opposed to the standard one. Finally, we conclude by providing some interesting future paths of research.

10.2 Previous Works

We present here two categories of work. The first one addresses parallel SAT algorithms while the second one addresses an original, relevance restriction strategy that aims at improving automated theorem provers by identifying relevant clauses in the formula w.r.t. the theorem to be proved. This is clearly related to one of the fundamental questions on parallel solving with clause sharing: how to measure the relevance (quality) of the clause to be shared w.r.t. the formula to be proved.

10.2.1 Parallel SAT

We can distinguish between two categories of works: the ones which predate the introduction of the modern DPLL architecture, and the ones based on this architecture.

PSATO [29] is based on the SATO (SAtisfiability Testing Optimized) sequential solver [30]. Like SATO, it uses a *trie* data structure to represent clauses. PSATO uses the notion of *guiding-paths* to divide the search space of a problem. These paths are represented by a set of unit clauses added to the original formula. The parallel exploration is organized in a master/slave model. The master organizes the work by addressing guiding-paths to workers that have no interaction with each other. The first worker to finish stops the system. The balancing of the work is organized by the master.

In [5], the input formula is dynamically divided into disjoint subformulas. Each subformula is solved by a sequential SAT solver running on a particular processor. The algorithm uses optimized data structures to modify Boolean formulas. Additionally, workload balancing algorithms are used to achieve a uniform distribution of load among the processors.

The first parallel SAT solver based on a modern DPLL is Gradsat [7] which extends the zChaff solver with a master/slave model, and implements *guiding-paths* to divide the search space of a problem. These paths are represented by a set of unit clauses added to the original formula. Additionally, learned clauses are exchanged between all clients if they are less than a predefined limit on the number of literals.

[4] uses an architecture similar to Gradsat. However, a client incorporates a foreign clause if it is not subsumed by the current guiding-path constraints. Practically, clause sharing is implemented by *mobile-agents*. This approach is supposed to scale well on computational grids.

Nagsat [12] is a parallel SAT solver which exploits the heavy-tailed distribution of random 3-SAT instances. It implements *nagging*, a notion taken from the DALI theorem prover. Nagging involves a master and a set of clients called *naggers*. In Nagsat, the master runs a standard DPLL algorithm with a static variable ordering. When a nagger becomes idle, it requests a *nagpoint* which corresponds to the current state of the master. Upon receiving a nagpoint, it applies a transformation (e.g., a change in the ordering of the remaining variables) and begins its own search on the corresponding subproblem.

MiraXT is designed for shared memory multiprocessor systems [19]. It uses a divide-and-conquer approach where threads share a unique clause database which represents the original and the learnt clauses. Therefore in this system all the learnt clauses are shared. When a new clause is learnt by a thread, it uses a lock to safely update the common database. Read access can be done in parallel.

pMiniSat uses a standard divide-and-conquer approach based on guiding-paths [8]. It exploits these paths to improve clause sharing. When considered with the knowledge of the guiding-path of a particular thread, a clause can become small and therefore highly relevant. This allows pMiniSat to extend the sharing of clauses since a large clause can become small in another search context.

pMiniSat and MiraXT were respectively ranked second and third in the parallel track of the 2008 SAT-Race.

10.2.2 Relevance Strategy

In [26], the authors consider the challenge of identifying the relevant clauses of a problem in order to improve the efficiency of a sequential automated theorem proving process. They define relevance relative to a given set of clauses S and one or more distinguished sets of support T. In their theorem proving application, the set T can be played by the negation of the theorem to be proved or the query to be answered in S.

The concept of relevance distance between two clauses of S is defined using various metrics based on properties of the paths connecting the clauses (clauses are connected if they contain the same literal). This concept is extended to define relevance distance between a clause and a set (or multiple sets) of support.

Informally, the relevance distance reflects how closely two clauses are related. The relevance distance to one or more support sets is used to compute a relevance set R, a subset of S that is unsatisfiable if and only if S is unsafisfiable. R is computed as the set of clauses of S at distance less than n from one or more support sets: if n is sufficiently large then R is unsatisfiable if S is unsatisfiable. If R is much smaller than S, a refutation from R may be obtainable in much less time than a refutation from S. R must be efficiently computable to achieve overall efficiency improvement. The authors define and characterize different relevance metrics.

10.3 Technical Background

In this section, we introduce the computational features of modern SAT solvers. This is followed by a full overview of the ManySAT parallel SAT solver. We finish the section with the presentation of the principles behind the AIMD feedback control-based algorithm usually applied to solve TCP congestion control problems. This technique is exploited in our framework to deal with the combinatorial explosion problem caused by the huge number of possible clauses to be shared.

10.3.1 Computational Features of Modern SAT Solvers

Most of the state-of-the-art SAT solvers are based on the Davis, Putnam, Logemann and Loveland procedure, commonly called DPLL [10]. DPLL is a backtrack search procedure; at each node of the search tree, a decision literal is chosen according to some branching heuristic. Its assignment to one of the two possible values (true or false) is followed by an inference step that deduces and propagates some forced unit literal assignments. The assigned literals (decision literal and the propagated ones) are labeled with the same decision level, starting from 1, and increased at each decision (or branching) until a model is found or a conflict (or a dead end) is encountered. In the first case, the formula is answered to be satisfiable, whereas in the second case, we backtrack to the last decision level and assign the remaining value to the last decision literal. After backtracking, some variables are unassigned, and the current decision level is decreased accordingly. The formula is answered to be unsatisfiable when backtracking to level 0 occurs.

In addition to this basic scheme, modern SAT solvers use additional components such as restart policy, conflict-driven clause learning and activity-based heuristics. Let us give some details on these last two important features. First, to learn from conflict, they maintain a central data structure, the implication graph, which records the partial assignment that is under construction, together with its implications. When a dead end occurs, a conflict clause (called asserting clause) is generated by resolution following a bottom-up traversal of the implication graph. The learning process stops when a conflict clause containing only one literal from the current decision level is generated. Such a conflict clause (or learnt clause) expresses that the literal is implied at a previous level. Modern SAT solvers backtrack to the implication level and assign that literal the value true. The activity of each variable encountered during the resolution process is increased. The variable with greatest activity is selected to be assigned next.

As the number of learnt clauses can grow exponentially, even in the sequential case, modern SAT solvers regularly reduce the database of learnt clauses. In the SAT solver MiniSat [11], such a reduction (called reduceDB) is achieved as follows. When the size of the learnt database exceeds a given upper bound (B), it is reduced by half. The set of deleted clauses corresponds to the less active ones. Initially, B is set to $\frac{1}{3} \times |\mathscr{F}|$, where $|\mathscr{F}|$ is the number of clauses in the original formula $\mathscr{F}$. At each restart, B is increased by 10%.

Let us mention a strong result by Pipatsrisawat and Darwiche [24], proving that the proof system of modern SAT solvers p-simulates general resolution. This result shows that modern SAT solvers as usually used in practice are capable of simulating any resolution proof (given the right branching and restarting heuristics, of course).

10.3.2 ManySAT: A Parallel SAT Solver

The design of ManySAT benefits from the main weaknesses of modern SAT solvers: their sensitivity to parameter tuning and their lack of robustness. ManySAT uses a portfolio of complementary sequential algorithms obtained through careful variations of the most important components of modern SAT solvers –two-watched-literal, unit propagation, activity-based decision heuristics, lemma deletion strategies, and clause learning. Additionally, each sequential algorithm shares clauses with each other to improve the overall performance of the whole system. This contrasts with most of the parallel SAT solvers generally designed using the divide-and-conquer paradigm (see Section 10.2.1). The first version of ManySAT parameterizes these techniques and runs them on four different cores. In the following, we detail the particular combinations of techniques used by the different cores. This version won the parallel track of the 2008 SAT-Race. Interested readers will find a full presentation of this solver in [16].

10.3.2.1 Restart Policies

Restart policies represent an important component of modern SAT solvers. They were introduced by [9] and [14] to eliminate the heavy-tailed phenomena [13] observed on many families of SAT instances. Indeed, heavy-tailed distributions observed on SAT demonstrate that on many instances a different variable ordering might lead to dramatic performance variation of a given SAT solver. Restarts aim at eliminating the long runs by increasing the probability of falling on the best variable orderings. In modern SAT solvers, restarts are not used to eliminate the heavy-tailed phenomena since after restarting SAT solvers dive into the part of the search space they just left, influenced by the activity-based ordering. In these solvers, restart policies are used to compress the assignment stack and improve the order of assumptions.

Different restart policies have been previously presented. Most of them are static, and their cutoff values follow different evolution schemes (e.g., arithmetic [14], geometric [28], Luby [20]). To ensure the completeness of the SAT solver, in all the restart policies the cutoff value in terms of the number of conflicts increases over time. The performance of the different policies clearly depends on the considered SAT instances. More generally, rapid restarts (e.g., Luby) perform well on industrial instances, while on hard instances slow restarts are more suitable. Generally, it is difficult to say in advance which policy should be used on which problem class [17].

ManySAT runs a set of complementary restart policies to define the restart cutoff x_i. It uses the well-known Luby policy [20], and a classical geometric policy, $x_i = 1.5 \times x_{i-1}$ with $x_1 = 100$ [11], where x_i represents the cutoff value in term of number of conflicts allowed for the restart i. In addition, it introduces two new policies. A very slow arithmetic one, $x_i = x_{i-1} + 16,000$ with $x_1 = 16,000$, and a new dynamic one based on the evolution of the average size of backjumps. This infor-

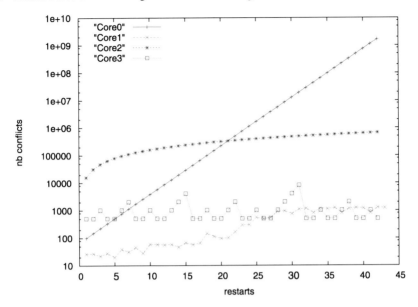

Fig. 10.1: Restart policies

mation is a good indicator of the decision errors made during search. Moreover, it can be seen as an interesting measure of the relative hardness of the instance. This new policy is designed in such a way that, for high (or low) fluctuation of the average size of backjumps (between the current and the previous restart), it delivers a low (or high) cutoff value. In other words, the cutoff value of the next restart depends on the average size of backjumps observed during the two previous consecutive runs. It is defined as $x_1 = 100$, $x_2 = 100$, and $x_{i+1} = \frac{\alpha}{y_i} \times |cos(1 - r_i)|$, $i \geq 2$, where $\alpha = 1200$, y_i represents the average size of backjumps at restart i, $r_i = \frac{y_{i-1}}{y_i}$ if $y_{i-1} < y_i$ and $r_i = \frac{y_i}{y_{i-1}}$ otherwise (see Table 10.1 column 3). The cutoff value x_i is minimal when the ratio between the average size of jumps between the two previous consecutive runs is equal to 1. This dynamic restart policy was introduced in ManySAT [15]. Other dynamic policies were presented in [27, 3, 25].

Figure 10.1 presents the previous restart policies: geometric, dynamic (plotted on a particular instance), arithmetic, and Luby with a unit of 512.

10.3.2.2 Heuristic and Polarity

In ManySAT, the first core uses increased random noise with its VSIDS heuristic [22] since its restart policy is the slowest one. Indeed, that core tends to intensify the search, and slightly increasing the random noise allows us to introduce more diversification.

Each time a variable is chosen, one needs to decide if such a variable might be assigned the value true (positive polarity) or false (negative polarity). Different kinds of polarity have been defined. For example, MiniSat usually chooses negative polarity, whereas RSat [23] uses progress saving. More precisely, each time a backtrack occurs, the polarities of the assigned variables between the conflict and the backjumping level are saved. If one of these variables is chosen again its saved polarity is preferred. In CDCL-based solvers, the chosen polarity might have a direct impact on the learnt clauses and on the performance of the solver.

The polarity policy of the core 0 is defined according to the number of occurrences of each literal in the learnt database. Each time a learnt clause is generated, the number of occurrences of each literal is increased by 1. Then to maintain a more constrained learnt database, the polarity of l is set to *true* when $\#occ(l)$ is greater than $\#occ(\neg l)$, and to *false* otherwise. For example, setting the polarity of l to *true* biases the occurrence of its negation $\neg l$ in the next learnt clauses.

This approach tends to balance the polarity of each literal in the learnt database. By doing so, it increases the number of possible resolvents between the learnt clauses. If the relevance of a given resolvent is defined as the number of steps needed to derive it, then a resolvent between two learnt clauses might lead to more relevant clauses in the database.

Since the restart strategy in core 0 tends to intensify the search, it is important to maintain a learnt database of better quality. However, for rapid restarts as in cores 1 and 3, progress saving is more suitable for saving the work. Core 2 applies a complementary polarity (*false* by default, as in MiniSat).

10.3.2.3 Learning

Learning is another important component crucial for the efficiency of modern SAT solvers. Most of the known solvers use similar CDCL approaches associated with the first UIP (Unique Implication Point) scheme.

In ManySAT a new learning scheme obtained using an extension of the classical implication graph is also used [2]. This new notion considers additional arcs, called inverse arcs. These are obtained by taking into account the satisfied clauses of the formula, which are usually ignored in classical conflict analysis. The new arcs present in this extended graph can show that even some decision literals admit a reason, something ignored when using classical implication graphs. As a result, the size of the backjumps is often increased.

This new learning scheme is integrated into the SAT solvers of cores 0 and 3.

10.3.2.4 Clause Sharing

In ManySAT, each core exchanges a learnt clause if its size is less or equal to 8. This decision is based on extensive tests with representative industrial instances. The communication between the solvers of the portfolio is organized through lockless queues which contain the lemmas that a particular core wants to exchange.

10.3.2.5 Summary

Table 10.1 summarizes the choices made for the different solvers of the ManySAT portfolio. For each solver (core), we mention the restart policy, the heuristic, the polarity, the learning scheme and the size of the shared clauses.

Strategies	Core 0	Core 1	Core 2	Core 3
Restart	Geometric $x_1 = 100$ $x_i = 1.5 \times x_{i-1}$	Dynamic (Fast) $x_1 = 100, x_2 = 100$ $x_i = f(y_{i-1}, y_i), i > 2$ if $y_{i-1} < y_i$ $f(y_{i-1}, y_i) =$ $\frac{\alpha}{y_i} \times \|cos(1 - \frac{y_{i-1}}{y_i})\|$ else $f(y_{i-1}, y_i) =$ $\frac{\alpha}{y_i} \times \|cos(1 - \frac{y_i}{y_{i-1}})\|$ $\alpha = 1200$	Arithmetic $x_1 = 16000$ $x_i = x_{i-1} + 16000$	Luby 512
Heuristic	VSIDS (3% rand.)	VSIDS (2% rand.)	VSIDS (2% rand.)	VSIDS (2% rand.)
Polarity	if $\#occ(l) > \#occ(\neg l)$ $l = true$ else $l = false$	Progress saving	false	Progress saving
Learning	CDCL-extended	CDCL	CDCL	CDCL-extended
Cl. sharing	size ≤ 8	size ≤ 8	size ≤ 8	size ≤ 8

Table 10.1: ManySAT: different strategies

10.3.3 AIMD Feedback Control-Based Algorithm

The Additive Increase/Multiplicative Decrease (AIMD) algorithm is a feedback control algorithm used in TCP congestion avoidance. AIMD guesses the communication bandwidth available between two communicating nodes. The algorithm performs successive probes, increasing the communication rate w linearly as long as no packet loss is observed, and decreasing it exponentially when a loss is encountered. More precisely, the evolution of w is defined by the following $AIMD(a, b)$ formula:

- $w = w - a \times w$, if a loss is detected
- $w = w + \frac{b}{w}$, otherwise

Different proposals have been made to prevent congestion in communication networks based on different values for a and b. Today, AIMD is the major component of TCP's congestion avoidance and control [18]. On probe of network bandwidth, increasing too quickly will overshoot practical limits (i.e., underlying capacities). On notice of congestion, decreasing too slowly will not be reactive enough.

In the context of clause sharing, our control policies aim at achieving a particular throughput or a particular throughput of maximum quality. Since any increase in the size limit can potentially generate a very large number of new clauses, AIMD's slow increase can help us to avoid a quick overshoot of the throughput. Similarly, in the case of overshooting, an aggressive decrease can help us to quickly reduce clause sharing by a very large amount.

10.4 Control-Based Clause Sharing in Parallel SAT Solving

10.4.1 Motivation

To motivate further interest in our proposed framework, we conducted a simple experiment using the standard MiniSat algorithm [11]. In Figure 10.2 we show the evolution of the percentage of learnt clauses of size less than or equal to 8 on a particular family of 16 industrial instances, *AProVE07_**. This limit represents the default static clause sharing limit used by the ManySAT parallel solver [16].

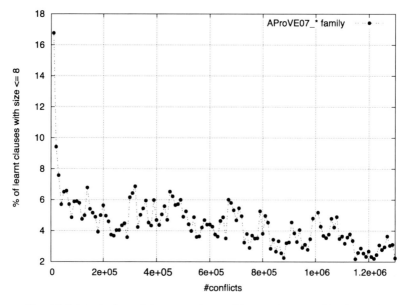

Fig. 10.2: Evolution of the percentage of learnt clauses with size ≤ 8

This percentage is computed every 10,000 conflicts, and as can be observed, it decreases over time[1]. Initially, nearly 17% of the learnt clauses could be exchanged, but as search goes on, this percentage falls below 4%. Our observation is very general and can be performed on different instances with similar or different clause size limits. It illustrates the second flaw reported above: the size of learnt clauses tends to increase over time. Consequently, in a parallel SAT setting, any static limit might lead to a halt of the clause sharing process. Therefore, if one wants to maintain a quantity of exchange over time, there does not exist an optimal *static* policy for that. This clearly shows the importance of a dynamic control clause sharing policy.

10.4.2 Throughput and Quality-Based Control Policies

In this section, we describe our dynamic control-based clause sharing policies, which control the exchange between any pair of processing units through dynamic pairwise size limits.

The first policy controls the throughput of clause sharing. Each unit considers the number of foreign clauses received from other units. If this number is below or above a predefined throughput threshold, the pairwise limits are all increased or decreased using an AIMD feedback algorithm. The second policy is an extension of the first one. It introduces a measure of the quality of foreign clauses. With this information, the increase or decrease of the pairwise limits becomes proportional to the underlying quality of the clauses shared by each unit. The first (or second) policy allows the system to maintain a throughput (or throughput of better quality).

We consider a parallel SAT solver with n different processing units. Each unit u_i corresponds to a SAT solver with clause learning capabilities. Each solver can work either on a subspace of the original instance, as in divide-and-conquer techniques, or on the full problem, as in ManySAT (see details in Sections 10.2 and 10.3.2). We assume that these different units communicate through shared memory (as in multicore architectures).

In our control strategy, we consider a control-time sequence as a set of steps t_k with $t_0 = 0$ and $t_k = t_{k-1} + \alpha$, where α is a constant representing the time window defined in terms of the number of conflicts. The step t_k of a given unit u_i corresponds to the conflict number $k \times \alpha$ encountered by the solver associated with u_i. In the sequel, when there is no ambiguity, we sometimes denote t_k simply by k. Then, each unit u_i can be defined as a sequence of states $S_i^k = (\mathscr{F}, \Delta_i^k, R_i^k)$, where $\mathscr{F}$ is a CNF formula, Δ_i^k is the set of its proper learnt clauses and R_i^k is the set of foreign clauses received from the other units between two consecutive steps $k - 1$ and k. The different units achieve pairwise exchange using pairwise limits. Between two consecutive steps $k - 1$ and k, a given unit u_i receives from all the other remaining units u_j, where $0 \le j < n$ and $j \ne i$, a set of learnt clauses $\Delta_{j \to i}^k$ of length less than

---

[1] The regular, small raises are the result of the cyclic reduction of the learnt base through reduceDB.

or equal to a size limit $e_{j\rightarrow i}^k$, i.e., $\Delta_{j\rightarrow i}^k = \{c \in \Delta_j^k / \ |c| \le e_{j\rightarrow i}^k\}$. Then, the set R_i^k can be formally defined as $\cup_{0 \le j < n, j \ne i} \Delta_{j\rightarrow i}^k$.

Using a fixed throughput threshold T of shared clauses, we describe our control-based policies, which allow each unit u_i to guide the evolution of the size limit $e_{j\rightarrow i}$ using an AIMD feedback mechanism.

10.4.2.1 Throughput-Based Control

As illustrated in Figure 10.3, at step k a given unit u_i checks whether the throughput is exceeded or not. If $|R_i^k| < T$ (or $|R_i^k| > T$) the size limit $e_{j\rightarrow i}^{k+1}$ is additively increased (or multiplicatively decreased). More formally, the upper bound $e_{j\rightarrow i}^{k+1}$ on the size of clauses that a solver j shares with the solver i between k and $k+1$ is changed using the following AIMD function:

$$aimdT(R_i^k)\{$$
$$\quad \forall j | 0 \le j < n, j \ne i$$
$$\quad e_{j\rightarrow i}^{k+1} = \begin{cases} e_{j\rightarrow i}^k + \frac{b}{e_{j\rightarrow i}^k}, if(|R_i^k| < T) \\ e_{j\rightarrow i}^k - a \times e_{j\rightarrow i}^k, if(|R_i^k| > T) \end{cases}$$
$$\}$$

where a and b are positive constants.

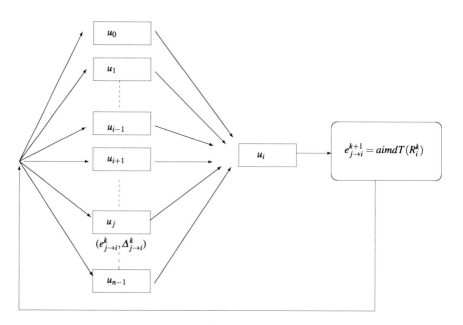

Fig. 10.3: Throughput-based control policy

10.4.2.2 Throughput- and Quality-Based Control

In this policy, to control the throughput of a given unit u_i, we introduce a quality measure $Q[^k_{j\to i}]$ (see Definitions 10.1 and 10.2) to estimate the relative quality of the clauses received by u_i from u_j. In the throughput- and quality-based control policy, the evolution of the size limit $e^k_{j\to i}$ is related to the estimated quality.

Our throughput- and quality-based control policy changes the upper bound $e^{k+1}_{j\to i}$ on the size of clauses that a solver j shares with the solver i between k and $k+1$ using the following AIMD function:

$$aimdTQ(R^k_i)\{$$
$$\forall j | 0 \leq j < n, j \neq i$$
$$e^{k+1}_{j\to i} = \begin{cases} e^k_{j\to i} + Q[^k_{j\to i}] \times \frac{b}{e^k_{j\to i}}, if(|R^k_i| < T) \\ e^k_{j\to i} - (1 - Q[^k_{j\to i}]) \times a \times e^k_{j\to i}, if(|R^k_i| > T) \end{cases}$$
$$\}$$

where a and b are positive constants.

As shown by the AIMD function of the throughput- and quality-based control policy, the adjustment of the size limit depends on the quality of the shared clauses. Indeed, as can be seen from the above formula, when the exchange quality between u_j and u_i ($Q[^k_{j\to i}]$) tends to 100% (or 0%), the increase in the limit size tends to be maximal (or minimal) while the decrease tends to be minimal (or maximal).

Our goal in this second policy is to maintain a throughput of good quality, the rationale being that the information recently received from a particular source is qualitatively linked to the information that could be received from it in the very near future.

In the following, we propose two quality-based measures. The first one estimates the relevance of a clause c received by u_i from u_j according to the activities of the literals of c in u_i, whereas the second one considers the truth values of its literals in the current assignment of u_i.

Activity-Based Quality

Our activity based-quality measure, Qa, is defined using the activity of the variables at the basis of the VSIDS heuristic [22], another important component of modern SAT solvers. The variables with greatest activity represent those involved in most of the (recent) conflicts. Indeed, when a conflict occurs, the activity of the variables whose literals appear in the clauses encountered during the generation of a learnt clause are updated. The most active variables are those related to the current part of the search space. Consequently, our quality measure exploits these activities to quantify the relevance of a clause learnt by unit u_j to the current state of a given unit

u_i. To define our quality measure, suppose that, at any time of the search process, we have $\mathscr{A}_i^{max}$, the current maximal activity of u_i's variables, and $\mathscr{A}_i(x)$, the current activity of a given variable x.

Definition 10.1 (Activity-Based Quality). Let c be a clause sent by u_i to u_j and $\mathscr{L}_{\mathscr{A}_i}(c) = \{x/x \in c | \mathscr{A}_i(x) \geq \frac{\mathscr{A}_i^{max}}{2}\}$ be the set of active literals of c with respect to unit u_i. We define $\mathscr{P}_{j \rightarrow i}^k = \{c/c \in \Delta_{j \rightarrow i}^k | \mathscr{L}_{\mathscr{A}_i}(c) | \geq f(|c|)\}$ to be the set of clauses received by u_i from u_j between steps $k-1$ and k with at least $f(|c|)$ active literals. We define the quality of clauses sent by u_j to u_i at a given step k as $Qa_{j \rightarrow i}^{[k} =$ $\frac{|\mathscr{P}_{j \rightarrow i}^k|+1}{|\Delta_{j \rightarrow i}^k|+1}$.

From the definition below, we can see that the quality of a given clause sent by u_j to u_i depends on the activity of its literals. In our experiments the lower bound $f(|c|)$ is set to $\frac{|c|}{3}$.

Proof-Based Quality

The best clause to share between u_j and u_i is the one leading to a conflict on u_i. In other words, the clause sent by u_j to u_i is falsified by the current assignment ρ_{u_i}. Such a clause induces conflict analysis and backjumping for unit u_i. The second most preferred shared clause is clearly one that causes unit propagation on u_i. Continuing the same reasoning, we claim that the most "relevant" clause contains a large number of falsified literals w.r.t. the current assignment ρ_{u_i}. Consequently, a clause with the largest number of falsified literals is clearly relevant. Interestingly enough, as the variable ordering used in modern SAT solvers is based on the activities of the variable, if a clause received by u_i shares a large portion of assigned literals, then it involves active literals. This means that the shared clause is closely related to the current portion of the search space of u_i.

The following definition describes our second quality measure, called proof-based quality, as its relevance is computed with respect to the current assignment.

Definition 10.2 (Proof-Based Quality). Let c be a clause sent by u_j to u_i, ρ_{u_i} the current assignment of u_i, and $\mathscr{L}_{\mathscr{P}_i}(c) = \{x/x \in c | \rho_{u_i}(x) \neq false\}$ be the set of unsigned or satisfied literals of c with respect to unit u_i. We define $\mathscr{P}_{j \rightarrow i}^k = \{c/c \in \Delta_{j \rightarrow i}^k | \mathscr{L}_{\mathscr{P}_i}(c) | \leq f(|c|)\}$ to be the set of clauses received by u_i from u_j between steps $k-1$ and k with at most $f(|c|)$ unsigned or satisfied literals. We define the quality of clauses sent by u_j to u_i at a given step k as $Qp_{j \rightarrow i}^{[k} =$ $\frac{|\mathscr{P}_{j \rightarrow i}^k|+1}{|\Delta_{j \rightarrow i}^k|+1}$.

From Definition 10.2, we can observe that a given clause is considered relevant if its ratio of unassigned and satisfied literals is less than a given upper bound $f(|c|)$. In our experiment, we consider $f(|c|) = \frac{|c|}{log(|c|)}$. The idea is that as the clause length increases, the number of falsified literals that a given clause might contain also increases. For example, using the ratio $\frac{|c|}{log(|c|)}$, unary and binary clauses are considered relevant, while a ternary clause is considered relevant if it contains one literal assigned to false. A clause of length 10, 20 or 30 is considered relevant only if it contains 6, 14 or 22 literals assigned to false, respectively.

10.5 Evaluation

10.5.1 Experiments

Our policies were implemented and tested on top of the ManySAT parallel SAT solver (see Section 10.3.2). Our tests were done on Intel Xeon Quadcore machines with 16 GB of RAM running at 2.3 Ghz. We used a timeout of 1,500 seconds for each problem. ManySAT was used with four DPLL strategies, each running on a particular core (unit). To alleviate the effects of unpredictable thread scheduling, each problem was solved three times and the average was taken.

Our dynamic clause sharing policies were added to ManySAT and compared to ManySAT with its default static policy *ManySAT e* = 8, which exchanges clauses up to size 8. Since each pairwise limit is read by one unit and updated by another, our proposal can be integrated without any lock.

In all our experiments, the default parameter settings are $a = 0.125$ and $b = 8$ for aimdT, aimdTQa, and aimdTQp, associated with a time window of $\alpha = 10,000$ conflicts. The throughput T is set to 5,000. Each unit u_i can then receive from the three remaining units $5,000 \times 3 = 15,000$ clauses in the worst case. Each pairwise limit $e_{j \to i}$ was initialized to 8. The results corresponding to these settings are presented in Tables 10.2 and 10.3 and Figures 10.4 and 10.6.

10.5.2 Industrial Problems

Tables 10.2, 10.3 and 10.4 present the results on the 100 industrial problems of the 2008 SAT-Race. The problem set contains families with several instances or individual instances. We present the family/instance name and the number of instances per family. For each method, we report the average CPU time (seconds) for each family and the number of problems solved (#S) before timeout. In the case of dynamic policies, we also provide $\bar{e}$, the average of the $e_{j \to i}$ observed during the computation. When the symbol "-" is reported, this means that the instance is not solved in the allowed time limit. The last row provides for each method the total number of problems solved, and the cumulated runtime (with 1,500 seconds added for each unsolved instance). For the dynamic policies, it also presents the average of the $\bar{e}$ values (without considering the unsolved instances).

family/instance	#inst	#S	ManySAT $e = 8$ time(s)	#S	ManySAT aimdT time(s)	$\bar{e}$
ibm_*	20	19	**204**	19	218	7
manol_*	10	10	120	10	**117**	8
mizh_*	10	6	762	7	**746**	6
post_*	10	9	325	9	**316**	7
velev_*	10	8	585	8	**448**	5
een_*	5	5	2	5	2	8
simon_*	5	5	111	5	**84**	10
bmc_*	4	4	7	4	7	7
gold_*	4	1	1160	1	**1103**	12
anbul_*	3	2	742	3	**211**	11
babic_*	3	3	2	3	2	8
schup_*	3	3	129	3	**120**	5
fuhs_*	2	2	90	2	**59**	11
grieu_*	2	1	783	1	**750**	8
narain_*	2	1	786	1	**776**	8
palac_*	2	2	20	2	**8**	3
jarvi-eq-atree-9	1	1	70	1	**69**	25
aloul-chnl11-13	1	0	-	0	-	-
marijn-philips	1	0	-	1	**1133**	34
maris-s03	1	1	11	1	11	10
vange-col	1	0	-	0	-	-
Total/(average)	100	83	10406	86	9180	(10.15)

Table 10.2: SAT-Race 2008, industrial problems: ManySAT $e = 8$ and aimdT

In Table 10.2, we present comparative results of the standard ManySAT with the static size limit set to 8 and the dynamic ManySAT aimdT using throughput-based control policy. We stress that the static policy ($e = 8$) was found optimal in preliminary testing, i.e., it corresponds to the static limit which provides the best performance on this set of problems. We can observe that the static policy solves 83 problems while the dynamic policy aimdT solves 86 problems. Except on the ibm_* family, the dynamic policy aimdT always exhibits a runtime better than or equivalent to the static one. Unsurprisingly, when the runtime is significant but not a drastic improvement over the static policy, the values of $\bar{e}$ are often close to 8, i.e., equivalent to the static size limit.

In Table 10.3, we report the experimental comparison between the dynamic throughput-based control ManySAT aimdT, left-hand side, and the quality-based control ManySAT aimdTQa (activity-based quality) and ManySAT aimdTQp (proof-based quality), on the right-hand side.

We can see that aimdT is faster than aimdTQa and aimdTQp. However, the quality-based policies solve more problems. The aimdT method solves 86 problems, while aimdTQa and aimdTQp solve 89 and 88 problems, respectively. We can explain this as follows. The quality-based policy intensifies the search by favoring the exchange of clauses related to the current exploration of each unit. This intensification leads to the resolution of more difficult problems. However, it increases the runtime on easier instances where a more diversified search is often more beneficial. On these last instances, a small overhead can be observed. It can be explained by the time needed to evaluate the quality of exchanged clauses. Overall these results are very good since our dynamic policies are able to outperform the best possible static tuning.

family/instance	#inst	ManySAT aimdT #S	time(s)	$\bar{e}$	ManySAT aimdTQa #S	time(s)	$\bar{e}$	ManySAT aimdTQp #S	time(s)	$\bar{e}$
ibm_*	20	19	218	7	19	286	6	**20**	**143**	9
manol_*	10	10	117	8	10	205	7	10	**115**	10
mizh_*	10	7	746	6	**10**	441	5	9	**428**	8
post_*	10	9	**316**	7	9	375	7	9	330	9
velev_*	10	8	**448**	5	8	517	7	8	544	10
een_*	5	5	2	8	5	2	7	5	2	7
simon_*	5	5	84	10	5	59	9	5	**17**	9
bmc_*	4	4	7	7	4	**6**	9	4	**6**	9
gold_*	4	1	1103	12	1	1159	12	2	**1092**	14
anbul_*	3	3	**211**	11	3	689	11	1	1047	12
babic_*	3	3	2	8	3	2	8	3	2	8
schup_*	3	3	**120**	5	3	160	5	3	143	6
fuhs_*	2	2	**59**	11	2	77	10	2	75	12
grieu_*	2	1	750	8	1	750	8	2	**217**	10
narain_*	2	1	**776**	8	1	792	8	1	794	8
palac_*	2	2	**8**	3	2	54	7	2	29	5
jarvi-eq-atree-9	1	1	69	25	1	**43**	17	1	57	14
aloul-chnl11-13	1	0	-	-	0	-	-	0	-	-
marijn-philips	1	1	1133	34	1	**1132**	29	0	-	-
maris-s03	1	1	11	10	1	11	8	1	27	12
vange-col	1	0	-	-	0	-	-	0	-	-
Total/(average)	100	86	9180	(10.15)	89	9760	(9.47)	88	9568	(9.55)

Table 10.3: Industrial problems: ManySAT aimdT, aimdTQa, and aimdTQp

Our hypothesis is that the introduction of quality measures to control the overall exchange might be beneficial in situations where each unit is flooded by the huge number of received clauses. To validate the hypothesis, in Table 10.4 we consider new parameters setting corresponding to the worst-case situation, where the amount of exchanged clauses is maximal. To simulate this situation, we first set the time window α to a small value of $1,000$ conflicts in order to increase the application frequency of our control-based policies. Secondly, the throughput T is set to $1,000 \times 3 = 3,000$ clauses, which corresponds to the maximal number of clauses that can be received by each unit u_i during a time window of α conflicts. Finally, to allow a fast arithmetic increase, we also set the value b to 64. Our goal behind these last settings is to quickly bring each unit towards a maximum amount of exchange, and then evaluate the effect of our quality-based measures.

We can observe that the results obtained in Table 10.4 prove that our hypothesis is correct. Indeed, aimdTQa and aimdTQp are better than aimdT both in CPU time and in terms of the number of solved problems. Clearly, aimdTQa (activity-based quality) obtained the best performance in the worst-case situation. We can also observe that in this intense situation of exchange, aimdTQa obtained the smallest average value of $\bar{e}$ (see that last row).

family/instance	#inst	ManySAT aimdT			ManySAT aimdTQa			ManySAT aimdTQp		
		#S	time(s)	$\bar{e}$	#S	time(s)	$\bar{e}$	#S	time(s)	$\bar{e}$
ibm_*	20	19	204	46	19	216	29	**20**	**154**	43
manol_*	10	10	113	48	10	113	26	10	**112**	39
mizh_*	10	9	**348**	45	9	504	24	**10**	596	91
post_*	10	9	364	43	9	**355**	24	9	360	24
velev_*	10	8	576	49	8	**488**	28	8	529	74
een_*	5	5	2	7	5	1	4	5	2	6
simon_*	5	5	168	62	5	**38**	31	5	77	50
bmc_*	4	4	9	62	4	10	37	4	9	34
gold_*	4	1	1174	72	1	**1164**	62	1	1169	81
anbul_*	3	1	1086	40	1	1093	48	2	**968**	102
babic_*	3	3	2	8	3	2	8	3	2	8
schup_*	3	3	134	39	3	**110**	26	3	142	32
fuhs_*	2	2	104	56	2	**87**	51	2	88	62
grieu_*	2	1	750	8	2	**4**	21	1	750	8
narain_*	2	1	788	8	2	460	28	1	769	8
palac_*	2	2	49	33	2	**4**	15	2	12	27
jarvi-eq-atree-9	1	1	66	94	1	60	84	1	67	70
aloul-chnl11-13	1	0	-	-	0	-	-	0	-	-
marijn-philips	1	0	-	-	1	1230	183	0	-	-
maris-s03	1	1	11	60	1	**8**	33	1	18	50
vange-col	1	0	-	-	0	-	-	0	-	-
Total/(average)	100	85	10448	43	88	8947	40	88	10324	44

Table 10.4: Industrial problems: ManySAT aimdT, aimdTQa, and aimdTQp for $\alpha = 1,000$, $T = 3,000$ and $b = 64$

10.5.3 Crafted Problems

We present here results on the crafted category (201 problems) of the 2007 SAT competition. These problems are handmade and many of them are designed to beat all the existing SAT solvers. They contain, for example, quasi-group instances, forced random SAT instances, counting, ordering and pebbling instances, and social golfer problems.

The scatter plot (in log scale) given in Figure 10.4 illustrates the comparative results of the static and dynamic throughput versions of ManySAT. The x-axis (or y-axis) corresponds to the CPU time tx (or ty) obtained by ManySAT $e = 8$ (or ManySAT aimdT). Each dot with (tx, ty) coordinates corresponds to a SAT instance. Dots below (or above) the diagonal indicate that ManySAT aimdT is faster (or slower) than ManySAT $e = 8$. The results clearly exhibit that the throughput-based policies outperform the static policy on the crafted category.

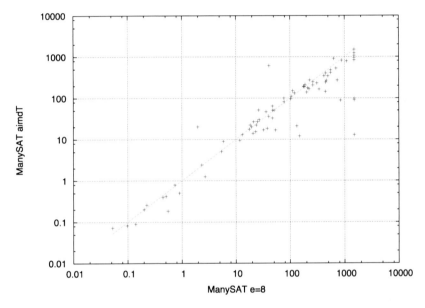

Fig. 10.4: SAT Competition 2007, crafted problems (aimd $e = 8$ and aimdT)

The above improvements are illustrated in Figure 10.5, which shows the time in seconds needed to solve a given number of instances (#*instances*). We can observe that both aimdT ($\bar{e} = 23.14$ with a peak at 94) and aimdTQa ($\bar{e} = 26.17$ with a peak at 102) solve seven more problems than the static policy. Also, the aimdTQp ($\bar{e} = 30.44$ with a peak at 150) solves six more problems than the static policy. The proof-based quality policy (aimdTQp), even if it solves one less problem than aimdT and aimdTQa, obtains better results in term of CPU times. As for the previous problem category, aimdT remains faster than aimdTQa, but slower than aimdTQp. The proof-based quality is better than the activity-based one. It seems that since by definition these problems do not have a structure which can be advantageously exploited by the activity-based heuristics (even in the sequential case), the proof-based quality is clearly more convenient as it measures the relevance of clauses according to their relationship to the current state of the search process.

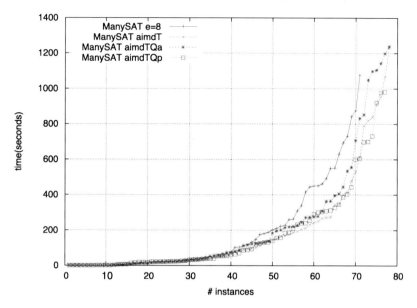

Fig. 10.5: SAT Competition 2007, crafted problems (aimd $e = 8$, aimdT, aimdTQa, aimdTQp)

10.5.4 The Dynamic of $\bar{e}$

It is interesting to consider the evolution of $\bar{e}$, the average of the $e_{j \to i}$ observed during the computation. In Figure 10.2 it was shown on the 16 instances of the *APRoVE07_** family that the size of learnt clauses was increasing over time.

We present in Figure 10.6 the evolution of $\bar{e}$ with ManySAT aimdT on the same family. The evolution is given for each unit (core0 to core3), and the average of the units is also presented (average). We can see that our dynamic policy overcomes the disappearing of "small" clauses by the incremental raising of the pairwise limits. It shows the typical sawtooth behavior that represents the probe for throughput T. This figure and our results on the industrial and crafted problems show that the evolution of the pairwise limits does not reach unacceptable levels if not bounded.

10.6 Conclusion

We have presented several dynamic clause sharing policies for parallel SAT solving. They use an AIMD feedback control-based algorithm to dynamically adjust the size of shared clauses between any pair of processing units. Our first policy maintains an overall number of exchanged clauses (throughput) whereas our second policy additionally exploits the relevance quality of shared clauses. These policies have

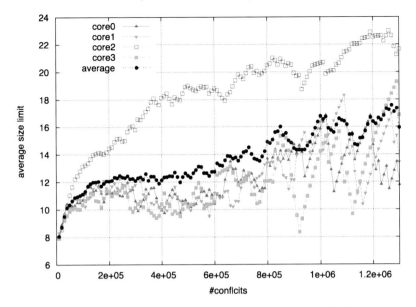

Fig. 10.6: ManySAT aimdT: $\bar{e}$ on the AProVE07_* family

been devised as an efficient answer to the various flaws of the classical static size limit policy. The experimental results comparing our proposed dynamic policies to the static policy show important improvements in the state-of the-art parallel SAT solver ManySAT. Our proposed framework offers interesting perspectives. For example, the design of new, relevant quality measures for clause sharing is of great importance. It could benefit sequential solvers by improving their learnt base reduction strategy and, as demonstrated by this work, have an important impact on parallel SAT solving.

In the context of Autonomous Search, our work demonstrates how control theory can be integrated during search. In general, feedback mechanisms such as the one presented here can be used to adjust the behavior of a search algorithm to the difficulty of the problem at hand. In future work, we are planning to integrate more feedback mechanisms to address the challenges of large-scale parallel SAT solving.

References

[1] Büning, H., Zhao, X. eds.: Proceedings of Theory and applications of satisfiability testing - 11th International Conference SAT 2008. Guangzhou, China, May 12–15, 2008. *Lecture Notes in Computer Science*, vol. 4996. Springer (2008)

[2] Audemard, G., Bordeaux, L., Hamadi, Y., Jabbour, S., Sais, L.: A generalized framework for conflict analysis. In: Proc. of SAT, pp. 21–27 (2008)

[3] Biere, A.: Adaptive restart strategies for conflict driven SAT solvers. In: Büning and Zhao [1], pp. 28–33 (2008)

[4] Blochinger, W., Sinz, C., Küchlin, W.: Parallel propositional satisfiability checking with distributed dynamic learning. Parallel Computing **29**(7), 969–994 (2003)

[5] Böhm, M., Speckenmeyer, E.: A fast parallel SAT-solver – efficient workload balancing. Ann. Math. Artif. Intell. **17**(3-4), 381–400 (1996)

[6] Chiu, D. M., Jain, R.: Analysis of the increase and decrease algorithms for congestion avoidance in computer networks. Comp. Networks **17**, 1–14 (1989)

[7] Chrabakh, W., Wolski, R.: GrADSAT: A parallel SAT solver for the grid. Tech. rep., UCSB CS TR N. 2003-05 (2003)

[8] Chu, G., Stuckey, P.: pMiniSat: A parallelization of MiniSat 2.0. Tech. rep., SAT-Race 2008, solver description (2008)

[9] Crawford, J. M., Baker, A. B.: Experimental results on the application of satisfiability algorithms to scheduling problems. In: AAAI, pp. 1092–1097 (1994)

[10] Davis, M., Logemann, G., Loveland, D. W.: A machine program for theorem-proving. Comm. of the ACM **5**(7), 394–397 (1962)

[11] Eén, N., Sörensson, N.: An extensible SAT-solver. In: Proc. of SAT'03, pp. 502–518 (2003)

[12] Forman, S. L., Segre, A. M.: Nagsat: A randomized, complete, parallel solver for 3-sat. In: Proc. of SAT 2002, pp. 236–243 (2002)

[13] Gomes, C. P., Selman, B., Crato, N., Kautz, H.: Heavy-tail phenomena in satisfiability and constraint satisfaction. Journal of Automated Reasoning pp. 67–100 (2000)

[14] Gomes, C. P., Selman, B., Kautz, H.: Boosting combinatorial search through randomization. In: Proceedings of the Fifteenth National Conference on Artificial Intelligence (AAAI'98), pp. 431–437. Madison, Wisconsin (1998)

[15] Hamadi, Y., Jabbour, S., Sais, L.: ManySAT: Solver description. Tech. Rep. MSR-TR-2008-83, Microsoft Research (2008)

[16] Hamadi, Y., Jabbour, S., Sais, L.: ManySAT: A parallel SAT solver. Journal on Satisfiability, Boolean Modeling and Computation **6**, 245–262 (2009)

[17] Huang, J.: The effect of restarts on the efficiency of clause learning. In: IJCAI, pp. 2318–2323 (2007)

[18] Jacobson, V.: Congestion avoidance and control. In: Proc. SIGCOMM '88, pp. 314–329 (1988)

[19] Lewis, M., Schubert, T., Becker, B.: Multithreaded SAT solving. In: Proc. ASP DAC'07 (2007)

[20] Luby, M., Sinclair, A., Zuckerman, D.: Optimal speedup of Las Vegas algorithms. Information Processing Letters **47**, 173–180 (1993)

[21] Marques-Silva, J., Sakallah, K. A.: GRASP – A New Search Algorithm for Satisfiability. In: Proceedings of IEEE/ACM International Conference on Computer-Aided Design, pp. 220–227 (1996)

[22] Moskewicz, M. W., Madigan, C. F., Zhao, Y., Zhang, L., Malik, S.: Chaff: Engineering an efficient SAT solver. In: Proceedings of the 38th Design Automation Conference (DAC'01), pp. 530–535 (2001)

[23] Pipatsrisawat, K., Darwiche, A.: A lightweight component caching scheme for satisfiability solvers. In: Proc. SAT'07, pp. 294–299 (2007)

[24] Pipatsrisawat, K., Darwiche, A.: On the power of clause-learning SAT solvers with restarts. In: Proceedings of the 15th International Conference on the Principles and Practice of Constraint Programming – CP 2009, pp. 654–668 (2009)

[25] Pipatsrisawat, K., Darwiche, A.: Width-based restart policies for clause-learning satisfiability solvers. In: O. Kullmann (ed.) SAT, *Lecture Notes in Computer Science*, vol. 5584, pp. 341–355. Springer (2009)

[26] Plaisted, D. A., Yahya, A. H.: A relevance restriction strategy for automated deduction. Artif. Intell. **144**(1-2), 59–93 (2003)

[27] Ryvchin, V., Strichman, O.: Local restarts. In: Büning and Zhao [1], pp. 271–276

[28] Walsh, T.: Search in a small world. In: Proceedings of the Sixteenth International Joint Conference on Artificial Intelligence – IJCAI 99, pp. 1172–1177 (1999)

[29] Zhang, H., Bonacina, M. P., Hsiang, J.: PSATO: a distributed propositional prover and its application to quasigroup problems. Journal of Symbolic Computation **21**, 543–560 (1996)

[30] Zhang, H., Stickel, M. E.: Implementing the Davis-Putnam algorithm by tries. Tech. rep., AI Center, SRI International, Menlo (1994)

Chapter 11
Learning Feature-Based Heuristic Functions

Marek Petrik and Shlomo Zilberstein

11.1 Introduction

Planning is the process of creating a sequence of actions that achieve some desired goals. Automated planning arguably plays a key role in both developing intelligent systems and solving many practical industrial problems. Typical planning problems are characterized by a structured state space, a set of possible actions, a description of the effects of each action, and an objective measure. In this chapter, we consider planning as an optimization problem, seeking plans that minimize the cost of reaching the goals or some other performance measure.

An important challenge in planning research is developing *general-purpose* planning systems that rely on *minimal human guidance*. General-purpose planners – also referred to as *domain-independent* – can be easily applied to a large class of problems, without relying on domain-specific assumptions. Such planners often take as input problems specified using PDDL [29], a logic-based specification language inspired by the well-known STRIPS formulation [27, 61].

Planning plays an important role in many research areas such as robotics, reinforcement learning, and operations research. Examples of planning problems in the various research fields are as follows:

Mission planning – How do we guide an autonomous spacecraft? These problems are often modeled using highly structured and discrete state and action

Marek Petrik
Department of Computer Science
University of Massachusetts Amherst
Amherst, MA, USA, e-mail: petrik@cs.umass.edu

Shlomo Zilberstein
Department of Computer Science
University of Massachusetts Amherst
Amherst, MA, USA, e-mail: shlomo@cs.umass.edu

Y. Hamadi et al. (eds.), *Autonomous Search*,
DOI 10.1007/978-3-642-21434-9_11,
© Springer-Verlag Berlin Heidelberg 2011

spaces. They have been addressed extensively by the classical planning community [30].

Inventory management – What is the best amount of goods to keep in stock to minimize holding costs and shortages? These problems are often modeled using large continuous state spaces and have been widely studied within operations research [57].

Robot control – How do we control complex robot manipulations? These stochastic planning problems have a shallow state space structure often with a small number of actions, and have been studied by the reinforcement learning community [66].

Interestingly, despite the diverse objectives in these research areas, many of the solution methods are based on heuristic search [30]. However, different research fields sometimes emphasize additional aspects of the problem that we do not address here. Reinforcement learning, for example, considers model learning to be an integral part of the planning process.

Planning problems can be solved in several different ways. One approach is to design a *domain-specific* solution method. Domain-specific methods are often very efficient but require significant effort to design. Another approach is to formulate the planning problem as a *generic optimization* problem, such as SAT, and solve it using general-purpose SAT solvers [40]. This approach requires little effort, but often does not exploit the problem structure. Yet another approach, which is the focus of this chapter, is to use *heuristic search* algorithms such as A* or Branch-and-Bound [6, 61]. Heuristic search algorithms represent a compromise between domain-specific methods and generic optimization. The domain properties are captured by a heuristic function which assigns an approximate utility to each state. Designing a heuristic function is usually easier than designing a domain-specific solver. The heuristic function can also reliably capture domain properties, and therefore heuristic search algorithms can be more efficient than generic optimization methods.

The efficiency of heuristic search depends largely on the accuracy of the heuristic function, which is often specifically designed for each planning problem. Designing a good heuristic function, while easier than designing a complete solver, often requires considerable effort and domain insight. Therefore, a truly autonomous planning system should not rely on handcrafted heuristic functions. To reduce the need for constructing heuristics manually, much of the research in autonomous planning concentrates on automating this process. The goal of automatic heuristic construction is to increase the applicability of heuristic search to new domains, not necessarily to improve the available heuristics in well-known domains.

The construction of heuristic functions often relies on *learning* from previous experience. While constructing a good heuristic function can be useful when solving a single problem, it is especially beneficial when a number of related problems must be solved. In that case, early search problems provide the necessary samples for the learning algorithm, and the time spent on constructing a good heuristic can be amortized over future search problems.

The focus of this chapter is on constructing heuristic functions within a planning framework. The specific context of construction within the planning system

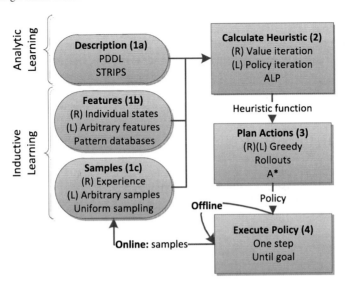

Fig. 11.1: Framework for learning heuristic functions. The numbers in parentheses are used to reference the individual components

is also crucial in designing the learning algorithms. Planning, despite being similar to generic optimization problems, such as SAT, usually encompasses a number of other components. A general framework for learning heuristics in planning systems is shown in Figure 11.1. The ovals represent inputs, the rectangles represent computational components, and the arrows represent information flow. The computational components list sample implementations and the inputs list sample information sources.

The inputs are formally defined in Section 11.3. Intuitively, the problem description (1a) denotes a precise model in some description language that can be easily manipulated. The description languages are often logic-based, such as STRIPS and PDDL. The features (1b) represent functions that assign a set of real values to each state. Finally, the samples (1c) represent simple sequences of states and actions that respect the transition model.

The methods for constructing heuristics can be divided into two broad categories based on the input they require: 1) analytic methods and 2) inductive methods [74]. *Analytic methods* use the formal description of the domain to derive a relaxed version of the problem, which in turn is used to derive a heuristic. For example, an integer linear program may be relaxed to a linear program [11, 5], which is much easier to solve. A STRIPS problem may be abstracted by simply ignoring selected literals [72]. When applicable, analytic methods typically work well, but the required description is sometimes unavailable.

Inductive methods rely on a provided set of features and transition samples of the domain. A heuristic function is calculated as a function of these *state features* based on *transition samples*. Inductive methods are easier to apply than analytic ones, since they rely on fewer assumptions. Moreover, the suitable structure needed to use abstraction is rarely present in stochastic domains [3].

Inductive heuristic learning methods can be further classified by the source of the samples into *online* and *offline* methods. Online methods interleave execution of a calculated plan with sample gathering. As a result, a new plan may be often recalculated during plan execution. Offline methods use a fixed number of samples gathered earlier, prior to plan execution. They are simpler to analyze and implement than online methods, but may perform worse due to fewer available samples.

To illustrate the planning framework, consider two common planning techniques that can learn heuristic functions: Real-Time Dynamic Programming (RTDP) [2] and Least Squares Policy Iteration (LSPI) [45]. Both methods are examples of inductive learning systems and do not rely on domain descriptions. The component implementations of RTDP and LSPI are marked with "(R)" and "(L)" respectively in Figure 11.1. The methods are described in more detail in Section 11.3.3.

RTDP is an online method that is often used to solve stochastic planning problems. While it is usually considered a type of heuristic search, it can also be seen as an inductive method for learning heuristic functions. Because features are independent for each state, it is possible to represent an arbitrary heuristic function as shown in Section 11.3.2. However, RTDP updates the heuristic function after obtaining each new sample during execution; thus it does not generalize to unseen states.

LSPI is an offline method that uses a fixed set of predetermined samples. It has been developed by the reinforcement learning community, which often refers to the heuristic function as the *value function*. The features may be arbitrary and therefore could limit the representable heuristic functions. However, unlike RTDP, LSPI also generalizes values to states that have not been visited.

Figure 11.1 represents a single planning system, but it is possible to combine multiple learning systems. Instead of the calculated heuristic function directly being used to create a policy, it can be used as an additional feature within another planning system. This is convenient, because it allows analytic methods to create good features for use with inductive learning.

The remainder of the chapter presents a unified framework for learning heuristic functions. In particular, it focuses on calculating heuristics – component (2) of the general framework – but it also describes other components and their interactions. Section 11.2 introduces the framework and the search algorithms used in components (3) and (4). Section 11.3 describes the existing inductive and analytic approaches to learning heuristic functions. The chapter focuses on inductive learning methods – the analytic methods are described because they can serve as a good source of features. Section 11.4 describes an inductive method for constructing heuristic functions using linear programming. The linear programming method is further analyzed in Section 11.5. Experimental results are presented in Section 11.6.

This chapter concentrates on using previous experience to create or improve heuristic functions. Previous experience can be used in a variety of ways. For example, some planners use samples to learn control rules that speed up the computation [30]. Early examples of such systems are SOAR [46] and PRODIGY [50]. Other notable approaches include learning state invariants [60], abstracting planning rules [62], and creating new planning rules [47]. In comparison, learning a heuristic function offers greater flexibility, but it is only relevant to planning systems that use heuristic search.

11.2 Search Framework

This section formally defines the search framework and describes common search algorithms. The section also discusses metrics of heuristic function quality that can be used to measure the required search effort. The search problem is defined as follows:

Definition 1 *A search problem is a tuple* $\mathscr{P} = (\mathscr{S}, \mathscr{A}, T, r)$, *where* $\mathscr{S}$ *is a finite set of states with a single initial state* σ *and a single goal state* τ, $\mathscr{A}$ *is a finite set of actions available in each state, and the function* $T(s, a) = s'$ *specifies the successor state* s' *of some state* $s \neq \tau$ *when action a is taken. The set finite of all successors of s is denoted by* $\mathscr{C}(s)$. *The function* $r(s, a) \in \mathbb{R}$ *specifies the reward. A discount factor* $0 < \gamma \leq 1$ *applies to future rewards.*

The discount factor γ is considered only for the sake of generality; the framework also applies to search problems without discounting (i.e., $\gamma = 1$). The assumption of a single goal or initial state is for notational convenience only. Problems with many goals or initial states can be easily mapped into this framework.

The solution of a search problem is a *policy* $\pi : \mathscr{S} \to \mathscr{A} \cup \{v\}$, which defines the action to take in each state. Here, v represents an undefined action and by definition $\pi(\tau) = v$. The policy is chosen from the set of all valid policies Π. In a deterministic search problem, a policy defines a *path* from a state s_1 to the goal. The path can be represented using the sequence

$$(\{s_i \in \mathscr{S}, a_i \in \mathscr{A} \cup \{v\}\})_{i=1}^n,$$

such that for $i < n$

$$a_i \in \mathscr{A} \qquad\qquad a_i = \pi(s_i)$$
$$t(s_i, a_i) = s_{i+1} \qquad\qquad s_n = \tau.$$

The *return*, or value, of such a sequence for a given policy π is

$$v_\pi(s_1) = \sum_{i=1}^{n-1} \gamma^i r(s_i, a_i).$$

The optimal value, denoted by $v^*(s_1)$, is defined as

$$v^*(s) = \max_{\pi \in \Pi} v_\pi(s).$$

The optimal value of a state is the maximal possible value over any goal-terminated sequence starting in that state. The objective is to find a policy π^* that defines a path for the initial state σ with the maximal *discounted cumulative* return $v_{\pi^*}(\sigma)$.

Notice that the optimal policy only needs to define actions for the states connecting the initial state σ to the goal state τ. For all other states the policy may have the value v. For simplicity, assume that there is always a path from σ to any state s and that there is a path from any state s to the goal τ. It is easy to modify the techniques for problems with dead ends and unreachable states. To guarantee the finiteness of the returns when $\gamma = 1$, assume that there are no cycles with positive cumulative rewards.

11.2.1 Heuristic Search

Heuristic methods calculate a policy for search problems from a heuristic function and an initial state. A *heuristic function* $h : \mathscr{S} \to \mathbb{R}$ returns for each state s an estimate of its optimal value $v^*(s)$. An important property of a heuristic function is admissibility, which guarantees optimality of some common algorithms. A heuristic function h is admissible when

$$h(s) \geq v^*(s) \text{ for all } s \in \mathscr{S}.$$

Notice we consider maximization problems instead of more traditional minimization problems to be consistent with the reinforcement learning literature. Therefore, an admissible heuristic is an *overestimate* of the optimal return in our case.

Admissibility is implied by *consistency*, which is defined as

$$h(s) \geq r(s,a) + \gamma h(T(s,a)) \text{ for all } s \in \mathscr{S}, a \in \mathscr{A}.$$

Consistency is often necessary for some heuristic search algorithms to be efficient.

The simplest heuristic search algorithm is *greedy best-first search*. This search simply returns the following policy:

$$\pi(s) = \underset{a \in \mathscr{A}}{\operatorname{argmax}}\{r(s,a) + \gamma h(s)\},$$

with ties broken arbitrarily. Greedy search does not require the initial state and defines a valid action for each state.

Many contemporary heuristic search methods are based on the A* algorithm, a simplified version of which is depicted in Algorithm 2. It is a best-first search, guided by a function $f : \mathscr{S} \to \mathbb{R}$, defined as

$$f(s) = g(s) + h(s).$$

Algorithm 2: A* heuristic search algorithm

Input: Heuristic function h, initial state σ

1   $\mathcal{O} \leftarrow s_0$;   // Initialize the open set
2   $s' \leftarrow \arg\max_{s \in \mathcal{O}} f(s)$;   // $f(s) = g(s) + h(s)$
3 **while** $s' \neq \tau$ **do**
4     $\mathcal{O} \leftarrow \mathcal{O} \setminus \{s'\}$;   // Remove state s' from the open set
5     $\mathcal{O} \leftarrow \mathcal{O} \cup \mathcal{C}(s')$;   // Append the one with the greatest g value
6     $s' \leftarrow \arg\max_{s \in \mathcal{O}} f(s)$;   // $f(s) = g(s) + h(s)$

The total reward (possibly negative) achieved for reaching s from the initial state σ is denoted by $g(s)$. States are chosen from an open set $\mathcal{O}$ according to their f value. The successors of the chosen states are consequently added to the open set. A* is guaranteed to find the optimal solution when the heuristic function is *admissible*.

A* is an optimally efficient search algorithm since it expands the minimal number of states needed to prove the optimality of the solution with a *consistent* heuristic function. When the heuristic function is inconsistent, the efficiency of A* could be improved using simple modifications such as PATHMAX [73].

Despite the theoretical optimality of A*, its applicability is often limited by the exponential growth of the open list $\mathcal{O}$. A* modifications have been proposed that address its large memory requirements. One common modification is Iterative Deepening A* (IDA*), which repeatedly searches the graph in a depth-first manner with a decreasing threshold on the value f of states [43]. Others have proposed modifications that do not guarantee optimality, such as weighted A* [55, 32, 68]. Because most of these algorithms retain the basic characteristics of A* and use the same heuristic function, the rest of the chapter focuses on A* as a general representative of these heuristic search methods.

The presented framework also applies to stochastic domains, formulated as Markov decision processes (MDPs) [58]. An MDP is a generalization of a search problem, in which the successor state is chosen stochastically after an action. A* is not directly applicable in stochastic domain, but some modifications have been proposed. The most commonly used heuristic search algorithm for MDPs is the greedy best-first search method described above. More sophisticated algorithms used in stochastic domains are RTDP and LAO*, which are discussed in Section 11.3.2. However, in some domains, even greedy search can be overly complicated [57].

11.2.2 Metrics of Heuristic Quality and Search Complexity

This section discusses the influence of heuristic function quality on search complexity. To start, it is necessary to define a measure of heuristic quality. The measure must accurately capture the time complexity of the search process and also must be easy to estimate and optimize. For example, the number of states that A* expands is a very precise measure but is hard to calculate. As it turns out, a precise and simple quality measure does not exist yet, but approximations can be used with good results.

The discussion in this section is restricted to admissible heuristic functions. This is a limitation when a suboptimal solution is sought and admissibility is not crucial. However, the analysis of inadmissible heuristics is hard and involves trading off solution quality with time complexity. Moreover, admissible heuristic functions can be used with suboptimal algorithms, such as weighted A* [55], to obtain approximate solutions faster.

Early analysis shows bounds on the number of expanded nodes in terms of a worst-case additive error [56, 28]:

$$\varepsilon = \max_{s \in \mathscr{S}} |h(s) - v^*(s)|.$$

The bounds generally show that even for a small value of ε, a very large number of nodes may be expanded unnecessarily. A more general analysis with regard to errors weighted by the heuristic function shows similar results [53]. In addition, most of these bounds assume a single optimal solution, which is often violated in practice.

Recent work shows tight bounds on the number of nodes expanded by A* in problems with multiple solutions [15]. In particular, the number of expanded nodes may be bounded in terms of

$$\varepsilon = \max_{s \in \mathscr{S}} \frac{|h(s) - v^*(s)|}{|v^*(s)|}.$$

This work assumes that there is a relatively small number of optimal solutions clustered together.

The existing bounds usually require a small number of goal states and states that are close to them. These assumptions have been questioned because many benchmark problems do no satisfy them [36, 35]. When the assumptions do not hold, even a good heuristic function according to the measures may lead A* to explore exponentially many states. In light of this evidence, the utility of the existing quality bounds is questionable.

Because a widely acceptable measure of the quality of heuristic functions does not yet exist, we focus on two objectives that have been studied in the literature and are relatively easy to calculate. They are

L_1, *the average-case error:* $\|h - v^*\|_1 = \sum_{s \in \mathscr{S}} |h(s) - v^*(s)|$ and
L_∞, *the worst-case error:* $\|h - v^*\|_\infty = \max_{s \in \mathscr{S}} |h(s) - v^*(s)|$.

11.3 Learning Heuristic Functions

This section gives an overview of existing methods for learning heuristic functions in planning domains. Most of these methods, however, can also be applied to search. As mentioned earlier, methods that use experience to learn control rules or other structures are beyond the scope of this chapter [74]. First, to classify the methods based on the inputs in Figure 11.1, we define the inputs more formally below.

Description (1a)

The domain description defines the structure of the transitions and rewards using a well-defined language. This description allows us to easily simulate the domain and to thoroughly analyze its properties. Domain descriptions in planning usually come in one of three main forms [30]: 1) *Set-theoretic representation* such as STRIPS, 2) *classical representation* such as PDDL, or 3) *state-variable representation* such as SAS+. Extensive description of these representations is beyond the scope of this chapter. The details can be found in [30].

Features (1b)

A feature is a function $\phi : \mathscr{S} \to \mathbb{R}$ that maps states to real values. One is typically given a set of features:

$$\Phi = \{\phi_1 \dots \phi_k\}.$$

The heuristic function is then constructed using a combination function $\bar{\theta}$:

$$h(s) = \bar{\theta}(\phi_1(s), \phi_2(s), \dots, \phi_k(s)). \tag{11.1}$$

The space of combination functions Θ is often restricted to linear functions because of their simplicity. Using linear θ, the heuristic function is expressed as

$$h(s) = \theta(\phi_1(s), \phi_2(s), \dots, \phi_k(s)) = \sum_{i=1}^{k} x_i \phi_i(s).$$

The actual combination function $\bar{\theta}$ may be either fixed for the domain or may be computed from other inputs.

One simple way to obtain features is state aggregation. In aggregation, the set of all states $\mathscr{S}$ is partitioned into $\mathscr{S}_1 \dots \mathscr{S}_m$. A feature ϕ_j is then defined as

$$\phi_j(s_i) = 1 \Leftrightarrow s_i \in \mathscr{S}_j.$$

A trivial example of an aggregation is when aggregates are singletons, as in $\mathscr{S}_i = \{s_i\}$. Such features allow us to express an arbitrary heuristic function, but do not generalize the heuristic values.

Samples (1c)

Samples are sequences of state-action pairs of one of the following types.

1. *Arbitrary* goal-terminated paths:

$$\Sigma_1 = (s_i^j, a_i^j)_{i=1}^{n_j}, \quad s_{n_j} = \tau.$$

This set also contains all shorter sample paths that start with a state later in the sequence, which are also arbitrary goal-terminated samples.

2. *Optimal* goal-terminated paths:

$$\Sigma_2 = (s_i^j, a_i^j)_{i=1}^{n_j}, \quad s_{n_j} = \tau.$$

We assume also that $\Sigma_2 \subseteq \Sigma_1$. As in Σ_1, all paths that start with a state later in the sequence are included in Σ_2.

3. *One-step* samples:

$$\Sigma_3 = (s_1^j, a_1^j, s_2^j).$$

We assume that Σ_3 also contains all transitions pairs from Σ_1 and Σ_2.

In general, the effort needed to gather an individual sample decreases in the order $\Sigma_2, \Sigma_1, \Sigma_3$. However, the utility of the samples also decreases in the same way.

Samples may come either from previously solved problem instances or from randomly selected state transitions. Samples that come from previously solved problems are in the form Σ_1 or Σ_2. Randomly gathered samples are usually in the form Σ_3 and have relatively low utility, but are much simpler to obtain. The availability of the various samples depends on the application. For example, it is rare to have optimal goal-terminated paths in online learning algorithms such as RTDP. Sampling is also more complicated in stochastic domains, such as Markov decision processes, because many samples may be required to determine the transition probabilities.

The inputs to the learning procedure determine its applicability and theoretical properties. Therefore, the procedures used in component (2) are classified according to their inputs as follows:

Inductive methods rely on state features (1b) or transition samples (1b) or both, but do not require a domain description. The methods usually cannot use the description even if it is provided. They are further classified as follows:

Feature-based methods only require features (1b), without any samples. The methods are typically very simple and require high-quality features in order to produce admissible heuristic functions. The methods are described in detail in Section 11.3.1.

Sample-based methods use samples (1c) to learn the heuristic function for states that have been directly sampled. The methods cannot take advantage of features to generalize to unseen states. The methods are described in detail in Section 11.3.2.

Sample and feature-based methods use samples (1c) to combine provided features (1b) to construct a heuristic value for states that have not been sampled. The features may be arbitrary, but the quality of the heuristic function

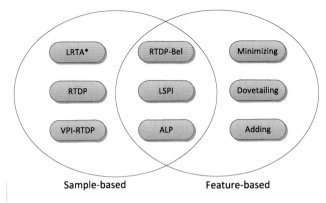

Fig. 11.2: Examples of inductive methods based on the inputs they use

highly depends on their quality. The methods are described in detail in Section 11.3.3.

Analytic methods rely on a domain description (1a), such as SAS+. These methods often cannot take advantage of features and samples, even if they are provided. They can be further classified as follows:

> *Abstraction-based methods* create a heuristic function without using samples or features. Instead, they simplify the problem using abstraction and solve it to obtain the heuristic function. The methods are described in detail in Section 11.3.4.
>
> *Other methods:* There is a broad range of analytic methods for many different problem formulations. A notable example is the simplification of mathematical programs by relaxations. For example, a linear program relaxation of an integer linear program defines an admissible heuristic function [70]. These relaxation methods have been also previously used in planning with success [5], but are not discussed in detail here.

The classification roughly follows Zimmerman et al. [74]. As mentioned above, the main focus of the chapter is on inductive methods. However, the inductive methods require features which may be hard to come by. The analytic methods, particularly those that are abstraction-based, can be a good source of the features.

Selected inductive heuristic learning methods are classified in Figure 11.2. The feature- and sample-based methods, lying at the intersection of the two sets, are the most complex and interesting to study. The intersection contains also manyreinforcement learning methods, which are inherently incompatible with A*, as discussed in Section 11.3.3. Note that the classification in the figure is with respect to the implementation of component (2) and it ignores differences in other components.

11.3.1 Feature-Based Methods

Feature-based methods are inherently simple because the heuristic function is computed statically regardless of the specific problem structure. As a result, they are useful only when the features ϕ_i represent admissible heuristic functions. The three major feature-based methods are as follows

1. Minimum of heuristic functions[1],
2. Dovetailing,
3. Sum of heuristic functions.

Feature-based methods must use a *fixed* combination function $\bar{\theta}$ in (11.1) since there are no other inputs available.

The most common feature-based method is to simply take the minimum value of the features. The combination function $\bar{\theta}$ is defined as

$$h(s) = \bar{\theta}(\phi_1(s), \ldots, \phi_k(s)) = \min\{\phi_1(s), \ldots, \phi_k(s)\}.$$

The main advantage of this method is that if the features ϕ_i are admissible heuristic functions, then the resulting heuristic function h is also admissible. *Dovetailing* is a refinement of minimization, which can reduce memory requirements by defining each feature only for a subset of the state space [12].

The minimization method is simple but usually provides only a small improvement over the provided heuristic functions. An alternative is to statically add the available features:

$$h'(s) = \bar{\theta}(\phi_1(s), \ldots, \phi_k(s)) = \sum_{i=1}^{k} \phi_i(s).$$

There are, however, additional requirements in this case to ensure that h' is admissible. It is not sufficient to use arbitrary admissible heuristic functions as features. Designing suitable features is often complicated, as has been shown for *additive pattern databases* [71].

11.3.2 Sample-Based Methods

Sample-based methods are sometimes regarded as heuristic search methods, not as methods for learning heuristic functions. Nevertheless, they fit very well into the framework in Figure 11.1. They use samples to learn a heuristic function for the states visited online, but can be easily applied in offline settings as well. The main research question and the differences between the algorithms relate to the selection of the samples, which is usually guided by some predetermined heuristic function.

The two main sample-based methods are LRTA* [44] and RTDP [2]. RTDP can be seen as a generalization of LRTA*, and therefore we focus on it and its many

---

[1] This would be a maximum in a cost minimization problem.

Algorithm 3: Real-Time Dynamic Programming (RTDP) [64]

Input: $\hat{h}$ – an initial heuristic
Input: Σ_1 – arbitrary goal terminated samples

1 $h(s) \leftarrow \hat{h}(s) \; \forall s$;   /* Component (1c) */
2 **while** *Convergence criterion is not met* **do**
3   *visited*.Clear() ;
4   Pick initial state s ;
5   **while** $s \neq \tau$ **do**
6   *visited*.Push(s);
7     $h(s) \leftarrow \max_{a \in \mathscr{A}} r(s,a) + \gamma h(s)$;   /* Component (2) */
8     $s \leftarrow$ ChooseNextState(s,h) ; /* Component (3), greedy in RTDP */
   /* Component (2) :   */
9   **while** $\neg$ *visited.IsEmpty* **do**
10    $s \leftarrow$ *visited*.Pop() ;
11    $h(s) \leftarrow \max_{a \in \mathscr{A}} r(s,a) + \gamma h(s)$;

---

refinements. The section also includes a comparison with a significant body of theoretical work on sampling algorithms in reinforcement learning, such as E^3 and R-Max.

RTDP, shown in Algorithm 3, refines the heuristic values of the individual states by backward propagation of heuristic values, essentially using value iteration [58]. The method builds a table of visited states with values $h(s)$. When a state s is expanded with children $s_1 \dots s_k$, it is stored in a table along with the value

$$h(s) \leftarrow \max_{a \in \mathscr{A}} r(s,a) + \gamma h(s).$$

When the function $h(s)$ is initialized to an upper bound on v^*, RTDP will converge at the limit to $h(s) = v^*(s)$ for all visited states.

Modifications of RTDP may significantly improve its performance. Some notable RTDP extensions are Labeled RTDP [8], HDP [7], Bounded RTDP [48], Focused RTDP [65], and VPI-RTDP [64]. The algorithms differ in how the next states are chosen in "ChooseNextState" and in the convergence criterion. The procedure "ChooseNextState" is modified from greedy to one that promotes more exploration in the right places. Bounded RTDP, Focused RTDP, and VPI-RTDP determine the exploration by additionally using a lower bound on the value of v^*. Note that h represents an upper bound on v^*.

The LAO* algorithm, described also in Section 11.2, can be seen as a sample-based method for learning heuristic functions. LAO* is based on policy iteration [58], unlike value iteration-based RTDP.

Many similar methods exist in reinforcement learning, such as R-Max [10], E^3 [41], UCT [42], UCRL2 [1]. Like RTDP, they modify sample selection in value iteration, but their motivation is quite different. In particular, the RL methods differ in the following ways: 1) they assume that the model is unknown, and 2) they have a well-defined measure of optimality, namely *regret*. Overall, RTDP focuses on the *optimization* part while the RL methods focus on the *learning* part. That is, while the RL methods aim to work almost as well as value iteration in *unknown* domains, RTDP methods attempt to improve on value iteration in *fully known* domains.

The difficulty with the purely sample-based methods is that the heuristic function is only calculated for sampled states. Generalization in RTDP can be achieved by using state features to represent the function $h(s)$. Regular RTDP can be seen as using features $\phi_1 \ldots \phi_n$, defined as

$$\phi_i(s_j) = \begin{cases} 1 & i = j \\ 0 & i \neq j \end{cases},$$

which are combined linearly. It is possible to extend RTDP to use arbitrary features, which would effectively turn it into Approximate Value Iteration (AVI). AVI is a common method in reinforcement learning but is in general unsuitable for constructing heuristic functions to be used with A*, as we discuss in Section 11.3.3.

11.3.3 Feature- and Sample-Based Methods

Feature- and sample-based methods use features to generalize from the visited state and samples to combine the features adaptively. These methods are the most general and the most complex of the inductive heuristic learning methods.

As mentioned above, the heuristic function h is constructed using a combination function θ from some predetermined set Θ:

$$h(s) = \bar{\theta}(\phi_1(s), \phi_2(s), \ldots, \phi_k(s)).$$

The objective is to determine the $\bar{\theta} \in \Theta$ that would result in the best heuristic function. That is, for some measure of heuristic quality $\omega : \mathbb{R}^{\mathscr{S}} \to \mathbb{R}$, the heuristic selection problem is

$$\bar{\theta} = \underset{\theta \in \Theta}{\text{argmin}}\, \omega\left(\theta(\phi_1(\cdot), \phi_2(\cdot), \ldots, \phi_k(\cdot))\right).$$

The objective function ω is chosen as discussed in Section 11.2.2.

The space of combination functions Θ is typically restricted to linear functions for simplicity. The heuristic function is then expressed as

$$h(s) = \theta(\phi_1(s), \phi_2(s), \ldots, \phi_k(s)) = \sum_{i=1}^{k} x_i \phi_i(s),$$

and the construction of the heuristic function becomes

$$\bar{\theta} = \underset{x \in \mathbb{R}^k}{\text{argmin}}\, \omega\left(\sum_{i=1}^{k} x_i \phi_i(\cdot)\right).$$

That is, the objective is to determine the weights x_i for all features. As described in detail below, these methods are closely related to linear value function approximation in reinforcement learning [57, 66]. The linear feature combination in the vector form is

$$h = \Phi x,$$

where Φ is a matrix that represents the features as explained in Section 11.4. The ordering of the states and actions is arbitrary but fixed.

The samples in feature- and sample-based methods may be incomplete. That is, there are potentially many transitions that are not sampled. The samples are also incomplete in classical planning when the search problem is fully known. The limited set of samples simplifies solving the problem. This is in contrast with reinforcement learning, in which more samples cannot be obtained. The presence of the model opens up the possibility of choosing the "right" samples.

Few feature- and sample-based methods have been proposed within classical planning. The symbolic RTDP method [25] enables RTDP to generalize beyond the visited states in some domains. Yoon et al. [72] propose a method that calculates a heuristic function from a linear combination of state features. The coefficients x are calculated using an iterative regression-like method. They also propose constructing state features from relaxed plans, obtained by deleting some of the preconditions in actions. Another feature- and sample-based method, RTDP-Bel, was proposed in the context of solving POMDPs [9]. RTDP-Bel endows RTDP with a specific set of features which allow generalization in the infinite belief space of a POMDP. The resulting heuristic function is not guaranteed to be admissible.

Reinforcement learning

The notion of a heuristic function in search is very similar to the notion of a value function in reinforcement learning [66]. A value function also maps states to real values that represent the expected sum of discounted rewards achievable from that state. In fact, the value function is an approximation of v^*. The objective of most methods in reinforcement learning is to learn the value function as a combination of provided features and a set of transition samples using approximate dynamic programming [57]. The policy is then calculated from the value function using greedy search.

Approximate Dynamic Programming (ADP) fits well into the category of feature- and sample-based methods. It has also been used in some interesting planning settings [17, 26]. Yet, there is significant incompatibility between ADP and classical planning applications. The policy in ADP is calculated using greedy best-first search, while in classical learning it is calculated using heuristic search. This results in two main difficulties with using ADP for learning heuristic functions: (1) ADP calculates functions that minimize the Bellman residual, which has little influence on the quality of the heuristic, and (2) they do not consider admissibility.

As Section 11.2.2 shows, there are difficulties with assessing the quality of a heuristic function used with A*. It is much simpler if the heuristic function is instead used with greedy search. Greedy search has fixed time complexity and the heuristic influences only the quality of the policy. The quality of the policy is bounded by the Bellman residual of h. The Bellman residual $\mathscr{B}(s)$ for a state s and a heuristic function h is defined as

$$\mathscr{B}(s) = h(s) - \max_{a \in \mathscr{A}} \left(h(T(s,a)) - r(s,a) \right).$$

The quality of the greedy solution is a function of [51]:

$$\frac{\max_{s \in \mathscr{S}} \mathscr{B}(s)}{1 - \gamma}.$$

The bounds requires a discount factor $\gamma < 1$, which is not the case in most planning problems.

Most approximate dynamic programming methods minimize a function of the Bellman residual to get a good greedy policy. A heuristic function with a small Bellman residual may be used with A*, but it may be very far away from the true heuristic function. In particular, adding a constant to the heuristic changes the Bellman residual by factor of $1 - \gamma$, which is very small when $\gamma \to 1$, but significantly increases the number of nodes expanded by A*. Figures 11.3 and 11.4 illustrate this issue on a simple test problem. The horizontal axes represent the state space and the vertical axes represent respectively the Bellman residual and the value of the corresponding state. The solid line is the optimal value function v^*. Notice that the dotted function has a very small Bellman residual, but a large true error. On the other hand, the dashed function has a large Bellman residual, but the error of the value is small.

Because of the above-mentioned issues, most reinforcement learning methods are unsuitable for calculating heuristic functions to be used with A*. One notable exception is approximate linear programming, which is described in detail in Section 11.4. Approximate linear programming does not minimize the Bellman residual but bounds the error of the heuristic function.

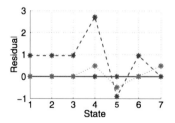

Fig. 11.3: Bellman residual of three heuristic functions for a simple chain problem

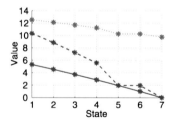

Fig. 11.4: Three value functions for a simple chain problem

A natural question in feature-based methods is about the source of the features. Unfortunately, there is no good answer yet and the methods are usually domain-specific. Often, existing heuristic functions for the domain are a good source of features. This is practical when obtaining heuristic functions in a domain is easy, such as when analytic methods are applicable.

11.3.4 Abstraction-Based Methods

Abstraction-based methods can quickly generate many relevant heuristic functions in domains with suitable structure. They work by solving a simplified version of the problem, and can be traced at least to work by Samuel [63]. There has been significant progress in recent years in automatic construction of heuristic functions using state abstraction in either deterministic [38] or stochastic domains [3]. Effective methods have been developed based on hierarchical heuristic search [37] and pattern databases [13]. These method are often able to solve challenging problems that were unsolvable before.

Abstraction-based approaches can be classified into two main types, as follows [69]:

Homomorphism. Additional transitions among states are allowed. This leads to a shorter path to the goal and an admissible heuristic function.

Embedding. Sets of states are grouped together and treated as a single state. This leads to a much smaller problem, which can be solved optimally. The solution of the simplified problem represents an admissible heuristic function.

A key advantage of abstraction-based methods is that they guarantee admissibility of the heuristic function.

Formally, when using abstraction on a search problem $\mathscr{P} = (\mathscr{S}, \mathscr{A}, T, r)$, a new search problem $\mathscr{P}' = (\mathscr{S}', \mathscr{A}', T', r')$ is defined, with a mapping function $b : \mathscr{S} -> \mathscr{S}'$. The optimal value functions v^* and $v^{*'}$ must satisfy, for all $s \in \mathscr{S}$,

$$v^{*'}(b(s)) \geq v^*(s).$$

That means that the optimal value function in $\mathscr{P}'$ represents an admissible heuristic function in $\mathscr{P}$.

Abstraction methods can also be defined formally as follows. An abstraction is called a *homomorphism* when

$$\mathscr{S} = \mathscr{S}'$$

$$b(s) = s \qquad\qquad\qquad \forall s \in \mathscr{S}$$

$$T(s,a) = s' \Rightarrow T'(s,a) = s' \qquad\qquad \forall s \in \mathscr{S}, a \in \mathscr{A}$$

An abstraction is called an *embedding* when

$$|\mathscr{S}'| \leq |\mathscr{S}|$$
$$T(s,a) = s' \Leftrightarrow T'(b(s),a) = b(s') \qquad\qquad \forall s \in \mathscr{S}, a \in \mathscr{A}.$$

Although abstraction can be defined for arbitrary search spaces, additional structure is required to make it efficient. Such structure can be provided by a description using a logic-based language.

Embedding provides two main advantages over homomorphism. First, assume that a search problem $\mathscr{P}$ is abstracted using *homomorphism* into $\mathscr{P}'$ and solved using blind search to obtain a heuristic h'. Then, the combined effort of blindly solving $\mathscr{P}'$ and solving $\mathscr{P}$ using A* with h' equals solving $\mathscr{P}$ using *blind search* [69]. In such settings, homomorphism provably does not provide any improvement. Second, a heuristic function computed using homomorphism may be difficult to store, while embedding provides a natural method for storing the heuristic function. Because the number of abstract states is typically small, the heuristic function is easy to store and reuse over multiple searches. As a result, homomorphism is useful only when the relaxed value may be calculated faster than blind search.

Despite difficulties with homomorphism, it has been used successfully in general-purpose planning by Yoon et al. [72]. While they calculate the heuristic function using a form of blind search, the policy is calculated from the heuristic function using a greedy search. As a result, Valtorta's theorem does not apply. The method leads to good results since the constructed heuristic function is sufficiently accurate.

Embedding is very useful in general-purpose planning, because it is easily applied to domains represented using logic relations [39, 37]. Such languages are discussed in Section 11.3. Embeddings in problems described by logic languages are created by simply ignoring selected predicates [19, 21] and are called *pattern databases* [13, 14, 18].

When designing a pattern database, it is crucial to properly choose the right abstraction. This is a challenging problem, but some recent progress has been made recently [16, 20, 33, 34, 35]. The existing methods use a local search in the space of potential abstraction to determine the best abstraction. They often have good empirical results in specific domains, but their behavior is not well understood and is very hard to analyze.

Because it is easy to obtain many pattern databases from different abstractions, it is desirable to be able to combine them into a single heuristic function. Research on *additive pattern databases* has tried to develop pattern databases that can be combined using the simple additive feature-based method [71], described in Section 11.3.1. Additive pattern databases guarantee that the sum of the heuristic functions produces an admissible heuristic function, but constructing them is nontrivial. Therefore, combining pattern databases using feature- and sample-based methods seems to be more promising.

11.4 Feature Combination as a Linear Program

This section describes in detail a method for learning heuristic functions based on approximate linear programming (ALP) [54]. ALP, a common method in reinforcement learning, uses a linear program to calculate the heuristic function. The ALP approach can be classified as feature- and sample-based and is related to methods described in Section 11.3.3. However, unlike most other feature- and sample-based methods, ALP is guaranteed to produce admissible heuristic functions.

Linear programming, the basis for the ALP formulation, is a mathematical optimization method in which the solution is constrained by a set of linear inequalities. The optimal solution minimizes a linear function of the set of constraints. A general form of a linear program is

$$\min_{x} c^{\mathsf{T}} x \qquad (11.2)$$
$$\text{s.t.} \quad Ax \geq b$$

where c is the objective function and $Ax \geq b$ are the constraints. Linear programs can express a large variety of optimization problems and can be solved efficiently [70].

The heuristic function in ALP is obtained as a *linear* combination of the features ϕ_i. The set of feasible heuristic functions forms a linear vector space $\mathcal{M}$, spanned by the columns of matrix Φ, which is defined as follows:

$$\Phi = \begin{pmatrix} \phi_1(s_1) & \phi_2(s_1) & \cdots \\ \phi_1(s_2) & \phi_2(s_2) & \cdots \\ & \vdots & \end{pmatrix}.$$

That is, each row of Φ defines the features for the corresponding state. The heuristic function may then be expressed as

$$h = \Phi x$$

for some vector x. Vector x represents the weights on the features, which are computed by solving the ALP problem. An important issue in the analysis is the representability of functions using the features, defined as follows.

Definition 2 *A heuristic function h is* representable in $\mathcal{M}$ *if $h \in \mathcal{M} \subseteq \mathbb{R}^{|\mathcal{S}|}$, i.e., there exists a z such that $h = \Phi z$.*

It is important to use a relatively small number of features in an ALP for two main reasons. First, it makes the linear programs easy to solve. Second, it allows us to use a small sample of all ALP constraints. This is important, since the total number of ALP constraints is greater than the total number of states in the search problem. Constraint sampling is explained in detail later in the section.

The remainder of the section describes multiple ALP formulations and their trade-offs. Section 11.4.1 describes two formulations of the constraints that ensure the admissibility of h. Section 11.4.2 introduces a new function θ that represents a *lower* bound on v^*. The difference between h and θ can be used to determine the accuracy of the heuristic function h. The difference is used when formulating the linear program, as described in Section 11.4.3.

11.4.1 Upper Bound Constraints

This section describes two sets of linear constraints that can ensure the admissibility of the heuristic function. The feasible set is represented by a set of linear inequalities.

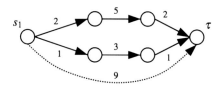

Fig. 11.5: Formulations ensuring admissibility

The two basic formulations are depicted in Figure 11.5. The first formulation, represented by the dotted line in the figure, is to simply bound the heuristic value by the value received in the sampled states. Formally, this is stated as

$$h(s_i^j) \geq \sum_{k=i}^{n_j} \gamma^{k-i} r(s_k^j, a_k^j) \quad \forall j \in \Sigma_2, \forall i \in 1 \dots n_j. \tag{11.3}$$

Clearly, this formulation ensures the admissibility of the heuristic function. Notice that this will possibly mean multiple inequalities for each state, but only the dominating ones need to be retained. Thus, let v_i denote the largest right-hand side for state s_i. The function must be restricted to the vector subspace spanned by the columns of Φ. For notational convenience, we formulate the problem in a way that is independent of the samples. Let h and v^* be column vectors with each row corresponding to a state. Then the inequality may be written as

$$h = \Phi x \geq v^*,$$

treating h as a vector. In general, only some of the inequalities are provided based on the available samples; with all samples, $v = v^*$. To simplify the notation, we denote this feasible set as H_1, and thus $h \in H_1$.

The second formulation is based on approximate linear programming, and is represented by the solid lines in Figure 11.5. In this case, the sample paths do not need to be terminated by the goal node. However, the heuristic function is actually required to be *consistent*, which is a stronger condition than admissibility. That is, for each observed sequence of two states, the difference between their heuristic values must be greater than the reward received. Formally,

$$h(s_i^j) \geq \gamma h(s_{i+1}^j) + r(s_i^j, a_i^j) \quad \forall j \in \Sigma_2, \forall i \in 1 \ldots (n_j - 1)$$

$$h(\tau) \geq 0.$$

$$(11.4)$$

In this case, we can define an action transition matrix T_a for action a. The matrix captures whether it is possible to move from the state defined by the row to the state defined by the column.

$$T_a(i, j) = 1 \Leftrightarrow t(s_i, a) = s_j.$$

A transition matrix T for all actions can then be created by vertically appending these matrices as follows:

$$T = \begin{pmatrix} T_{a_1} \\ \vdots \end{pmatrix}.$$

Similarly, we define a vector r_a of all the rewards for action a such that $r_a(i) = r(s_i, a)$. The vector r of all the rewards for all the actions can then be created by appending the vectors:

$$r = \begin{pmatrix} r_{a_1} \\ \vdots \end{pmatrix}.$$

The constraints on the heuristic function in matrix form become

$$h \geq \gamma T_a h + r_a \quad \forall a \in \mathscr{A},$$

together with the constraint $h(\tau) \geq 0$. To include the basis Φ to which the heuristic function is constrained, the problem is formulated as

$$(I - \gamma T_a)\Phi x \geq r_a \quad \forall a \in \mathscr{A}$$

$$h(\tau) \geq 0,$$

where I is the identity matrix. To simplify the notation, we denote the feasible set as H_2, and thus $h \in H_2$.

The formulation in (11.4) ensures that the resulting heuristic function will be admissible.

Proposition 1 *Given a complete set of samples, the heuristic function $h \in H_2$ is admissible. That is, for all $s \in \mathscr{S}$, $h(s) \geq v^*(s)$.*

Proof. By induction on the length of the path from state s to the goal with maximal value. The base case follows from the definition. For the inductive case, let the optimal path to the goal from state s take action a, breaking ties arbitrarily. Let $s_1 = t(s, a)$. From the inductive hypothesis and subpath optimality, we have that $h(s_1) \geq v^*(s_1)$. Then

$$h(s) \geq \gamma h(s_1) + r(s, a) \geq \gamma v^*(s_1) + r(s, a) = v^*(s).$$

Therefore, for a finite state space, the function h is an admissible heuristic function. □

In addition to being admissible, given incomplete samples, the heuristic function obtained from (11.4) is guaranteed not to be lower than the lowest heuristic value feasible in (11.3), as the following proposition states.

Proposition 2 *Let Σ_2 be a set of samples that does not necessarily cover all states. If h is infeasible in (11.3), then h is also infeasible in (11.4).*

The proof of this proposition is simple and relies on the fact that if an inequality is added for every segment of a path that connects it to the goal, then the value in this state cannot be less than the sum of the transition rewards.

Proposition 2 shows that given a fixed set of samples, (11.4) guarantees admissibility whenever (11.3) does. However, as we show below, it may also lead to a greater approximation error. We therefore analyze a hybrid formulation, weighted by a constant α:

$$\forall j \in \Sigma_2, \forall i \in 1\ldots(n_j-1):$$
$$h(s_i^j) \geq \alpha\gamma h(s_{i+1}^j) + \alpha r(s_i^j, a_i^j) + (1-\alpha)v^*(s_i^j) \tag{11.5}$$

Here $v^*(s_i^j)$ is the value of state s_i^j in sequence j. When it is not available, an arbitrary lower bound may be used. For $\alpha = 0$, this formulation is equivalent to (11.3), and for $\alpha = 1$, the formulation is equivalent to (11.4). We denote the feasible set as H_3, and thus $h \in H_3$. The key property of this formulation is stated in the following lemma, which is used later in the chapter to establish approximation bounds, and is straightforward to prove.

Lemma 1 *The optimal value function v^* is a feasible solution of (11.5) for an arbitrary α.*

11.4.2 Lower Bound Constraints

This section shows how to obtain a lower bound on the value of each state. This is important because it allows us to evaluate the difference between the heuristic value and the true value of each state. The lower bounds on the values of some selected states are obtained from the optimal solutions.

The formulation we consider is similar to that of (11.3):

$$\theta(s_i^j) \leq \sum_{k=i}^{n_j} \gamma^{k-i} r(s_k^j, a_k^j) \quad \forall j \in \Sigma_1, \forall i \in 1\ldots n_j. \tag{11.6}$$

That is, the bounds are on the values of states that were solved optimally and on any nodes that are on the path connecting the start state with the goal state. These bounds can also be written in matrix notation, as in (11.3):

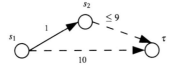

Fig. 11.6: Lower bound formulations, where the dotted lines represent paths of arbitrary length. The numbers next to the edges represent the rewards received

$$\theta = \Phi y \geq v^*.$$

We denote this feasible set by G_1, and thus $\theta \in G_1$. Additional bounds may be introduced as well. Given an admissible heuristic function, bounds can be deduced for any state that is expanded, even when it is not on an optimal path. While these bounds may not be tight in many cases, they will only increase the probability that the function θ is a lower bound. Notice that these constraints are sampled in the same manner as the constraints that ensure feasibility.

Proposition 3 *When the set of samples is complete and θ satisfies (11.6),*

$$\theta(s) \leq v^*(s) \quad \forall s \in \mathcal{S}.$$

The proof of this proposition is straightforward.

In addition to the formulation above, a variation of (11.4) can also be considered. For this, assume that every state is reachable from the initial state σ. Then, the bounds can be written for $\forall j \in \Sigma_1, \forall i \in 1 \ldots (n_j - 1)$ as

$$\theta(s_{i+1}^j) \leq \gamma\theta(s_i^j) - r(s_i^j, a_i^j) \tag{11.7}$$
$$\theta(\sigma) \leq v^*(\sigma).$$

Unlike in (11.4), these constraints alone do not guarantee that the function θ will be a lower bound on the optimal value of the states. Figure 11.6 depicts a situation in which these bounds are satisfied, but there is a feasible solution that is not an upper bound. However, as in (11.4), the bounds may be formulated as

$$(I - \gamma T_a)\Phi y \geq r_a \quad \forall a \in \mathcal{A}$$
$$\theta(\sigma) \leq v^*(\sigma).$$

We denote this feasible set by G_2, and thus $\theta \in G_2$.

11.4.3 Linear Program Formulation

Given the above, we are ready to formulate the linear program for an admissible heuristic function. As discussed in Section 11.2.2, two simple metrics used for judging the quality of a heuristic function are the L_1 norm and L_∞ norm. Linear program formulations for each of the norms follow.

The linear program that minimizes the L_1 norm of the heuristic function error is the following:

$$\min_h \mathbf{1}^\top h \qquad (11.8)$$
$$\text{s.t.} \quad h \in H_3.$$

The formulation corresponds exactly to approximate linear programming when $\alpha = 1$. It is easy to show that the optimal solution of (11.8) minimizes $\|h - v^*\|_1$ [22]. A linear program that minimizes the L_∞ norm of the heuristic function error is the following:

$$\min_{\delta, h} \delta$$
$$\text{s.t.} \quad h(s) - \theta(s) \leq \delta \quad \forall s \in S \qquad (11.9)$$
$$h \in H_3 \quad \theta \in G_1.$$

It is easy to show that (11.9) minimizes $\|h - v^*\|_\infty$. This is because from the definition $h(s) \geq \theta(s)$ for all s. In addition, even when the linear program is constructed from the samples only, this inequality holds. Notice that the number of constraints $h(s) - \theta(s) \leq \delta$ is too large, because one constraint is needed for each state. Therefore, in practice these constraints as well as the remaining states will be sampled. In particular, we use those states s for which $v^*(s)$ is known. While it is possible to use G_2 instead of G_1, that somewhat complicates the analysis. We summarize below the main reasons why the formulation in (11.9) is more suitable than the one in (11.8).

11.5 Approximation Bounds

We showed above how to formulate the linear programs for optimizing the heuristic function. It is however important whether these linear programs are feasible and whether their solutions are close to the best heuristic that can be represented using the features in basis Φ. In this section, we extend the analysis used in approximate linear programming to show new conditions for obtaining a good heuristic function.

We are interested in bounding the maximal approximation error $\|h - v^*\|_\infty$. This bound limits the maximal error in any state, and can be used as a rough measure of the extra search effort required to find the optimal solution. Alternatively, given that $\|h - v^*\|_\infty \leq \varepsilon$, the greedily constructed solution with this heuristic will have an approximation error of at most $m\varepsilon$, where m is the number of steps required to reach the goal. This makes it possible to solve the problem without search. For simplicity, we do not address here the issues related to limited sample availability, which have been previously analyzed [22, 23, 4, 31].

The approximation bound for the solution of (11.8) with the constraints in (11.4) comes from approximate linear programming [22]. In the following, we use $\mathbf{1}$ to denote the vector of all 1s. Assuming $\mathbf{1}$ is representable in $\mathcal{M}$, the bound is

$$\|v^* - h\|_c \leq \frac{2}{1-\gamma} \min_x \|v^* - \Phi x\|_\infty,$$

where $\|\cdot\|_c$ is an L_1 error bound weighted by a vector c, elements of which sum to 1. The approximation bound contains the multiplicative factors, because even when Φx is close to v^* it may not satisfy the required feasibility conditions. This bound only ensures that the sum of the errors is small, but errors in some of the states may still be very large. The bound can be directly translated to an L_∞ bound, assuming that $c = \mathbf{1}$, that is, a vector of all 1s. The bound is as follows:

$$\|v^* - h\|_\infty \leq |\mathcal{S}| \frac{2}{1-\gamma} \min_x \|v^* - \Phi x\|_\infty.$$

The potential problem with this formulation is that it may be very loose when (1) the number of states is large, since it depends on the number of states $|\mathcal{S}|$, or (2) the discount factor γ is close to 1 or is 1.

We show below how to address these problems using the alternative formulation of (11.9) and taking advantage of the additional structure of the approximation space.

Lemma 2 *Assume that $\mathbf{1}$ is representable in $\mathcal{M}$. Then there exists a heuristic function $\hat{h}$ that is feasible in (11.5) and satisfies*

$$\|\hat{h} - v^*\|_\infty \leq \frac{2}{1-\gamma\alpha} \min_x \|v^* - \Phi x\|_\infty.$$

Proof. Let the closest representable heuristic function be $\tilde{h}$, defined as

$$\varepsilon = \|\tilde{h} - v^*\|_\infty = \min_x \|v^* - \Phi x\|_\infty.$$

This function may not satisfy the inequalities (11.5). We show that it is possible to construct an $\hat{h}$ that satisfies the inequalities. From the assumption, we have that

$$v^* - \varepsilon\mathbf{1} \leq \tilde{h} \leq v^* + \varepsilon\mathbf{1}$$

and, for $a \in \mathcal{A}$,

$$0 \leq T_a\mathbf{1} \leq \gamma\alpha\mathbf{1},$$

directly from the definition of the inequalities (11.5). We use $\mathbf{0}$ to denote a zero vector of the appropriate size. Now let $\hat{h} = \tilde{h} + d\mathbf{1}$ for some d. An appropriate value of d to make $\hat{h}$ feasible can be derived using Lemma 1 as follows

$$\begin{aligned}
\hat{h} &= \tilde{h} + d\mathbf{1} \\
&\geq v^* - \varepsilon\mathbf{1} + d\mathbf{1} \\
&\geq Tv^* + r + (d-\varepsilon)\mathbf{1}
\end{aligned}$$

$$\geq T(\tilde{h} - \varepsilon\mathbf{1}) + r + (d - \varepsilon)\mathbf{1}$$
$$\geq T\tilde{h} - \varepsilon\gamma\alpha\mathbf{1} + r + (d - \varepsilon)\mathbf{1}$$
$$= T\tilde{h} + r + (d - (\gamma\alpha + 1)\varepsilon)\mathbf{1}$$
$$= T(\hat{h} - d\mathbf{1}) + r + (d - (\gamma\alpha + 1)\varepsilon)\mathbf{1}$$
$$\geq T\hat{h} + r + ((1 - \gamma\alpha)d - (1 + \gamma\alpha)\varepsilon)\mathbf{1}.$$

Therefore, $\hat{h} \geq T\hat{h} + r$ if

$$((1 - \gamma\alpha)d - (1 + \gamma\alpha)\varepsilon)\mathbf{1} \geq \mathbf{0}$$
$$d \geq \frac{1 + \gamma\alpha}{1 - \gamma\alpha}\varepsilon.$$

Since $d \geq \varepsilon$, also $\hat{h}(\tau) \geq 0$ is satisfied. Finally,

$$\|\hat{d} - v^*\|_\infty \leq \|\tilde{h} - v^*\|_\infty + \|d\mathbf{1}\|_\infty \leq \frac{2}{1 - \gamma\alpha}\varepsilon.$$

□

Using similar analysis, the following lemma can be shown.

Lemma 3 *Assume* **1** *is representable in* $\mathcal{M}$. *Then there exists a lower bound* $\hat{\theta}$ *that is feasible in* (11.6), *such that*

$$\|\hat{\theta} - v^*\|_\infty \leq 2\min_x \|v^* - \Phi x\|_\infty.$$

This lemma can be proved simply by subtracting $\varepsilon\mathbf{1}$ from the θ that is closest to v^*. The above lemmas lead to the following theorem with respect to the formulation in (11.9).

Theorem 1 *Assume that* **1** *is representable in* $\mathcal{M}$, *and let* $\hat{h}, \hat{\theta}, \delta$ *be an optimal solution of* (11.9). *Then*

$$\delta = \|\hat{h} - v^*\|_\infty \leq \left(2 + \frac{2}{1 - \gamma\alpha}\right)\min_x \|v^* - \Phi x\|_\infty.$$

Proof. Assume that that the solution δ does not satisfy the inequality. Then, using Lemmas 3 and 2, it is possible to construct a solution $\hat{h}, \hat{\theta}, \hat{\delta}$. This leads to a contradiction, because $\hat{\delta} < \delta$. □

Therefore, by solving (11.9) instead of (11.8), the error becomes independent of the number of states. This is a significant difference, since the approach is proposed for problems with a very large number of states.

Even when (11.9) is solved, the approximation error depends on the factor $1/(1 - \gamma\alpha)$. For $\gamma = \alpha = 1$, the bound is infinite. In fact, the approximate linear program may become infeasible in this case, unless the approximation basis Φ satisfies some requirements. In the following, we show which requirements are necessary to ensure that there will always be a feasible solution.

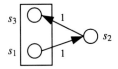

Fig. 11.7: An approximation with loose bounds

To illustrate this problem with the approximation, consider the following simple example with states $\mathscr{S} = \{s_1, s_2, s_3\}$ and a single action $\mathscr{A} = \{a\}$. The goal is the state $\tau = s_3$, and thus there is no transition from this state. The transitions are $t(s_i, a) = s_{i+1}$, for $i = 1, 2$. The rewards are also $r(s_i, a) = 1$ for $i = 1, 2$. Now, let the approximation basis be

$$\Phi = \begin{pmatrix} 1 & 0 & 1 \\ 0 & 1 & 0 \end{pmatrix}^{\mathsf{T}}.$$

This example is depicted in Figure 11.7, in which the rectangle represents the aggregated states in which the heuristic function is constant. The bounds of (11.4) in this example are

$$h(s_1) \geq \gamma h(s_2) + 1$$
$$h(s_2) \geq \gamma h(s_3) + 1$$
$$h(s_3) \geq 0.$$

The approximation basis Φ requires that $h(s_1) = h(s_3)$. Thus we get that

$$h(s_1) \geq \gamma h(s_2) + 1 \geq \gamma^2 h(s_3) + \gamma + 1 = \gamma^2 h(s_1) + \gamma + 1.$$

As a result, despite the fact that $v^*(s_1) = 2$, the heuristic function is $h(s_2) = (1+\gamma)/(1-\gamma^2)$. This is very imprecise for high values of γ. A similar problem was addressed in standard approximate linear programming by introducing so-called Lyapunov vectors. We build on this idea to define conditions that enable us to use (11.5) with high γ and α.

Definition 3 (Lyapunov vector hierarchy) *Let $u^1 \ldots u^k \geq 0$ be a set of vectors, and T and r be partitioned into T_i and r_i respectively. This set of vectors is called a Lyapunov vector hierarchy if there exist $\beta_i < 1$ such that*

$$T_i u^i \leq \beta_i u^i$$
$$T_j u^i \leq 0 \quad \forall j < i.$$

The second condition requires that no states in partition j transit to a state with positive u^i.

An example of such a hierarchy would be an abstraction, depicted in Figure 11.8. Let the state space $\mathscr{S}$ be partitioned into l subsets $\mathscr{S}_i$, with $i \in 1 \ldots l$. Assume that the transitions satisfy

$$\forall a \in \mathscr{A}: \quad t(s,a) = s' \wedge s \in \mathscr{S}_i \wedge s' \in \mathscr{S}_j \Rightarrow j < i.$$

That is, there is an ordering of the partitions consistent with the transitions. Let u^i be a vector of the size of the state space, defined as

$$u^i(s) = 1 \Leftrightarrow s \in \mathscr{S}_i,$$

and zero otherwise. It is easy to show that these vectors satisfy the requirements of Definition 3. When the approximation basis Φ can be shown to contain such u^i, it is, as we show below, possible to use the formulation with $\gamma = \alpha = 1$ with low approximation error.

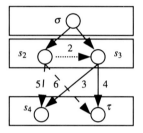

Fig. 11.8: An example of the Lyapunov hierarchy. The dotted line represents a constraint that needs to be removed and replaced with the dashed ones

Lemma 4 *Assume that there exists a Lyapunov hierarchy $u^1 \ldots u^l$ such that each u^i is representable in $\mathscr{M}$. Then there exists a heuristic function $\hat{h}$ in Φ that is feasible in (11.5), such that*

$$\|\hat{h} - v^*\|_\infty \le \left(\prod_{i=1}^{l} \frac{(1 + \alpha\gamma)\max_{s \in \mathscr{S}} u^i(s)}{(1 - \alpha\gamma\beta_i)\min_{s \in \mathscr{S}} u^i_i(s)} \right) 2\min_x \|v^* - \Phi x\|_\infty,$$

where u^i_i is the vector u^i restricted to states in partition i.

Proof. First, let

$$\varepsilon = 2\|\tilde{h}_1 - v^*\|_\infty = 2\min_x \|\Phi x - v^*\|_\infty.$$

Construct $\tilde{h} = \tilde{h}_1 + \varepsilon \mathbf{1}$ such that

$$v^* \le \tilde{h} \le v^* + \varepsilon.$$

The proof follows by induction on the size l of the Lyapunov hierarchy. Assume that the inequalities are satisfied for all $i' < i$, with the error ε and the property that the current $\tilde{h} \ge v^*$. Then let $\hat{h} = \tilde{h} + d\mathbf{1}$ for some d. Then, using Lemma 1, we have

$$\hat{h} = \tilde{h} + du^i \ge v^* + du^i \ge T_i v^* + r_d u^i$$

$$\geq T_i \tilde{h} - \gamma \alpha \varepsilon \mathbf{1} + r + du^i$$
$$\geq T_i (\hat{h} - du^i) - \gamma \alpha \varepsilon \mathbf{1} + r + du^i$$
$$\geq T_i \hat{h} + r - \alpha \beta_i \gamma du^i + du^i - \gamma \alpha \varepsilon e$$

To satisfy $\hat{h} \geq T_i \hat{h} + r_i$, set d to

$$\alpha \beta_i \gamma du^i + du^i \geq \gamma \alpha \varepsilon \mathbf{1}$$
$$d \geq \frac{\gamma \alpha}{1 - \alpha \beta_i \gamma} \frac{1}{\min_{s \in \mathscr{S}} u_i^i(s)} \varepsilon.$$

Therefore the total approximation error for $\hat{h}$ is

$$\|\hat{h} - v^*\|_\infty \leq \frac{\gamma \alpha}{1 - \alpha \beta_i \gamma} \frac{\max_{s \in \mathscr{S}} u^i(s)}{\min_{s \in \mathscr{S}} u_i^i(s)} \varepsilon.$$

The lemma follows because $d \geq 0$ and $u^i \geq 0$, and thus the condition $\tilde{h} \geq v^*$ is not violated. In the end, all the constraints are satisfied from the definition of the Lyapunov hierarchy. $\square$

The bound on the approximation error of the optimal solution of (11.9) may be then restated as follows.

Theorem 2 *Assume that there exists a Lyapunov hierarchy $u^1 \ldots u^l$, and for each u^i there exists z_i such that $u^i = \Phi z_i$. Then for the optimal solution $\hat{h}, \delta$ of (11.9),*

$$\delta = \|\hat{h} - v^*\|_\infty \leq \left(1 + \prod_{i=1}^{l} \frac{(1 + \alpha \gamma) \max_{s \in \mathscr{S}} u^i(s)}{(1 - \alpha \gamma \beta_i) \min_{s \in \mathscr{S}} u_i^i(s)}\right) 2\varepsilon,$$

where $\varepsilon = \min_x \|v^* - \Phi x\|_\infty$.

The proof follows from Lemma 4 similarly to Theorem 1. The theorem shows that even when $\gamma = 1$, it is possible to guarantee the feasibility of (11.9) by including the Lyapunov hierarchy in the basis.

A simple instance of a Lyapunov hierarchy is a set of features that depends on the number of steps to the goal. Therefore, the basis Φ must contain a vector u^i such that $u^i(j) = 1$ if the number of steps to get to s_j is i, and 0 otherwise. This is practical in problems in which the number of steps to the goal is known in any state. Assuming this simplified condition, Theorem 2 may be restated as follows:

$$\|\hat{h} - v^*\|_\infty \leq \left(1 + \prod_{i=1}^{l} \frac{1 + \alpha \gamma}{1 - \alpha \gamma \beta_i}\right) 2 \min_x \|v^* - \Phi x\|_\infty.$$

This, however, indicates exponential growth in error with the size of the hierarchy with $\gamma = \alpha = 1$. It is possible to construct an example in which this bound is tight. We have not observed such behavior in the experiments, and it is likely that finer error bounds could be established. As a result of the analysis above, if the basis contains the Lyapunov hierarchy, the approximation error is finite even for $\gamma = 1$.

In some problems it is hard to construct a basis that contains a Lyapunov hierarchy. An alternative approach is to include only constraints that respect the Lyapunov hierarchy present in the basis. These may include multistep constraints, as indicated in Figure 11.8. As a result, only a subset of the constraints is added, but this may improve the approximation error significantly. Another option is to define features that depend on actions, and not only on states. This is a nontrivial extension, however, and we leave the details to future work. Finally, when all the rewards are negative and the basis contains only a Lyapunov hierarchy, then it can be shown that no constraints need to be removed.

11.6 Empirical Results

We evaluate the approach on the sliding eight tile puzzle problem -a classic search problem [59]. The purpose of these experiments is to demonstrate the applicability of the ALP approach. We used the eight tile puzzle particularly because it has been studied extensively and can be solved relatively quickly. This allows us to evaluate the quality of the heuristic functions we obtain in different settings. Since all the experiments we describe took less than a few seconds, scaling to large problems with many sample plans is very promising. Scalability mostly relies on the ability to efficiently solve large linear programs -an area that has seen significant progress over the years. In all instances, we use the formulation in (11.9) with different values of α.

The heuristic construction method relies on a set of features available for the domain. Good features are crucial for obtaining a useful heuristic function, since they must be able to differentiate between states based on their heuristic value. In addition, the set of features must be limited to facilitate generalization. Notice that although the features are crucial, they are in general much easier to select than a good admissible heuristic function. We consider the following basis choices:

1. Manhattan distance of each tile from its goal position, including the empty tile. This results in nine features that range in values from 0 to 4. This basis does not satisfy the Lyapunov property condition. The minimal admissible heuristic function from these features will assign value -1 to the feature that corresponds to each tile, except the empty tile, the feature for which is assigned 0.
2. Abstraction based on the sum of the Manhattan distances of all pieces. For example, feature 7 will be 1 if the sum of the Manhattan distances of the pieces is 7, and 0 otherwise. This basis satisfies the Lyapunov hierarchy condition as defined above, since all the rewards in the domain are negative.
3. The first feature is the Nilsson sequence score [52] and the second feature is the total Manhattan distance minus 3. The Nilsson sequence score is obtained by checking around the non-central square in turn, allotting 2 for every tile not

followed by its proper successor and 0 for every other tile, except that a piece in the center scores 1. This value is not an admissible heuristic.

First, we evaluate the approach in terms of the number of samples that are required to learn a good heuristic function. We first collect samples from a blind search. To avoid expanding too many states, we start with initial states that are close to the goal. This is done by starting with the goal state and performing a sequence of 20 random actions. Typical results obtained in these experiments are shown in Figure 11.9, all performed with $\alpha = 1$. The column labeled "States" shows the total number of node-action pairs, not necessarily unique, expanded and used to learn the heuristic function. The samples are gathered from solving the problem optimally for two states. The results show that relatively few nodes are required to obtain a heuristic function that is admissible, and very close to the optimal heuristic function with the given features. Observe that the heuristic function was obtained with no prior knowledge of the domain and without any a priori heuristic function. Very similar results were obtained with the second basis.

x_0	x_1	x_2	x_3	x_4	x_5	x_6	x_7	x_8	States
0	-1	-1	-1	-1	0	0	-1	0	958
0	-1	-1	-1	-1	-1	0	-1	0	70264
0	-1	-1	0	-1	-1	-1	-1	-1	63
0	-1	-1	-1	-1	-1	-1	-1	-1	162

Fig. 11.9: Weights calculated for individual features using the first basis choice. Column x_i corresponds to the weight assigned to the feature associated with tile i, where 0 is the empty tile. The top two rows are based on data from blind search, and the bottom two on data from search based on the heuristic from the previous row

Next, we compare the two formulations for the upper bounds, (11.3) and (11.4), with regard to their approximation error. Notice that a big advantage of the formulation depicted by (11.4) is the ability to use transitions from states that are not on a path to the goal. The data is based on 100 goal-terminated searches and 1,000 additional randomly chosen states. The results are shown in Figure 11.10. Here, δ is the objective value of (11.9), which is the maximal overestimation of the heuristic function in the given samples. Similarly, δ' is the maximal overestimation obtained based on 1,000 state samples independently of the linear program. The approximate fraction of states in which the heuristic function is admissible is denoted by p. These results demonstrate the trade-off that α offers. A lower value of α generally leads to a better heuristic function, but at the expense of admissibility. The bounds with regard to the sampled constraints, as presented in [22], do not distinguish between the two formulations. A deeper analysis of this will be an important part of future work. Interestingly, the results show that the Nilsson sequence score is not admissible, but it becomes admissible when divided by 4.

α	1	0.99	0.9	0.8	0
x_1	0	-0.26	-1	-1	-1
x_2	-0.25	-0.25	-0.59	-0.6	-0.6
x_3	0	0	2.52	2.6	2.6
δ	17	16.49	13.65	13.6	13.6
p	1	1	0.97	0.96	0.95
δ'	19	17	15	14.4	14.4

Fig. 11.10: The discovered heuristic functions as a function of α in the third basis choice, where the x_i are the weights on the corresponding features in the order in which they are defined

The ALP approach has also been applied to Tetris, a popular benchmark problem in reinforcement learning [67]. Because Tetris is an inherently stochastic problem, we used the Regret algorithm [49] to solve it using a deterministic method. The Regret algorithm first generates samples of the uncertain component of the problem and then treats it as a deterministic problem. It is crucial in Tetris to have a basis that contains a Lyapunov hierarchy, since the rewards are positive. Since we always solve the problem for a fixed number of steps forward, such a hierarchy can be defined based on the number of steps remaining, as in Figure 11.8. Our initial experiments with Tetris produced promising results. However, to outperform the state-of-the-art methods -such as approximate linear programming [24] and cross-entropy methods [67]- we need to scale up our implementation to handle millions of samples. This is mostly a technical challenge that can be addressed using methods developed by [24].

11.7 Conclusion

This chapter presents a unified framework for learning heuristic functions in planning settings. Analytic methods use the structure of the domain to calculate heuristic functions, while inductive methods rely on state features and properties. The framework also encompasses most of the methods used in reinforcement learning, which are often unsuitable for use with heuristic search algorithms.

The ALP-based approach is a feature- and sample-based method that learns *admissible* heuristic functions. When applied naively, these techniques may lead to formulations that have no feasible solutions. We show that by guaranteeing certain properties, it is possible to ensure that the approximation is finite and -in some cases- accurate, with tight error bounds. This work lays the foundation for further understanding of how sampling techniques can produce good heuristic functions for complex planning problems.

Learning heuristic functions automatically is a long-standing challenge in artificial intelligence. Despite the existing shortcomings, the ALP approach has several important advantages. The formulation is very general and it places only modest assumptions on the features. It requires little domain knowledge. It works well with samples of plans -both optimal and non-optimal ones. And, most importantly, it makes it possible to compute guarantees on the admissibility of the learned heuristic function.

Acknowledgements This work was supported by the Air Force Office of Scientific Research under Grant No. FA9550-08-1-0171.

References

[1] Auer, P., Jaksch, T., Ortner, R.: Near-optimal regret bounds for reinforcement learning. In: Advances in Neural Information Processing Systems (2009)

[2] Barto, A., Bradtke, S. J., Singh, S. P.: Learning to act using real-time dynamic programming. Artificial Intelligence **72**(1), 81–138 (1995)

[3] Beliaeva, N., Zilberstein, S.: Generating admissible heuristics by abstraction for search in stochastic domains. In: Abstraction, Reformulation and Approximation, pp. 14–29. Springer Berlin / Heidelberg (2005)

[4] Ben-Tal, A., Nemirovski, A.: Selected topics in robust optimization. Mathematical Programming, Series B **112**, 125–158 (2008)

[5] Benton, J., van den Briel, M., Kambhampati, S.: A hybrid linear programming and relaxed plan heuristic for partial satisfaction planning problems. In: International Conference on Automated Planning and Scheduling (ICAPS) (2007)

[6] Bonet, B., Geffner, H.: Planning as heuristic search. Artificial Intelligence **129**(1-2), 5–33 (2001)

[7] Bonet, B., Geffner, H.: Faster heuristic search algorithms for planning under uncertainty and full feedback. In: International Joint Conference on Artificial Intelligence (2003)

[8] Bonet, B., Geffner, H.: Labeled RTDP: Improving the convergence of real-time dynamic programming. In: International Conference on Autonomous Planning (ICAPS) (2003)

[9] Bonet, B., Geffner, H.: Solving POMDPs: RTDP-Bel vs. point-based algorithms. In: International Joint Conference on Artificial Intelligence (IJCAI) (2009)

[10] Brafman, R. I., Tennenholtz, M.: R-MAX -a general polynomial time algorithm for near-optimal reinforcement learning. Journal of Machine Learning Research **3**, 213–231 (2002)

[11] Bylander, T.: A linear programming heuristic for optimal planning. In: National Conference on Artificial Intelligence, pp. 694–699 (1997)

[12] Culberson, J. C., Schaeffer, J.: Efficiently searching the 15-puzzle. Tech. rep., Department of Computer Science, University of Alberta (1994)

[13] Culberson, J. C., Schaeffer, J.: Searching with pattern databases. In: Advances in Artifical Intelligence, pp. 402–416. Springer Berlin / Heidelberg (1996)

[14] Culberson, J. C., Schaeffer, J.: Pattern databases. Computational Intelligence **14**(3), 318–334 (1998)
[15] Dinh, H., Russell, A., Su, Y.: On the value of good advice: The complexity of A* search with accurate heuristics. In: AAAI (2007)
[16] Drager, K., Fingbeiner, B., Podelski, A.: Directed model checking with distance-preserving abstractions. In: International SPIN Workshop, LNCS, vol. 3925, pp. 19–34 (2006)
[17] Dzeroski, S., de Raedt, L., Driessens, K.: Relational reinforcement learning. Machine Learning **43**, 7–52 (2001)
[18] Edelkamp, S.: Planning with pattern databases. In: ECP (2001)
[19] Edelkamp, S.: Symbolic pattern databases in heuristic search planning. In: AIPS (2002)
[20] Edelkamp, S.: Automated creation of pattern database search heuristics. In: Workshop on Model Checking and Artificial Intelligence (2006)
[21] Edelkamp, S.: Symbolic shortest paths planning. In: International Conference on Automated Planning and Scheduling (ICAPS) (2007)
[22] de Farias, D. P.: The linear programming approach to approximate dynamic programming: Theory and application. Ph.D. thesis, Stanford University (2002)
[23] de Farias, D. P., van Roy, B.: On constraint sampling in the linear programming approach to approximate dynamic programming. Mathematics of Operations Research **29**(3), 462–478 (2004)
[24] Farias, V., van Roy, B.: Probabilistic and Randomized Methods for Design Under Uncertainty, chap. 6: Tetris: A Study of Randomized Constraint Sampling. Springer-Verlag (2006)
[25] Feng, Z., Hansen, E. A., Zilberstein, S.: Symbolic generalization for on-line planning. In: Uncertainty in Artificial Intelligence (UAI), pp. 209–216 (2003)
[26] Fern, A., Yoon, S., Givan, R.: Approximate policy iteration with a policy language bias: Solving relational Markov decision processes. Journal of Artificial Intelligence Research (JAIR) **25**, 85–118 (2006)
[27] Fikes, R. E., Nilsson, N. J.: STRIPS: A new approach to the application of theorem proving to problem solving. Artificial Intelligence **2**(189-208), 189–208 (1971)
[28] Gaschnig, J.: Ph.D. thesis, Carnegie-Mellon University (1979)
[29] Gerevini, A., Long, D.: Plan constraints and preferences in PPDL3. Tech. rep., Dipartimento di Elettronica per l'Automazione, Universita degli Studi di Brescia (2005)
[30] Ghallab, M., Nau, D., Traverso, P.: Automated Planning: Theory and Practice. Morgan Kaufmann (2004)
[31] Goldfarb, D., Iyengar, G.: Robust convex quadratically constrained programs. Mathematical Programming **97**, 495–515 (2003)
[32] Hansen, E. A., Zhou, R.: Anytime heuristic search. Journal of Artificial Intelligence Research **28**, 267–297 (2007)
[33] Haslum, P., Bonet, B., Geffner, H.: New admissible heuristics for domain-independent planning. In: National Conference on AI (2005)

[34] Haslum, P., Botea, A., Helmert, M., Bonet, B., Koenig, S.: Domain-independent construction of pattern database heuristics for cost-optimal planning. In: National Conference on Artificial Intelligence (2007)

[35] Helmert, M., Mattmuller, R.: Accuracy of admissible heuristic functions in selected planning domains. In: National Conference on Artificial Intelligence (2008)

[36] Helmert, M., Roger, G.: How good is almost perfect. In: National Conference on AI (2008)

[37] Holte, R. C., Grajkowski, J., Tanner, B.: Hierarchical heuristic search revisited. In: Abstraction, Reformulation and Approximation, pp. 121–133. Springer Berlin / Heidelberg (2005)

[38] Holte, R. C., Mkadmi, T., Zimmer, R., MacDonald, A.: Speeding up problem solving by abstraction: a graph oriented approach. Artificial Intelligence **85**, 321–361 (1996)

[39] Holte, R. C., Perez, M., Zimmer, R., MacDonald, A.: Hierarchical A*: Searching abstraction hierarchies efficiently. In: National Conference on Artificial Intelligence (AAAI), pp. 530–535 (1996)

[40] Kautz, H. A., Selman, B.: Pushing the envelope: Planning, propositional logic, and stochastic search. In: National Conference on Artificial Intelligence (AAAI) (1996)

[41] Kearns, M., Singh, S.: Near-polynomial reinforcement learning in polynomial time. Machine Learning **49**, 209–232 (2002)

[42] Kocsis, L., Szepesvári, C.: Bandit based Monte-Carlo planning. In: European Conference on Machine Learning (ECML) (2006)

[43] Korf, R.: Depth-first iterative deepening: An optimal admissible tree search. Artificial Intelligence **27**(1), 97–109 (1985)

[44] Korf, R. E.: Real-time heuristic search. In: National Conference on AI (AAAI) (1988)

[45] Lagoudakis, M. G., Parr, R.: Least-squares policy iteration. Journal of Machine Learning Research **4**, 1107–1149 (2003)

[46] Laird, J. E., Rosenbloom, P. S., Newell, A.: Chunking in SOAR: The anatomy of of a general learning mechanism. Machine Learning **1**, 11–46 (1986)

[47] Leckie, C., Zuckerman, I.: Inductive learning of search control rules for planning. Artificial Intelligence **101**(1-2), 63–98 (1998)

[48] McMahan, H. B., Likhachev, M., Gordon, G. J.: Bounded real-time dynamic programming: RTDP with monotone upper bounds and performance guarantees. In: International Conference on Machine Learning (ICML) (2005)

[49] Mercier, L., van Hentenryck, P.: Performance analysis of online anticipatory algorithms for large multistage stochastic integer programs. In: International Joint Conference on AI, pp. 1979–1985 (2007)

[50] Minton, S., Knoblock, C., Kuokka, D. R., Gil, Y., Joseph, R. L., Carbonell, J. G.: PRODIGY 2.0: The manual and tutorial. Tech. rep., Carnegie Mellon University (1989)

[51] Munos, R.: Error bounds for approximate policy iteration. In: International Conference on Machine Learning, pp. 560–567 (2003)

[52] Nilsson, N.: Problem-Solving Methods in Artificial Intelligence. McGraw Hill (1971)

[53] Pearl, J.: Heuristics: Intelligent search strategies for computer problem solving. Addison-Wesley, Reading, MA (1984)

[54] Petrik, M., Zilberstein, S.: Learning heuristic functions through approximate linear programming. In: International Conference on Automated Planning and Scheduling (ICAPS), pp. 248–255 (2008)

[55] Pohl, I.: Heuristic search viewed as path finding in a graph. Artificial Intelligence 1, 193–204 (1970)

[56] Pohl, I.: Practical and theoretical considerations in heuristic search algorithms. Machine Intelligence 8, 55–72 (1977)

[57] Powell, W. B.: Approximate Dynamic Programming. Wiley-Interscience (2007)

[58] Puterman, M. L.: Markov decision processes: Discrete stochastic dynamic programming. John Wiley & Sons, Inc. (2005)

[59] Reinefeld, A.: Complete solution of the eight-puzzle and the benefit of node ordering in IDA*. In: International Joint Conference on AI, pp. 248–253 (1993)

[60] Rintanen, J.: An iterative algorithm for synthesizing invariants. In: National Conference on Artificial Intelligence (AAAI) (2000)

[61] Russell, S., Norvig, P.: Artificial Intelligence A Modern Approach, 2nd edn. Prentice Hall (2003)

[62] Sacerdott, E.: Planning in a hierarchy of abstraction spaces. Artificial Intelligence 5(2), 115–135 (1974)

[63] Samuel, A.: Some studies in machine learning using the game of checkers. IBM Journal of Research and Development 3(3), 210–229 (1959)

[64] Sanner, S., Goetschalckx, R., Driessens, K., Shani, G.: Bayesian real-time dynamic programming. In: Intenational Joint Conference on Artificial Intelligence (IJCAI) (2009)

[65] Smith, T., Simmons, R. G.: Focused real-time dynamic programming. In: National Proceedings in Artificial Intelligence (AAAI) (2006)

[66] Sutton, R.S., Barto, A.: Reinforcement Learning. MIT Press (1998)

[67] Szita, I., Lorincz, A.: Learning Tetris using the noisy cross-entropy method. Neural Computation 18(12), 2936–2941 (2006)

[68] Thayer, J. T., Ruml, W.: Faster than weighted A*: An optimistic approach to bounded suboptimal search. In: International Conference on Automated Planning and Scheduling (2008)

[69] Valtorta, M.: A result on the computational complexity of heuristic estimates for the A* algorithm. Information Sciences 34, 48–59 (1984)

[70] Vanderbei, R. J.: Linear Programming: Foundations and Extensions, 2nd edn. Springer (2001)

[71] Yang, F., Coulberson, J., Holte, R., Zahavi, U., Felner, A.: A general theory of assitive state space abstraction. Journal of Artificial Intelligence Research 32, 631–662 (2008)

[72] Yoon, S., Fern, A., Givan, R.: Learning control knowledge for forward search planning. Journal of Machine Learning Research 9, 638–718 (2008)

[73] Zhang, Z., Sturtevant, N. R., Holte, R., Schaeffer, J., Felner, A.: A* search with inconsistent heuristics. In: International Joint Conference on Artificial Intelligence (IJCAI) (2009)
[74] Zimmerman, T., Kambhampati, S.: Learning-assisted automated planning. AI Magazine **24**(2), 73–96 (2003)